Lecture Notes of the Institute for Computer Sciences, Social Informatics and Telecommunications Engineering   134

For further volumes:
http://www.springer.com/series/8197

Gianni A. Di Caro · Guy Theraulaz (Eds.)

Bio-Inspired Models of Network, Information, and Computing Systems

7th International ICST Conference,
BIONETICS 2012
Lugano, Switzerland, December 10–11, 2012
Revised Selected Papers

 Springer

Editors
Gianni A. Di Caro
Dalle Molle Institute
   for Artificial Intelligence (IDSIA)
Manno-Lugano
Switzerland

Guy Theraulaz
Research Center on Animal Cognition
Université Paul Sabatier
Toulouse
France

ISSN 1867-8211   ISSN 1867-822X (electronic)
ISBN 978-3-319-06943-2   ISBN 978-3-319-06944-9 (eBook)
DOI 10.1007/978-3-319-06944-9
Springer Cham Heidelberg New York Dordrecht London

Library of Congress Control Number: 2014942656

Springer is part of Springer Science+Business Media (www.springer.com)

Preface

The field of *bio-inspired computing and modeling* has witnessed a rapid growth in the last 25 years. A relatively large and still increasing number of conferences and workshops, journals and special issues are devoted each year to different aspects related to the understanding of mechanisms in the natural world and to their reverse-engineering for the design of novel algorithms and hardware systems. Results from bio-inspired research find their successful application in a wide range of different fields, including, among others, optimization and operations research, networking, robotics, bioinformatics, and data mining.

While most conferences in the bio-inspired domain focus on a specific field of application (e.g., bio-inspired robot systems) or on specific families of bio-inspired approaches (e.g., swarm intelligence), the BIONETICS series of conferences avoid focusing on specific bio-inspired application domains or strategies, aiming instead at being truly *interdisciplinary*. BIONETICS is conceived as a general forum for researchers and practitioners from different disciplines and application fields who seek the understanding of fundamental principles and design strategies in biological systems, and leverage on this understanding to build bio-inspired systems for problem-solving and engineering applications.

The 2012 edition of BIONETICS, held in Lugano, Switzerland, during December 10–11, 2012, was the seventh in this series of conferences. The first edition was held in 2006, in Cavalese, in Italy, and was then held in Budapest, Hungary (2007), Hyogo, Japan (2008), Avignon, France (2009), Boston, MA, USA (2010), and York, UK (2011). The papers in this volume were orally presented at the conference, and represent a carefully selected sample of state-of-the-art research in the domain of bio-inspired modeling and applications. The conference featured 40 submissions, 23 of which were accepted as full papers, corresponding to a 57% acceptance rate, after a very careful blind-review process with three reviewers per paper. These submission numbers are similar to those of previous editions, showing a stable interest in the conference (apart from the 2010 edition, which had a surge of submissions), in spite of the fierce competition with the continually increasing number of conferences addressing related topics.

In line with the truly interdisciplinary philosophy of the conference, the papers included in this volume cover a wide range of topics and application domains: bio-inspired approaches for telecommunications, robotics and learning issues, modeling at the molecular scale, bioinformatics, optimization, and bio-inspired approaches to various problems such as social networking, game theory, and language analysis. Overall, these works provide a high-quality sample of the diversity in bio-inspired research, as well as its wide potential for use in real-world applications and as a tool for gaining a deeper understanding of biological and social processes.

Many people helped to make BIONETICS 2012 a successful event. In particular, I would like to thank Guy Theraulaz (general co-chair), Enrico Natalizio (publicity chair), Eduardo Feo (publication chair), Jawad Nagi (Web chair), and the BIONETICS Steering Committee (Imrich Chlamtac, Iacopo Carreras, Falko Dressler, Tadashi Nakano, Tatsuya Suda, Jun Suzuki). A special thank goes to the four invited keynote speakers: Nicola Gatti, Jürgen Schmidhuber, Franco Zambonelli, and Anthony Weiss, who presented an exciting and insightful overview of different applications and techniques of bio-inspired computing. Finally, I also want to express my gratitude to the European Alliance for Innovation (EAI), and in particular to Ms. Dina Shakirova and Ms. Elisa Mendini who provided full management support, and to the Scuola Universitaria Professionale della Svizzera Italiana (SUPSI), which sponsored the event.

June 2014   Gianni A. Di Caro

Organization

Steering Committee

Imrich Chlamtac	Create-Net and University of Trento, Italy (Commitee Chair)
Iacopo Carreras	Create-Net, Italy
Falko Dressler	University of Innsbruck, Austria
Tadashi Nakano	Osaka University, Japan
Tatsuya Suda	University of California, Irvine, USA
Jun Suzuki	University of Massachusetts, Boston, USA

Organizing Committee

General Chair

Gianni A. Di Caro — IDSIA, USI/SUPSI, Lugano, Switzerland

Program Co-chair

Guy Theraulaz — Université Paul Sabatier, Toulouse, France

Workshops and Special Sessions Chair

Luca Gambardella — IDSIA, USI/SUPSI, Lugano, Switzerland

Publicity Chair

Enrico Natalizio — Inria Lille, France

Publication Chair

Eduardo Feo — IDSIA, USI/SUPSI, Lugano, Switzerland

Web Chair

Jawad Nagi — IDSIA, USI/SUPSI, Lugano, Switzerland

Technical Program Committee

Ajith Abraham	Machine Intelligence Research Labs, Seattle, USA
Ozgur B. Akan	Middle East Technical University, Turkey
Andrew Adamatzky	University of the West of England, Bristol, UK

Contents

Molecular Scale and Bioinformatics

Optimization

Bio-Inspired Modeling

Telecommunications

A Hybrid Exact-ACO Algorithm for the Joint Scheduling, Power and Cluster Assignment in Cooperative Wireless Networks

Fabio D'Andreagiovanni[1,2](✉)

[1] Department of Optimization, Zuse-Institut Berlin (ZIB), 14195 Berlin, Germany
[2] Department of Computer, Control and Management Engineering,
Sapienza Università di Roma, 00185 Roma, Italy
d.andreagiovanni@zib.de
http://www.dis.uniroma1.it/~fdag/

Abstract. Base station cooperation (BSC) has recently arisen as a promising way to increase the capacity of a wireless network. Implementing BSC adds a new design dimension to the classical wireless network design problem: how to define the subset of base stations (clusters) that coordinate to serve a user. Though the problem of forming clusters has been extensively discussed from a technical point of view, there is still a lack of effective optimization models for its representation and algorithms for its solution. In this work, we make a further step towards filling such gap: (1) we generalize the classical network design problem by adding cooperation as an additional decision dimension; (2) we develop a strong formulation for the resulting problem; (3) we define a new hybrid solution algorithm that combines exact large neighborhood search and ant colony optimization. Finally, we assess the performance of our new model and algorithm on a set of realistic instances of a WiMAX network.

Keywords: Cooperative wireless networks · Binary linear programming · Exact local search · Ant colony optimization · Hybridization

1 Introduction

Wireless communications have experienced an astonishing expansion over the last years and network operators have consequently faced the challenge of providing higher capacity with the same limited amount of radio resources. In this context, the capacity increase has been mainly pursued through interference avoidance: by limiting interference, a higher number of users can indeed be served, thus increasing network capacity. Interference avoidance can be first obtained by a careful setting of the power emissions of the base stations constituting the network. In recent years, it was showed that an additional reduction

This work was partially supported by the *German Federal Ministry of Education and Research* (BMBF), project *ROBUKOM* [4,27], grant 03MS616E. The Author thanks Antonella Nardin for fruitful discussions.

G.A. Di Caro and G. Theraulaz (Eds.): BIONETICS 2012, LNICST 134, pp. 3–17, 2014.
DOI: 10.1007/978-3-319-06944-9_1, © Institute for Computer Sciences, Social Informatics and Telecommunications Engineering 2014

in interference can be obtained by allowing *cooperation* among the base stations: the central feature of cooperation is that service can be provided by a number of base stations forming a *cluster*, in contrast to classical systems where service is provided by a single base station. These new wireless systems are called *cooperative wireless networks* (CWN). The advantage of a CWN is clear: a user that requires a single powerful signal can be instead served by a (small) number of signals of lower power, that reduce interference towards other users. Remarkable gain in capacity are thus possible in wireless technologies supporting cooperation, such as LTE [33] and WiMAX [2]. On the other hand, cooperation has a cost and must be kept low: cooperating base stations must indeed coordinate and need to exchange a non-negligible amount of information. For a detailed discussion of the potentialities and advantages of adopting cooperation in wireless networks, we refer the reader to the recent work [21].

Cooperation introduces an additional decision dimension in the problem of designing a wireless network, thus making an already complicated problem even more complicated. In this context, Mathematical Optimization has arisen as a very effective methodology to improve the quality of design, as proven in successful industrial collaborations with major wireless providers [12,29,31]. Optimization is furthermore able to overcome the limitations of the trial-and-error approach commonly adopted by network engineers, by pursuing a much more efficient usage of the scarce radio resources of a network.

Tough CWNs have attracted a lot of attention from a technical and theoretical point of view, there is still a lack of effective optimization models and algorithms for their design. Moreover, available models and algorithms do not treasure previous experience about how tackling specific computational issues that affect the general design problem. The main objective of our work is to make further steps towards closing such gap. Specifically, our original contributions are: (1) a new binary linear model for the design of cooperative wireless networks, that jointly optimizes activated service links, power emissions and cooperative clusters; (2) a strengthening procedure that generates a stronger optimization model, in which sources of numerical issues are completely eliminated; (3) a hybrid solution algorithm, based on the combination of a special exact large neighborhood search called RINS [11] with ant colony optimization [20] (note that tough generic heuristics and pure and hybrid nature-inspired metaheuristics are not a novelty in (wireless) network design - see e.g., [3,5,9,14] - to the best of our knowledge such combination has not yet been investigated). This algorithm is built to effectively exploit the precious information coming from the stronger model and can rapidly find solutions of very satisfying quality.

To our best knowledge, our model is the first that considers the joint optimization of activated service links, power emissions and cooperative clusters. Models available in literature have indeed focused attention on more specific versions of the problem. In [34], a set covering formulation for selecting clusters is introduced and solved by a greedy heuristic. A weakness of this work is that the feasible clusters must be explicitly listed and their number is in general exponential. In [22], a linear binary formulation for the problem of forming cooperative clusters and assigning users to base stations is proposed and a limited computational experience on small-sized instances is presented. In [24] a non-linear model for the joint optimal clustering and beamforming in CWN is investigated. Given the intrinsic hardness of the resulting non-linear problem, the Authors propose an iterative heuristic approach that makes use of mean square errors. In contrast to our work, all these works do not consider the computational issues associated with the coefficients representing signal propagation and thus their models are weak and their solution approaches may lead to coverage errors, as explained in [8,12,26].

The remainder of this paper is organized as follows: in Sect. 2, we introduce our new model for the design of cooperative wireless networks and discuss its strength; in Sect. 3 we present our hybrid metaheuristic; in Sect. 4 we present preliminary computational results over a set of realistic WiMAX instances and, finally, in Sect. 5, we derive some conclusions.

2 Cooperative Wireless Network Design

We consider the design of a wireless network made up of a set B of base stations (BSs) that provide a telecommunication service to a set T of user terminals (UTs). A UT $t \in T$ is said to be *covered* (or *served*), if it receives the service within a minimum level of quality. The BSs and the UTs are characterized by a number of parameters (e.g., geographical location, power emission, frequency used for transmitting). The *Wireless Network Design Problem (WND)* consists in setting the values of the base station parameters, with the aim of maximizing a revenue function associated with coverage (e.g., number of covered UTs). For an exhaustive introduction to the WND, we refer the reader to [12,13,26].

An optimization model for the WND typically focuses attention only on a subset of the parameters characterizing a BS. In particular, the majority of the models considers the power emission and the assignment of served UTs to BSs as the main decision variables. These are indeed two critical decisions that must be taken by a network administrator, as indicated in several real studies (e.g., [1,6,7,12,29,31]). Other parameters that are commonly considered are the frequency and the transmission scheme used to serve a terminal (e.g., [10,17,30]). Different versions of the WND are discussed in [13], where a hierarchy of WND problems is also presented.

In this work, we consider a generalization of the *Scheduling and Power Assignment Problem* (SPAP), a version of the WND that is known to be HP-Hard [29]. In the SPAP, two decisions must be taken: (1) setting the power emission of each BS and (2) assigning served UTs to activated cluster of BSs (note that this corresponds to identify a subset of service links BS-UT that can be scheduled simultaneously without interference, so we use the term *scheduling*). Since we address the problem of designing a cooperative network, in our generalization we also include a third decision: for each served UT, we want to select a subset of BSs forming the cluster that serves the UT. We call the resulting generalization by the name *Scheduling, Power and Cluster Assignment Problem* (SPCAP).

To model these three decisions, we introduce three types of decision variables:

1. A non-negative discrete variable $p_b \in \mathcal{P} = \{P_1, \ldots, P_{|\mathcal{P}|}\}$, with $P_{|\mathcal{P}|} = P_{\max}$ and $P_l > P_{l-1} > 0$, for $l = 2, \ldots, |\mathcal{P}|$, representing the power emission of a BS b. Such variable can be represented as the linear combination of the power values P_l multiplied by binary variables: to this end, for each $b \in B$ we introduce one binary variable z_{bl} (*power variable*) that is equal to 1 if b emits at power P_l and 0 otherwise. Additionally, we must express the fact that each BS may emit only at a single power level. If we denote the set of feasible power levels by L, the definition of the binary power variables is: $p_b = \sum_{l \in L} P_l z_{bl}, \forall b \in B$, with $\sum_{l \in L} z_{bl} \leq 1$. We remark that a BS b such that $\sum_{l \in L} z_{bl} = 0$ has null power (i.e., $p_b = 0$) and therefore is not deployed in the network. Our model thus also decides the localization of the BSs to be deployed, though this feature is not explicitly pointed out by defining a specific decision variable.

2. A binary *service assignment variable* $x_t \in \{0, 1\}, \forall t \in T$, that is equal to 1 if UT $t \in T$ is served and to 0 otherwise.

3. A binary *cluster assignment variable* $y_{tb} \in \{0, 1\}, \forall t \in T, b \in B$, that is equal to 1 if BS b belongs to the cluster that serves UT t and to 0 otherwise.

Every UT $t \in T$ picks up signals from each BS $b \in B$ in the network and the power P_{tb} that t gets from b is proportional to the emitted power p_b by a factor $a_{tb} \in [0, 1]$, i.e. $P_{tb} = a_{tb} \cdot p_b$. The factor a_{tb} is an attenuation coefficient that summarizes the reduction in power experienced by a signal propagating from b to t [31]. The BSs whose signals are picked up by t can be subdivided into *useful* and *interfering*, depending on whether they contribute either to guarantee or to destroy the service. Given a cluster of BSs $C_t \subseteq B$ that cooperate to serve t, the BSs in C_t provide useful signals to t, while all the BSs that do not belong to C_t are interferers. For a given service cluster C_t, a UT t is considered served if the ratio of the sum of the useful powers to the sum of the interfering powers (*signal-to-interference ratio* or *SIR*) is above a threshold $\delta_t > 0$, which depends on the desired quality of service [31]:

$$\frac{\sum_{b \in C_t} a_{tb}\, p_b}{N + \sum_{b \in B \setminus C_t} a_{tb}\, p_b} \geq \delta_t. \tag{1}$$

Note that in the denominator, we highlight the presence of the system noise $N > 0$ among the interfering terms. By simple linear algebra operations, we can reorganize the ratio (1) into the following inequality, commonly called *SIR inequality*:

$$\sum_{b \in C_t} a_{tb}\, p_b - \delta_t \sum_{b \in B \setminus C_t} a_{tb}\, p_b \geq \delta_t\, N. \tag{2}$$

We now explain how to modify the basic SIR inequality (2), in order to build our optimization problem. As first step, we need to introduce the service assignment variable x_t into (2), as we want to decide which UTs $t \in T$ are served:

$$\sum_{b \in C_t} a_{tb}\, p_b - \delta_t \sum_{b \in B \setminus C_t} a_{tb}\, p_b + M\,(1 - x_t) \geq \delta_t\, N. \tag{3}$$

This inequality contains a large positive constant M (the so-called *big-M* coefficient [12,32]), that in combination with the binary variable x_t either activates or deactivates the constraint. It is straightforward to check that if $x_t = 1$, then (3) reduces to (2) and the SIR inequality *must be satisfied*. If instead $x_t = 0$ and M is sufficiently large (for example, $M = +\delta_t \sum_{b \in B \setminus C_t} a_{tb}\, P_{\max} + \delta_t\, N$), then (3) *becomes redundant*, as it is satisfied by any power configuration of the BSs. As second step, we need to introduce the cluster assignment variables y_{tb} into (3), as we want to decide which BSs $b \in B$ form the cluster C_t serving t:

$$\sum_{b \in B} a_{tb}\, p_b\, y_{tb} - \delta_t \sum_{b \in B} a_{tb}\, p_b\, (1 - y_{tb}) + M\,(1 - x_t) \geq \delta_t\, N. \tag{4}$$

In this case, we extend the summation of the inequalities to all the BSs $b \in B$ and the service cluster C_t is made up of the BSs b such that $y_{tb} = 1$. The BSs that are not in the cluster act as interferers and this is expressed by including the complement $1 - y_{tb}$ of the cluster variables into the interfering summation. By $p_b = \sum_{l \in L} P_l \cdot z_{bl}$ and by simply reorganizing the terms, we finally obtain:

$$(1 + \delta_h) \sum_{b \in B} \sum_{l \in L} a_{tb}\, P_l\, z_{bl}\, y_{tb} - \delta_t \sum_{b \in B} \sum_{l \in L} a_{tb}\, P_l\, z_{bl} + M\,(1 - x_t) \geq \delta_t\, N. \tag{5}$$

This last inequality constitutes the core of our model and is called *SIR constraint*. It is actually non-linear as it includes the product $z_{bl}\, y_{tb}$ of binary variables. However, this does not constitute an issue, as this is a special non-linearity that can be linearized in a standard way, by introducing a new binary variable $v_{tbl} = z_{bl}\, y_{tb}$ for each triple $(t, b, l) : t \in T, b \in B, l \in L$, and three new constraints: (c1) $v_{tbl} \leq z_{bl}$; (c2) $v_{tbl} \leq y_{tb}$, (c3) $v_{tbl} \geq z_{bl} + y_{tb} - 1$ [23]. If $z_{bl} = 0$ or $y_{tb} = 0$, then by any of (c1), (c2) we have $v_{tbl} = 0$. If instead $z_{bl} = y_{tb} = 1$, then by (c3) we have $v_{tbl} = 1$. After these considerations, we can finally state the following pure binary *linear* formulation for the SPCAP:

$$\max \sum_{t \in T} r_t\, x_t - \sum_{t \in T} c_t \left(\sum_{b \in B} y_{tb} - x_t \right) \tag{big-M SPCAP}$$

$$(1 + \delta_h) \sum_{b \in B} \sum_{l \in L} a_{tb}\, P_l\, v_{tbl} - \delta_t \sum_{b \in B} \sum_{l \in L} a_{tb}\, P_l\, z_{bl} + M\,(1 - x_t) \geq \delta_t\, N$$

$$t \in T, \tag{6}$$

$$\sum_{l \in L} z_{bl} \leq 1 \qquad\qquad\qquad\qquad b \in B, \qquad\qquad (7)$$

$$v_{tbl} \leq z_{bl} \qquad\qquad\qquad\qquad t \in T, b \in B, l \in L, \qquad (8)$$

$$v_{tbl} \leq y_{tb} \qquad\qquad\qquad\qquad t \in T, b \in B, l \in L, \qquad (9)$$

$$v_{tbl} \geq z_{bl} + y_{tb} - 1 \qquad\qquad\quad t \in T, b \in B, l \in L, \qquad (10)$$

$$x_t \in \{0,1\} \qquad\qquad\qquad\qquad t \in T, \qquad\qquad\qquad (11)$$

$$z_{bl} \in \{0,1\} \qquad\qquad\qquad\qquad b \in B, l \in L, \qquad\qquad (12)$$

$$y_{tb} \in \{0,1\} \qquad\qquad\qquad\qquad t \in T, b \in B, \qquad\qquad (13)$$

$$v_{tbl} \in \{0,1\} \qquad\qquad\qquad\qquad t \in T, b \in B, l \in L. \qquad (14)$$

The objective function maximizes the difference between the total revenue obtained by serving UTs (each served UT $t \in T$ grants a revenue $r_t > 0$) and the total cost incurred to establish cooperation between BSs (for each UT t, the cooperation cost $c_t > 0$ arises for each cooperating BS when at least two BSs are cooperating; no cost arises when the service is provided by a single BS). The *linear* SIR constraints (6) express the coverage conditions under cooperation. The constraints (7) ensure that each BS emits at a single power level, while the constraints (8)–(10) operate the linearization of the product $z_{bl}\, y_{tb}$. Finally, (11)–(14) define the decision variables of the problem.

Strengthening the Formulation for the PSCAP. The formulation (big-M SPCAP) constitutes a natural optimization model for the SPCAP. However, because of the presence of the big-M coefficients, it is known to be really weak, i.e. its linear relaxation produces very low quality bounds [32]. Moreover, as the fading coefficients typically vary in a very wide range, the coefficient matrix may be very ill-conditioned and this leads to heavy numerical instability in the solution process. As a consequence, the effectiveness of standard solution algorithms provided by state-of-the-art commercial solvers, such as IBM ILOG Cplex [25], may be greatly reduced and solutions may even contain coverage errors, as pointed out in several works (e.g., [8,12,13,15,26,29]).

In order to tackle these computational issues, we extend a very effective strengthening approach that we proposed in [12,13,16]: the basic idea is to exploit the *generalized upper bound (GUB)* constraints $\sum_{l \in L} v_{tbl} \leq 1$, implied by the other GUB constraints $\sum_{l \in L} z_{bl} \leq 1$ (as $v_{tbl} \leq z_{bl}$), to replace the SIR constraints (6) with a set of *GUB cover inequalities*. For an exhaustive introduction to the cover inequalities and to their GUB version, we refer the reader to the book [32] and to [35]. Here, we briefly recall the well-known main results about the general *cover inequalities*: a *knapsack constraint* $\sum_{j \in J} a_j x_j \leq b$ with a_j, $b \in \mathbb{R}_+$ and $x_j \in \{0,1\}$, $\forall j \in J$, can be replaced with a (in general exponential) number of cover inequalities $\sum_{j \in C} x_j \leq |C| - 1$, where C is a *cover*. A cover is a subset of indices $C \subseteq J$ such that the summation of the corresponding coefficients $a_j, j \in C$ violates the knapsack constraint, i.e. $\sum_{j \in C} a_j > b$. The cover inequalities thus represent combinations of binary variables x_j that cannot be activated at the same time (therefore, we can activate at most $|C| - 1$ variables in each cover C). The GUB cover inequalities are a stronger version of the simple cover inequalities, that can be defined when there are additional constraints of the form $\sum_{j \in K \subseteq J} x_j \leq 1$ (GUB constraints), that allows to activate at most one variable in the subset K.

By applying the definition of [35] and reasoning similarly to [12], we can define the general form of the GUB cover inequalities (GCIs) needed to replace the SIR constraint (6):

$$x_t + \sum_{i=1}^{|\Delta|} \sum_{l=1}^{\lambda_i} v_{tbl} + \sum_{i=1}^{|\Gamma|} \sum_{l=q_i}^{|L|} z_{bl} \leq |\Delta| + |\Gamma|, \tag{15}$$

with $t \in T$, $\Delta \subseteq B$, $\lambda = (\lambda_1, \ldots, \lambda_{|\Delta|}) \in L^{|\Delta|}$, $\Gamma \subseteq B \backslash \Delta$, $(q_1, \ldots, q_{|\Gamma|}) \in L^I(t, \Delta, \lambda, \Gamma)$, with $L^I(t, \Delta, \lambda, \Gamma) \subseteq L^{|\Gamma|}$ representing the subset of interfering levels of BSs in Γ that destroy the service of t provided by the cluster Δ of BSs, emitting with power levels $\lambda = (\lambda_1, \ldots, \lambda_{|\Delta|})$. Intuitively, for fixed UT, subset of serving BSs and subset of interfering BSs, a GCI is built by fixing a power setting of the serving BSs and defining a power setting of the interfering BSs that deny the coverage of the considered UT.

If we replace the big-M SIR constraints (6) with the GCIs (15) in the big-M formulation (big-M SPCAP), we obtain what we call a *Power-Indexed* formulation (PI-SPCAP). The *Power-Indexed* formulation (PI-SPCAP) does not contain big-M and fading coefficients and is very strong and completely stable. On the other hand, it generally contains an exponential number of constraints and must be solved through a typical *branch-and-cut* approach [32]: initially we define a starting formulation containing only a suitable subset of GCIs (15), then we insert additional GCIs if needed, through the solution of an auxiliary separation problem (we refer the reader to [12] for details about the separation of the GCIs of a Power-Indexed formulation). In our case, the starting formulation contains the GCIs:

$$x_t + \sum_{l=1}^{\lambda} v_{t\beta l} + \sum_{l=q}^{|L|} z_{bl} \leq 2. \tag{16}$$

Such GCIs are obtained by considering a relaxed version of the SIR constraints (6), which contain only a single-server and a single-interferer (i.e., there are no summations over multiple serving and interfering BSs). This relaxation comes from the practical observation that it is common to have a server and an interferer BSs that are sensibly stronger than all the other and coverage of the user just depends on the power configuration of them [12, 15]. Computational experience also shows that the relaxed SIR constraints provide a good approximation of (big-M SPCAP) [12]. The GCIs (16) of the relaxed SIR constraints can be used to define a relaxed Power-Indexed formulation denoted by (PI^0-SPCAP), that constitutes a very good starting point for a branch-and-cut algorithm used to solve (PI-SPCAP), as reported in [12].

Following the features of the improved ANT algorithm proposed in [28], in our ANT algorithm presented in the next section, we make use of two lower bounds for the SPCAP: (1) the one obtained by solving the linear relaxation of (PI-SPCAP), that we will denote by *PI-bound*; (2) the one obtained by solving the linear relaxation of (big-M SPCAP), strengthened with the GCIs of (PI^0-SPCAP), that we will denote by *BM-bound*; We remark that *PI-bound* is consistently better than *BM-bound*, as it comes from a stronger formulation. However, its computation takes more time than that of *BM-bound*, as it requires to separate additional GCIs (we recall that (PI-SPCAP) initially corresponds to (PI^0-SPCAP) and its solution generally requires to generate additional GCIs).

3   A Hybrid Exact-ACO Algorithm for the SPCAP

Ant Colony Optimization (ACO) is a metaheuristic approach to combinatorial optimization problems that was originally proposed by Dorigo and colleagues, in a series of works from the 1990s (e.g., [20]), and it was later extended to integer and continuous optimization problems (e.g., [19]). For an overview of the theory and applications of ACO, we refer the reader to [5,18,19]. It is now common knowledge that the algorithm draws its inspiration from the foraging behaviour of ants. The basic idea of an ACO algorithm is to define a loop where a number of feasible solutions are iteratively built in parallel, exploiting the information about the quality of solutions built in previous executions of the loop. The general structure of an algorithm can be depicted as follows:

UNTIL an arrest condition is not satisfied DO   (Gen-ACO)

1. Ant-based solution construction
2. Daemon actions
3. Pheromone trail update

We now describe the details of each phase presented above for our hybrid exact-ACO algorithm for the SPCAP. Our approach is hybrid since the canonical ACO step 1 is followed by a daemon action phase, where we exactly explore a large neighborhood of the generated feasible solutions, by exactly solving a Mixed-Integer Program, as explained in Subsect. 3.2.

3.1   Ant-Based Solution Construction

In step 1 of the loop, $m \in \mathbb{Z}_+$ computational agents called *ants* are defined and each ant iteratively builds a feasible solution for the optimization problem. At every iteration, the ant is in a so-called *state*, associated with a *partial solution* to the problem, and can complete the solution by selecting a *move* among a set of feasible ones. The move is probabilistically chosen on the basis of its associated pheromone trail values. For a more detailed description of the elements and actions of step 1, we refer the reader to the paper by Maniezzo [28]. The paper proposes an improved ANT algorithm (ANTS), which we take as reference for our work. We were particularly attracted by the improvements proposed in ANTS, as they are based on the attempt of better exploiting the precious information that comes from a *strong* Linear Programming formulations of the original discrete optimization problem. Moreover, ANTS also uses a reduced number of parameters and adopts mathematical operations of higher computational efficiency (e.g., multiplications instead of powers with real exponents).

Before describing the structure and the behaviour of our ants, we make some preliminary considerations. Our formulation for the SPCAP is based on four types of variables: (1) power variables z_{bl}; (2) cluster variables y_{tb}; (3) service variables x_t; (4) linearization variables v_{tbl}. Once that the power variables and the cluster variables are fixed: (i) the value of the linearization variables is immediately determined, because of constraints (8)–(10), and (ii) the objective function can be easily computed, as the served UTs can be identified by simply checking which SIR inequalities (6) are satisfied. As a consequence, in the ant-construction phase we can limit our attention to the cluster and service variables and we introduce the following two concepts of power and cluster states.

Definition 1. *Power state (PS): let $L_0 = L \cup \{0\}$ be the set of power levels plus the null power level. A* power state *represents the activation of a subset of BSs on some power level $l \in L_0$ and excludes that the same BS is activated on two power levels. Formally: $PS \subseteq B \times L_0 : \not\exists (b_1, l_1), (b_2, l_2) \in PS : b_1 = b_2$.*

We say that a power state PS is *complete* when it specifies the power configuration of every BS in B (thus $|PS| = |B|$). Otherwise the PS is called *partial* and we have $|PS| < |B|$. Furthermore, for a given power state PS, we denote by $B(PS)$ the subset of BSs whose power is fixed in PS (we call such BSs *configured*), i.e. $B(PS) = \{b \in B : \exists (b, l) \in PS\}$.

Definition 2. *Cluster state (CS): let $\bar{B} = B$ and let $T \times \bar{B}$ be the set of couples (t, b) representing the* non-*assignment of BS b to the cluster serving UT t. A* cluster state *represents the assignment or non-assignment of a subset of BSs to the clusters serving a subset of UTs and excludes that the same BS is at the same time assigned and non-assigned to the cluster of a UT. Formally: $CS \subseteq T \times B \cup T \times \bar{B} : \not\exists (t_1, b_1), (t_2, b_2) \in CS : b_1 \in B, b_2 \in \bar{B}$ and $t_1 = t_2, b_1 = b_2 = b$.*

We say that a cluster state CS is *complete* when it specifies the cluster assignment or non-assignment of every BS in B to every UT in T (thus $|CS| = |T||B|$). Otherwise the CS is called *partial* and we have $|CS| < |T||B|$). Moreover, for a given cluster state CS and UT t, we denote by $B_t(CS)$ the subset of BSs that are either assigned or not assigned to the service cluster of t in CS, i.e. $B_t(CS) = \{b \in B \cup \bar{B} : \exists (t, b) \in CS\}$.

In our ANT algorithm, we decided to first establish the value of the power variables and then the cluster variables. So an ant first passes through a sequence of partial power states, till a complete one is reached (power construction phase). Then it passes through a sequence of partial cluster states, till a complete one is reached (cluster construction phase). More formally, in the power phase, an ant moves from a partial power state PS_i to a partial power state PS_j such that:

$$PS_j = PS_i \cup \{(b, l)\} \text{ with } (b, l) \in B \times L_0 : b \notin B(PS_i).$$

Note that by the definition of power state, the added couple (b, l) may not contain a BS whose power is already fixed in a previous power state.

In the cluster phase, an ant moves from a partial cluster state CS_i to a partial cluster state CS_j such that:

$$CS_j = CS_i \cup \{(t, b)\} \quad \text{with } \{(t, b)\} \in T \times B \cup T \times \bar{B}.$$

Note that by the definition of cluster state, the added couple (t, b) may not contain a BS that has been already either assigned or not assigned to the same UT. Moreover, note that the definitions of power and cluster state can be immediately traced back to a sequence of fixing of the decision variables, thus relaxing the concept of state and reducing a move to the fixing of a decision variable j after the fixing of another decision variable i, as it is done in [28].

Every move adds a single new element to the partial solution. Once that the construction phases terminate, the value of the decision variables (z, y) is fully established and, as previously noted, we can immediately derive the value of the other variables (x, v), thus obtaining a complete feasible solution (x, z, y, v) for the SPCAP.

The probability that an ant k moves from a power (cluster) state i to a more complete power (cluster) state j, chosen among a set F of feasible power (cluster) states, is defined by the improved formula of [28]:

$$p_{ij}^k = \frac{\alpha \, \tau_{ij} + (1 - \alpha) \, \eta_{ij}}{\sum_{f \in F} \alpha \, \tau_{if} + (1 - \alpha) \, \eta_{if}},$$

where $\alpha \in [0, 1]$ is the parameter establishing the relative importance of trail and attractiveness. Of course, the probability of infeasible moves is set to zero. As discussed in [28], the trail values τ_{ij} and the attractiveness values η_{ij} should be provided by suitable lower bounds of the considered optimization problem. In our particular case: (1) τ_{ij} is derived from the values of the variables in the solution associated with the bound *PI-bound*, provided by the linear relaxation of the Power-Indexed formulation (PI-SPCAP) (see the next subsection for the specific setting of τ_{ij}); (2) η_{ij} is equal to the optimal solution of the linear relaxation of the big-M formulation (big-M SPCAP) strengthened with the GCIs of (PI^0-SPCAP) and including additional constraints to fix the value of the variables fixed in the considered state j. We denote the latter bound by *strongBM-bound* and we recall that this can be quickly computed and its computation becomes faster and faster as we move towards a complete state (the number of fixed variables indeed increases move after move).

As previously explained, once that an ant has finished its construction, we have a vector (z, y) that can be used to derive the value of the other variables (x, v) and define a complete feasible solution (x, z, y, v) for the SPCAP.

3.2   Daemon Actions: Relaxation Induced Neighborhood Search

We refine the quality of the feasible solutions found through the ant-construction phase by an *exact local search* in a *large neighborhood*, made for each feasible solution generated by the ants. Specifically, we adopt a modified *relaxation induced neighborhood search* (RINS) (see [11] for a detailed discussion of the method). The main steps of RINS are (1) defining a neighborhood by exploiting information about some continuous relaxation of the discrete optimization problem, and (2) exploring the neighborhood through a (Mixed) Integer Programming problem, that is optimally solved through an effective commercial solver.

Let S^{ANT} be a feasible solution to the SPCAP built by an ant and let S^{PI} be the optimal solution to the linear relaxation of (PI-SPCAP). Additionally, let S_j^{ANT}, S_j^{PI} denote the j-th component of the vectors. Our modified RINS *(mod-RINS)* solves a sub-problem of the big-M formulation (big-M SPCAP) strengthened with the GCIs of (PI^0-SPCAP) where:

1. we fix the variables whose value in S^{ANT} and S^{PI} differs of at most $\epsilon > 0$, (i.e., $S_j = 0$ if $S_j^{ANT} = 0 \cap S^{PI} \leq \epsilon$, $S_j = 1$ if $S_j^{ANT} = 1 \cap S^{PI} \geq 1 - \epsilon$;
2. set an objective cutoff based on the value of S^{ANT};
3. impose a solution time limit of T;

The time limit is set as the problem may be in general difficult to solve, so the exploration of the feasible set may need to be truncated. Note that in point 1 we generalize the fixing rule of RINS, in which $\epsilon = 0$.

3.3   Pheromone Trail Update

At the end of each construction phase t of the ants, the pheromone trails $\tau_{ij}(t-1)$ are updated according to the following improved formula (see [28] for a detailed discussion of its elements):

$$\tau_{ij}(t) = \tau_{ij}(t-1) + \sum_{k=1}^{m} \tau_{ij}^k \quad \text{with } \tau_{ij}^k = \tau_{ij}(0) \cdot \left(1 - \frac{z_{curr}^k - LB}{\bar{z} - LB}\right), \quad (17)$$

where, to set the values $\tau_{ij}(0)$ and LB, we solve the linear relaxation of (PI-SPCAP) and then we set $\tau_{ij}(0)$ equal to the values of the corresponding optimal decision variables and LB equal to the optimal value of the relaxation. Additionally, z_{curr}^k is the value of the solution built by ant k and $\bar{z}$ is the moving average of the values of the last ψ feasible solutions built. As noticed in [28], formula (17) substitutes a very sensible parameter, the pheromone evaporation factor, with the moving average ψ whose setting is much less critical.

The overall structure of our original hybrid exact-ACO algorithm is presented in Algorithm 1. The algorithm includes an outer loop repeated r times. At each execution of the loop, an inner loop defines m ants to build the solutions. Once that an ant finishes to build its solution, mod-RINS is applied in an attempt at finding an improvement. Pheromone trail updates are done at the end of each execution of the inner loop.

Algorithm 1. Hybrid exact-ACO for the SPCAP.

1. Compute the linear relaxation of (PI-SPCAP) and use it to initialize the values $\tau_{ij}(0)$.
2. FOR $t := 1$ TO r DO
   a) FOR $\mu := 1$ TO m DO
      i. build a complete power state;
      ii. build a complete cluster state;
      iii. derive a complete feasible solution to the SPCAP;
      iv. apply mod-RINS to the feasible solution.
      END FOR
   b) Update $\tau_{ij}(t)$ according to (17).
   END FOR

4 Computational Experiments

We tested the performance of our hybrid algorithm on a set of 15 realistic instances of increasing size, defined in collaboration with the Technical Strategy & Innovations Unit of British Telecom Italia (BT Italia). The experiments were made on a machine with a 1.80 GHz Intel Core 2 Duo processor and 2 GB of RAM and using the commercial solver IBM ILOG Cplex 11.1. All the instances refer to a WiMAX Network [2] and lead to the definition of very large and hard to solve (big-M SPCAP) formulations. Even when strengthened with the GCIs of (PI^0-SPCAP), (big-M SPCAP) continues to constitute a very hard problem and the simple identification of feasible solutions may be a hard task even for Cplex. In particular, for most instances it was not possible to find feasible solutions within one hour of computations and, when solutions were found, they were anyway of low value (up to the 35 % of covered UTs). Our heuristic algorithm was instead able to find good quality solutions.

After a series of preliminary tests, we found that a good setting of the parameters of the heuristic is: $\alpha = 0.5$ (balance between attractiveness and trail level), $m = |B|/2$ (number of ants equal to half the number of BSs), $\psi = m = |B|/2$ (width of the moving average equal to the number of ants), $\epsilon = 0.01$ (tolerance of fixing in mod-RINS), $T = 10$ min (time limit in mod-RINS). Moreover, the construction loop was executed 50 times. In Table 1, for each instance we report its ID and size and the number $|T^*|$ of covered UTs in the best solution found by mod-RINS (showing also the percentage coverage $Cov\%$) and in the corresponding ant solution. We also report the maximum size of a cluster in the best solution. The solutions found by the hybrid algorithm have a much higher value than those found by Cplex directly applied to (big-M SPCAP) and guarantee a good level of coverage ranging from 45 to 80 %. Moreover, we note that the execution of mod-RINS is able to increase the value of the ant solution from 5 to 13 %. Finally, it is interesting to note that the size of the clusters keeps in general low, presenting a maximum dimension of 5. We consider the overall performance highly satisfying, considering that our real aim is to use the solutions

Table 1. Experimental results

ID	$|T|$	$|B|$	$|T^*|$ (ACO)	$|T^*|$ (ACO+RINS)	Cov %	Max size cluster
I1	100	9	55	60	0.60	3
I2	100	12	62	67	0.67	2
I3	121	9	52	55	0.45	2
I4	121	15	72	80	0.66	4
I5	150	12	94	106	0.71	3
I6	150	15	96	106	0.71	3
I7	150	18	103	112	0.75	3
I8	169	12	80	87	0.51	2
I9	169	15	103	116	0.69	4
I10	169	18	121	133	0.79	2
I11	196	15	136	144	0.73	3
I12	196	21	140	156	0.80	5
I13	225	9	125	142	0.63	3
I14	225	15	142	149	0.66	3
I15	225	18	152	163	0.72	4

generated by the algorithm to favour a warm start in an exact cutting-plane algorithm applied to the Power-Indexed formulation (PI-SPCAP). Moreover, we are confident that refinements of the components of the heuristic and further tuning of the procedure can lead to the generation of solutions of higher quality.

5 Conclusions

Cooperative wireless networks have recently attracted a lot of attention, since cooperation among base stations may lead to remarkable increases in the capacity of a network and enhance the service experience of the users. Though base station cooperation has been extensively discussed from a theoretical and technical point of view, there is still a lack of effective optimization models and algorithms for its evaluation and implementation. To make a further step towards filling such gap, in this work we have presented a new model and solution algorithm for the problem of designing a cooperative wireless network. In particular, we have generalized the classical model for wireless network design, in order to include cluster definition and assignment. We have then showed how to strengthen the model, through the use of a special class of valid inequalities, the GUB cover inequalities, that eliminate all the sources of numerical problems. Finally, we have defined a hybrid heuristic based on the combination of ant colony optimization and relaxation induced neighborhood search, that exploits the important information provided by the relaxation of a strong formulation. Computational experiments on a set of realistic instances showed that our heuristic can find solutions of good quality, which could be used for a warm start in an exact branch-and-cut algorithm. Future work will consist in refining the components of the heuristic (for example, by better integrating the power and cluster state moves) and in integrating the heuristic with a branch-and-cut algorithm, in order to find solutions of higher value and whose quality is precisely assessed.

References

1. Amaldi, E., Capone, A., Malucelli, F., Mannino, C.: Optimization problems and models for planning cellular networks. In: Resende, M., Pardalos, P. (eds.) Handbook of Optimization in Telecommunication, pp. 917–939. Springer, Heidelberg (2006)
2. Andrews, J.G., Ghosh, A., Muhamed, R.: Fundamentals of WiMAX. Prentice Hall, Upper Saddle River (2007)
3. Atzori, L., D'Andreagiovanni, F., Mannino, C., Onali, T.: An algorithm for routing optimization in DiffServ-aware MPLS networks. DIS Technical report 4–2010, Sapienza Università di Roma, Roma (2010)
4. Bley, A., D'Andreagiovanni, F., Hanemann, A.: Robustness in communication networks: scenarios and mathematical approaches. In: Proceedings of the 12th ITG Symposium on Photonic Networks, pp. 1–8. VDE Verlag, Berlin (2011)
5. Blum, C., Puchinger, J., Raidl, G.R., Roli, A.: Hybrid metaheuristics in combinatorial optimization: A survey. Appl. Soft Comp. **11**(6), 4135–4151 (2011)
6. Büsing, C., D'Andreagiovanni, F.: New results about multi-band uncertainty in robust optimization. In: Klasing, R. (ed.) SEA 2012. LNCS, vol. 7276, pp. 63–74. Springer, Heidelberg (2012)
7. Büsing, C., D'Andreagiovanni, F.: A new theoretical framework for Robust Optimization under multi-band uncertainty. In: Helber, S. et al. (eds.) Operations Research Proceedings 2012, pp. 115–121. Springer, Heidelberg (2014)
8. Capone, A., Chen, L., Gualandi, S., Yuan, D.: A new computational approach for maximum link activation in wireless networks under the SINR model. IEEE Trans. Wireless Comm. **10**(5), 1368–1372 (2011)
9. Choi, Y.S., Kim, K.S., Kim, N.: The Displacement of Base Station in Mobile Communication with Genetic Approach. EURASIP J. Wireless Comm. Net. (2008). doi:10.1155/2008/580761
10. Classen, G., Koster, A.M.C.A., Schmeink, A.: A robust optimisation model and cutting planes for the planning of energy-efficient wireless networks. Comput. OR **40**(1), 80–90 (2013)
11. Danna, E., Rothberg, E., Le Pape, C.: Exploring relaxation induced neighborhoods to improve MIP solutions. Math. Program. **102**, 71–90 (2005)
12. D'Andreagiovanni, F., Mannino, C., Sassano, A.: GUB covers and power-indexed formulations for wireless network design. Manage. Sci. **59**(1), 142–156 (2013)
13. D'Andreagiovanni, F.: Pure 0–1 programming approaches to wireless network design. Ph.D. Thesis. 4OR-Q. J. Oper. Res. **10**(2), 211–212 (2012)
14. D'Andreagiovanni, F.: On improving the capacity of solving large-scale wireless network design problems by genetic algorithms. In: Chio, C., et al. (eds.) EvoApplications 2011, Part II. LNCS, vol. 6625, pp. 11–20. Springer, Heidelberg (2011)
15. D'Andreagiovanni, F., Mannino, C., Sassano, A.: Negative cycle separation in wireless network design. In: Pahl, J., Reiners, T., Voß, S. (eds.) INOC 2011. LNCS, vol. 6701, pp. 51–56. Springer, Heidelberg (2011)

16. D'Andreagiovanni, F., Mannino, C., Sassano, A.: A power-indexed formulation for wireless network design. DIS Technical report 14–2009, Sapienza Università di Roma, Roma (2009)

17. D'Andreagiovanni, F., Mannino, C.: A MILP formulation for WiMAX Network Planning. DIS DIS Technical report. 02–2008, Sapienza Università di Roma, Roma (2008)

18. Dorigo, M., Blum, C.: Ant colony optimization theory: a survey. Theoret. Comp. Sci. **344**(2–3), 243–278 (2005)

19. Dorigo, M., Di Caro, G., Gambardella, L.M.: Ant algorithms for discrete optimization. Artif. Life **5**(2), 137–172 (1999)

20. Dorigo, M., Maniezzo, V., Colorni, A.: Ant system: optimization by a colony of cooperating agents. IEEE Trans. Syst. Man Cybern. B **26**(1), 29–41 (1996)

21. Gesbert, D., Hanly, S., Huang, H., Shamai Shitz, S., Simeone, O., Yu, W.: Multi-Cell MIMO cooperative networks: a new look at interference. IEEE J. Sel. Areas Comm. **28**(9), 1380–1408 (2010)

22. Giovanidis, A., D'Andreagiovanni, F., Krolikowski, J., Tanzil, V.H., Brueck, S.: A 0–1 Program for Minimum Clustering in Downlink Base Station Cooperation. ZIB-Report 11–19, Zuse-Institut Berlin (2011)

23. Hammer, P.L., Rudeanu, S.: Boolean Methods in Operations Research and Related Areas. Springer, Heidelberg (1968)

24. Hong, M., Sun, R., Baligh, H., Luo, Z.: Joint Base Station Clustering and Beamformer Design for Partial Coordinated Transmission in Heterogeneous Networks. Technical report (2012). arXiv:1203.6390

25. Ibm, ILOG Cplex. http://www-01.ibm.com/software/integration/optimization/cplex-optimizer

26. Kennington, J., Olinick, E., Rajan, D.: Wireless Network Design: Optimization Models and Solution Procedures. Springer, Heidelberg (2010)

27. Koster, A.M.C.A., Helmberg, C., Bley, A., Grötschel, M., Bauschert, T.: BMBF Project ROBUKOM: robust communication networks. In: Proceedings of the ITG Workshop "Visions of Future Generation Networks" - EuroView2012, pp. 1–2. VDE Verlag, Berlin (2012)

28. Maniezzo, V.: Exact and approximate nondeterministic tree-search procedures for the quadratic assignment problem. INFORMS J. Comp. **11**(4), 358–369 (1999)

29. Mannino, C., Rossi, F., Smriglio, S.: The network packing problem in terrestrial broadcasting. Oper. Res. **54**(6), 611–626 (2006)

30. Montemanni, R., Smith, D.H., Allen, S.M.: An improved algorithm to determine lower bounds for the fixed spectrum frequency assignment problem. Europ. J. Oper. Res. **156**(3), 736–751 (2004)

31. Nawrocki, M., Aghvami, H., Dohler, M.: Understanding UMTS Radio Network Modelling, Planning and Automated Optimisation: Theory and Practice. John Wiley and Sons, Hoboken (2006)

32. Nehmauser, G., Wolsey, L.: Integer and Combinatorial Optimization. John Wiley and Sons, Hoboken (1988)

33. Sesia, S., Toufik, I., Baker, M.: LTE-The UMTS Long Term Evolution: From Theory to Practice. John Wiley and Sons, Hoboken (2009)

34. Weber, R., Garavaglia, A., Schulist, M., Brueck, S., Dekorsy, A.: Self-Organizing Adaptive Clustering for Cooperative Multipoint Transmission. In: Proceedings of the IEEE VTC2011-Spring, pp. 1–5 (2011)

35. Wolsey, L.: Valid inequalities for 0–1 knapsacks and mips with generalised upper bound constraints. Discrete Appl. Math. **29**(2–3), 251–261 (1990)

Proposal and Evaluation of Attractor Perturbation-Based Rate Control for Stable End-to-End Delay

Midori Waki$^{(\boxtimes)}$, Naoki Wakamiya, and Masayuki Murata

Graduate School of Information Science and Technology,
Osaka University, 1-5 Yamadaoka, Suita 565-0871, Japan
{m-waki,wakamiya,murata}@ist.osaka-u.ac.jp

Abstract. Due to the best-effort nature of the Internet, delay and delay jitter observed by a session always fluctuate, even if it generates CBR (Constant Bit Rate) traffic. Buffering at a host and packet scheduling at routers would solve the problem to some extent, but they require prior knowledge of delay variation and traffic characteristics. In this paper, we propose a novel rate control mechanism to achieve and maintain the target delay in the dynamically changing environment. Our proposal does not filter or conceal fluctuation, but it exploits fluctuation to accomplish the goal by using the attractor perturbation model derived from biological behavior. Through simulation experiments, we confirmed that our proposal could achieve and maintain the target delay when background traffic changed.

Keywords: Attractor perturbation · Rate control · End-to-end delay

1 Introduction

When there are multiple sessions sharing the same physical network resources, delay, delay jitter, and packet loss observed by a session always fluctuate, regardless of adopted protocol or characteristics of generated traffic. Since the origin of fluctuation includes changes in the number of sessions and the amount of traffic, the shadowing and fading of a wireless channel, rerouting of paths and others, that cannot be predicted or controlled by an individual session, researchers had made an effort to suppress fluctuations especially for delay-sensitive applications such as IPTV (Internet Protocol TeleVision) and video streaming.

Delay fluctuation is generally managed by a playout buffer at a receiver [1,2]. A playout buffer defers video playout so that it can deposit the sufficient number of packets at the beginning and then provides a video player with buffered packets. As such, as far as packets arrive at a receiver before a buffer becomes empty, a video can be presented to a user without interruptions. However, delay and delay jitter are not predictable. Therefore, it is very likely that a buffer runs out of packets and a user experiences freezes. Increasing the number of packets

G.A. Di Caro and G. Theraulaz (Eds.): BIONETICS 2012, LNICST 134, pp. 18–32, 2014.
DOI: 10.1007/978-3-319-06944-9_2, © Institute for Computer Sciences, Social Informatics and Telecommunications Engineering 2014

to buffer merely degrades the interactivity and timeliness of an application. For delay-sensitive applications, researchers proposed methods to control and reduce delay jitter by developing an intelligent packet scheduling algorithm at routers [3–7] and by multipath routing [8]. In [3], comparative analysis shows that packet scheduling at routers can reduce delay jitter even when buffering at a receiver cannot prevent freezes. However, it requires all intermediate routers from a server to a receiver to be equipped with the algorithm. On the contrary, a multipath routing method still relies on prior knowledge of delay variation, which is unpredictable in general. As a mechanism adopted at end systems, many rate control algorithms have been studied [9,10]. They infer the network state by observing, for example, delay, delay jitter, and packet loss and regulate the sending rate to avoid network congestion. Although they can reduce the packet loss probability, they do not take into account the delay sensitivity of interactive applications.

As long as the network condition, such as the degree of congestion, can easily be predicted or estimated, it is trivial to control delay, delay jitter, and packet loss. However, the ever-increasing size, complexity, and dynamics of an information network prevent a control mechanism revealing the network condition even with active and aggressive probing. Going back to the simplest paradigm, given a complex system, what an end system can do is to apply a force and see how it reacts. Only if there exists the clear relationship between them, one can obtain the desired result by putting the appropriate force to a system. An answer can be found in biology, which has the long history of investigating and understanding complex systems, i.e. living organisms. The relationship is modeled by a mathematical expression, called an attractor perturbation model [11,12]. It is derived from the relationship between fluctuations inherent in a biological system and its response against an external force. Biological systems are always exposed to internal and external fluctuation or noise caused by, for example, thermal fluctuation and phenotypic fluctuation. As a result, size, metabolic concentrations, and gene expression differ among cells cultured in the same medium and individuals are all different. Furthermore, gene expression of a cell dynamically changes to adapt to the surrounding conditions such as temperature, pH, and concentrations of chemical substances. Therefore, a cell is not always the same. It is considered that such fluctuation or diversity is a source of flexibility and adaptability of biological systems to environmental changes.

The attractor perturbation model explains how biological systems respond to environmental changes, which act as a force to trigger biological responses. Based on the model, given a change in the external force, the average of a measurable variable, such as the concentration of metabolic substances and the number of cells, shifts by the amount in proportional to the degree that a biological system fluctuates, i.e. the variance of the measurable variable. That is, more a biological system fluctuates, more it responds to the environmental change and alters its behavior.

Fluctuation is intrinsic to an information network as well. Then, the attractor perturbation model may hold in an information network and we can develop a control mechanism based on the model. When we regard a network as a biological system and injected traffic as an external force imposed on a system, we can estimate how a network responds to a change in the injected traffic. More specifically, by adopting the end-to-end delay as a measurable variable of the attractor perturbation model, we can derive the appropriate amount of increase or decrease of the sending rate to achieve the desired end-to-end delay from the observed variance of delay. For example, assume that the measured end-to-end delay is larger than the desired delay. When the variance is large, it is enough to slightly decrease the sending rate to push down the delay to the desired level. On the contrary, aggressive rate control is required in a network with small fluctuation, which implies that a network is stable. With such a control mechanism based on the attractor perturbation model, efficient and effective rate adaptation can be accomplished without detailed information about a network system or tailored facilities.

The remainder of this paper is organized as follows. In Sect. 2, we briefly introduce the attractor perturbation model. In Sect. 3, we verify the attractor perturbation principle in an information network. In Sect. 4, we propose a novel rate control mechanism to achieve the stable end-to-end delay based on the attractor perturbation model. Then we conduct simulation experiments and evaluate the proposal in Sect. 5. Finally, Sect. 6 concludes the paper.

2   Attractor Perturbation Model

The attractor perturbation model represents the general relationship between inherent fluctuation and response in biological systems [11]. The following is a mathematical expression of the attractor perturbation model.

$$\langle w \rangle_{a+\Delta a} - \langle w \rangle_a = b \Delta a \sigma_a^2 \tag{1}$$

where $\langle w \rangle_a$ and σ_a^2 are the average and variance of measurable quantity w, e.g. protein concentration, under the force a, e.g. genetic mutation, respectively. Δa is a small change in the force and b is a constant coefficient. The equation indicates that a shift in the average of a measurable variable against a change in the force is proportional to the variance of the measurable variable. From Eq. (1), one can derive the following equation.

$$\langle w \rangle_{a+\Delta a} = \langle w \rangle_a + b \Delta a \sigma_a^2 \tag{2}$$

Equation (2) gives an estimate of an effect of increasing the force a to $a + \Delta a$ when the current average is $\langle w \rangle_a$ and the variance is σ_a^2. From a viewpoint of control of the force, we can consider the following equation.

$$\Delta a = \frac{\langle w \rangle_{a+\Delta a} - \langle w \rangle_a}{b \sigma_a^2} \tag{3}$$

The equation gives the amount of change in the force, i.e. Δa, or the amount of force, i.e. $a + \Delta a$, to obtain the shifted average $\langle w \rangle_{a+\Delta a}$ from the current conditions $\langle w \rangle_a$, a, and σ_a^2. This brings a basic idea of our proposal.

3 Attractor Perturbation Concept in Network

In this section, we verify that the attractor perturbation principle holds for a network system. We regard the end-to-end delay as the variable w and the rate of injected traffic as the external force a, and confirm the linear relationship between fluctuation and response, i.e. the variance of delay and the shift in the average delay.

3.1 Analytical Verification of Attractor Perturbation Concept in M/D/1 Model

First in this section, we prove the attractor perturbation principle in an M/D/1 queuing system assuming Poisson arrival of fixed-length packets. In the following, λ is the arrival rate, μ is the service rate, and $\rho = \lambda/\mu < 1$ is the traffic intensity or the load.

In [13], the author analyzes the mean time spent in an M/G/1 system, where the service time has a general distribution with mean $E(X)$. The first and second moment of time spent in an M/G/1 system are denoted by $E(T)$ and $E(T^2)$.

$$E(T) = \frac{\lambda}{2(1-\rho)}E(X^2) + E(X) \tag{4}$$

$$E(T^2) = \frac{\lambda}{3(1-\rho)} + \frac{\lambda^2}{2(1-\rho)^2}\{E(X^2)\}^2 + \frac{E(X^2)}{1-\rho} \tag{5}$$

Since the service time in an M/D/1 is constant, by substituting $E(X) = 1/\mu$ and $E(X^2) = 1/\mu^2$ into the above equations, we can obtain the mean $d(\lambda)$ and variance $\sigma^2(\lambda)$ of time spent in an M/D/1 system as functions of the arrival rate λ.

$$d(\lambda) = \frac{2\mu - \lambda}{2\mu(\mu - \lambda)} \tag{6}$$

$$\begin{aligned}\sigma^2(\lambda) &= E(T^2) - \{E(T)\}^2 \\ &= \frac{\lambda(4\mu - \lambda)}{12\mu^2(\mu - \lambda)^2}\end{aligned} \tag{7}$$

By differentiating $d(\lambda)$ with respect to λ we obtain

$$d'(\lambda) = \frac{1}{2(\mu - \lambda)^2} \tag{8}$$

Assuming that $\Delta\lambda$ is small, we further obtain the following relationship.

$$\frac{d(\lambda + \Delta\lambda) - d(\lambda)}{\Delta\lambda} = d'(\lambda) \tag{9}$$

$$d(\lambda + \Delta\lambda) - d(\lambda) = d'(\lambda)\Delta\lambda$$
$$= \frac{1}{2(\mu - \lambda)^2}\Delta\lambda$$
$$= \frac{6}{\rho(4 - \rho)}\sigma^2(\lambda)\Delta\lambda \tag{10}$$
$$= b(\rho)\sigma^2(\lambda)\Delta\lambda \tag{11}$$

Therefore, the shift in the mean time spent in an M/D/1 queueing system is given as a product of the variance $\sigma^2(\lambda)$, the change $\Delta\lambda$ of arrival rate, and the coefficient $b(\rho)$. The coefficient $b(\rho)$ is depicted in Fig. 1. Whereas $b(\rho)$ exponentially decreases in the region of $\rho < 0.5$, it can be represented by a constant in the region of $\rho \geq 0.5$. We consider that rate adaptation is necessary especially when a network is moderately or highly loaded. Therefore, we can conclude that the attractor perturbation model is applicable to rate control in a moderately congested network system. In the next section, we verify the attractor perturbation concept in a packet-based network by simulation experiments using ns-2 [14].

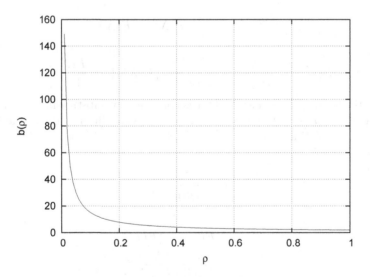

Fig. 1. variation of $b(\rho)$

3.2   Simulation-Based Verification of Linearity Between Fluctuation and Response

Figure 2 illustrates topology that we used for simulation. The dumbbell network models a bottleneck link of a network of arbitrary topology, which affects the end-to-end delay the most on a path. Two senders S_1 and S_2 are connected with two receivers D_1 and D_2, respectively, through routers E_0 and E_1. All links are full-duplex. The bandwidth and the propagation delay of a link between routers E_0 and E_1 are 15 Mbps and 5 ms, respectively. Those of the other links are 1 Gbps and 1 ms.

A drop-tail FIFO buffer with the capacity of 1000 packets is deployed on each router. A CBR session called "session 1" is established between nodes S_1 and D_1. We observe the one-way end-to-end delay on session 1 while changing the sending rate of UDP datagrams of a 1000-bytes payload. As background traffic, another UDP session, where the inter-arrival time of datagrams follows the exponential distribution and the payload size of a datagram is 1000 bytes, is set between nodes S_2 and D_2. It is called "session 2".

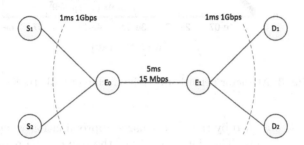

Fig. 2. Network topology used in simulation experiments

We observe the average $\langle w \rangle_a$ and variance σ_a^2 of one-way end-to-end delay of session 1 at the sending rate a Mbps. We prepared 10 traffic patterns of session 2 whose sending rate is 9 Mbps. For each of the pattern, we conducted 44 simulation experiments by increasing the sending rate a from 0.1 Mbps to 4.5 Mbps by 0.1 Mbps, i.e. $\Delta a = 0.1$. Then, from averages and variance obtained from 440 simulation experiments, we derive 430 pairs of σ_a^2 and $\langle w \rangle_{a+0.1} - \langle w \rangle_a$, i.e. $\sigma_{1.0}^2$ and $\langle w \rangle_{1.1} - \langle w \rangle_{1.0}$.

If the attractor perturbation principle holds, there exists the linear relationship between $\Delta a \cdot \sigma_a^2$ and $\langle w \rangle_{a+\Delta a} - \langle w \rangle_a$ as Eq. (1) indicates. 430 pairs of $0.1\sigma_a^2$ and $\langle w \rangle_{a+0.1} - \langle w \rangle_a$ are plotted on Fig. 3 as crosses. The figure shows the positive correlation between $0.1\sigma_a^2$ and $\langle w \rangle_{a+0.1} - \langle w \rangle_a$ and we can confirm the attractor perturbation principle in a packet-based network. When the sending rate of CBR session is low, there is little chance for packets to experience buffering at routers. As a result, the variance becomes small and the small increase of sending rate does not affect the delay much. Therefore, when the variance is small, the shift in the average delay becomes small as well. On the contrary, as the sending rate increases, the number of packets buffered at routers begins to fluctuate. It leads to both of the larger delay and the variance. Consequently, we observe the linear relationship between the variance and the shift in delay.

The proportional constant of the relationship between $0.1\sigma_a^2$ and $\langle w \rangle_{a+0.1} - \langle w \rangle_a$ corresponds to the coefficient b of Eq. (1). In Fig. 3, we show an approximate

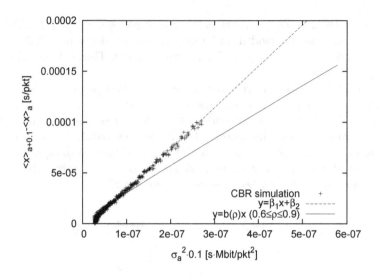

Fig. 3. Attractor perturbation relationship of CBR traffic

line $y = \beta_1 x + \beta_2$ obtained by the least squares approximation where x is $0.1\sigma_a^2$ and y is $\langle w \rangle_{a+0.1} - \langle w \rangle_a$. The slope, i.e. β_1, of the line can be regarded as the coefficient b, and its value is 407.63. The load ρ at the variance σ_a^2 is calculated by λ/μ, where μ is the service rate of the bottleneck link and λ is the arrival rate when the variance is σ_a^2. Therefore, ρ depends on σ_a^2 and the x-axis can be mapped to ρ. In Fig. 3, $y = b(\rho)x$ in the range of $0.6 < \rho < 0.9$ is depicted. To compare the analytical result of an M/D/1 system discussed in the previous section, we convert $b(\rho)$ to $\frac{750}{\rho(4-\rho)}$ by $\Delta\lambda = \frac{\Delta a \times 10^6}{8000}$ in Eq. (10). Although it is not a linear function due to the variation of ρ, the slope $b(\rho)$ is about 300 on average in the range of $0.6 < \rho < 0.9$. As shown in Fig. 3, there is a difference in slope between the analytical result and the simulation result. For the same variance, the shift $\langle w \rangle_{a+0.1} - \langle w \rangle_a$ is larger in the simulation than in the analysis. Given the variance σ_a^2, the load on a network in the case of the analysis, which can be derived from Eq. (7), is smaller than that of the simulation, which can be derived as $(a+9)/15$ considering that the amount of background traffic is 9 Mbps and the capacity of the bottleneck link is 15 Mbps. In general, when the sending rate increases, the end-to-end delay becomes larger in a congested network than in an unloaded network. Consequently, the growth rate or the slope is larger in the simulation than in the analysis. In Sect. 5, we used three alternatives of coefficient b, that is, 407.63, 300, and $b(\rho)$, to evaluate its influence.

4   Attractor Perturbation-Based Rate Control Mechanism

In this section, we propose a novel rate control mechanism based on the attractor perturbation model. We regard the delay as the measurable variable x and the sending rate as the force a and derive the appropriate sending rate to accomplish the target delay under the fluctuating environment.

We consider an application which communicates at least for several minutes. An application specifies the target one-way delay T s, the maximum sending rate a_{max} Mbps, and the minimum sending rate a_{min} Mbps. RTP/UDP and RTCP/UDP are employed and the sending rate is adjusted by adapting a transmission interval of RTP packets. Figure 4 illustrates how packets are exchanged and the sending rate is adjusted.

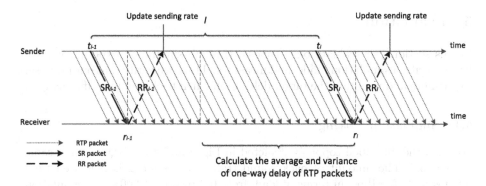

Fig. 4. Outline of proposal

At the beginning of a session, a sender sends RTP packets at the minimum rate a_{min} Mbps. Besides this, to obtain average delay d_i and variance v_i^2 a sender sends Sender Report (SR) packets at regular intervals of I s. The i-th SR packet emitted at time t_i s carries the information t_{i-1}, i.e. the instance when the $(i-1)$-th SR packet was sent, in its header. A receiver sends back a Receiver Report (RR) packet in response to a SR packet.

Now consider that a receiver receives the i-th SR packet at r_i s. It calculates the average d_{i-1} and variance v_{i-1}^2 of one-way delay of RTP packets received from $t_{i-1} + 2(r_i - t_i)$ s to r_i s (see Fig. 4). Since packets received from r_{i-1} s to $t_{i-1} + 2(r_i - t_i)$ s are sent before the rate adaptation initiated by reception of the $(i-1)$-th RR packet by a sender, they are excluded from the calculation. Then, the receiver sends a RR packet carrying the derived average and variance in an extended header.

On receiving the RR packet, the sender first calculates the amount Δa Mbps of rate change by substituting the received values, the target delay T, and the coefficient b to the following equation.

$$\Delta a = \frac{T - d_{i-1}}{bv_{i-1}^2} \qquad (12)$$

Next, the sender updates the sending rate to a_{new} Mbps, which is derived from the following equation.

$$a_{new} = \min\{a_{max}, \max(a_{min}, a + \Delta a)\} \qquad (13)$$

If a sender does not receive any of the $(i - n)$-th RR packets ($n \in 1, 2, 3$) by $t_i + 1 + I$ s, i.e. an instance to send the i-th SR packet, it considers that a network is considerably congested. Then, the sender reduces the sending rate by half and quits sending the i-th SR packet at $t_i + 1 + I$ s. After additional I s, if the sender receives any RR packets until then, it sends the i-th SR packet carrying $t_i - 1 + I$ s in an extended header to a receiver. On receiving the SR packet, the receiver calculates the average and variance of delay of RTP packets received from $t_{i-1} + I + 2(r_i - t_i)$ s to r_i s and sends a RR packet to the sender.

5   Evaluation

In this section, we evaluate our proposal through simulation experiments. We first describe a simulation model and measures. Then, we verify that our proposal can achieve and maintain the target delay even when background traffic changes.

5.1   Simulation Setting

We used the dumbbell topology depicted in Fig. 2 and set a UDP session same as Sect. 3.2. The amount of background traffic on session 2 is increased from 9 Mbps to 10.5 Mbps, in terms of load, from 0.6 to 0.7, at 200 s in a simulation run of 400 s. We employ our proposal on session 1 established between nodes S_1 and D_1. The size of a RTP packet including RTP, UDP, and IP headers is set at 1000 bytes. The sizes of a SR packet and a RR packet including an IP header are 64 and 72 bytes, respectively. The maximum sending rate a_{max} and the minimum sending rate a_{min} of our proposal are 15.0 Mbps and 0.1 Mbps, respectively. The interval I of SR packets is 10 s. The target delay is set at 8.2 ms, which is the one-way delay observed in the simulation experiments of the case of $\rho = 0.8$ in Sect. 3.2. Parameters used in evaluations are summarized in Table 1.

In order to evaluate the influence of the coefficient b, we conduct simulation experiments with $b = 300, 407.63$, and function $b(\rho)$. $b(\rho)$ enables dynamic adaptation of b with respect to the load condition. In the case of $b(\rho)$, we assume that a sender node can know the current load ρ of a network to derive the appropriate rate change Δa, whereas it is not possible to have the accurate and up-to-date information about the load condition of a network in an actual situation. More specifically, at t_i s, when a sender sends the i-th SR packet, the average load ρ_i on the bottleneck link between t_{i-1} s and t_i s is given and substituted into $b(\rho_i)$.

For comparison purposes, we additionally conduct simulation experiments for the cases of CBR traffic. In those cases, session 1 generates CBR traffic of 3 Mbps or 0.8 Mbps using RTP and RTCP. Note that a pair of SR and RR packets is sent every 10 s, but it is not used for rate control. We denote a case of CBR traffic with sending rate of 3 Mbps as CBR 3 Mbps and that of 0.8 Mbps as CBR 0.8 Mbps.

Table 1. Parameter setting

Parameter	Value
a_{min}	0.1 [Mbps]
a_{max}	15 [Mbps]
Interval I of SR packets	10 [s]
T	8.2 [ms]
b	300, 407.63

5.2 Evaluation Measures

To evaluate how our proposal achieves and maintains the target delay, we introduce the mean square error, the coefficient of variation, and the delay jitter defined in the following. We consider the first control interval after the initial transient state as the 0-th interval.

Mean Square Error. We evaluate the closeness to the target delay by the mean square error. First, we calculate the average delay T_i of successfully received RTP packets that are sent in the i-th control interval from t_i s to t_{i+1} s. Note that T_i is not equal to d_i, which is the average delay defined in Sect. 4. Then, we obtain the mean square error M as follows.

$$M = \frac{1}{n+1} \sum_{i=0}^{n} (T_i - T)^2 \tag{14}$$

Here T is the target delay, n is the number of SR packets sent in the whole simulation time. Therefore, T_n is the average delay of RTP packets that are sent from t_n s to the end of the simulation. A small M means that the average delay is close to the target delay in most of cases.

Coefficient of Variation. We evaluate the stability of the average delay by the coefficient of variation. We calculate the mean $\bar{T}$ and the standard deviation σ^2 of the average delay in the simulation as below.

$$\bar{T} = \frac{1}{j+1} \sum_{i=0}^{n} T_i \tag{15}$$

$$\sigma^2 = \sqrt{\frac{1}{n+1} \sum_{i=0}^{n} (T_i - \bar{T})^2} \tag{16}$$

Then we obtain the coefficient C of variation as follows.

$$C = \frac{\sigma^2}{\bar{T}} \tag{17}$$

A small C means that the average delay is kept constant and stable.

Delay Jitter. We define the delay jitter J as follows.

$$J = \max_{0 \leq i \leq n} \{|T_i - T|\} \tag{18}$$

The delay jitter is the maximum difference between the target delay T and the average delay T_i.

5.3   Evaluation Results

First we show an example of temporal variations in Figs. 5 and 6. In Fig. 5, variations of average delay T_i against the simulation time are depicted. In Fig. 6, variations of averaged sending rate per control interval are depicted. All results in the figures are obtained from simulation experiments with the identical background traffic pattern.

As shown in Fig. 5, CBR 3.0 Mbps results in the average delay close to the target delay at the beginning, but the delay becomes larger after the increase of background traffic. On the contrary, the average delay of CBR 0.8 Mbps is as low as the target delay from 200 s, whereas it is smaller than the target delay in the first half on the simulation run. Regarding our proposal, independently of the setting of coefficient b, the average end-to-end delay stays close to the target delay except for the period right after the sudden load increase. In the case of $b = 300$ for example, a sender node tries to decrease the sending rate on reception of a RR packet from a receiver node at 201 s. However, the decrease is only 0.33 Mbps at that time as shown in Fig. 6. It is because the delay and variance informed by the RR packet are derived from RTP packets sent before the load increase. At the next timing of rate control at 211 s, delay and variance have grown much

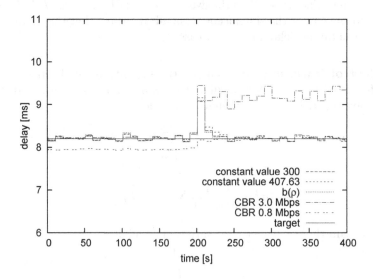

Fig. 5. Comparison of average delay

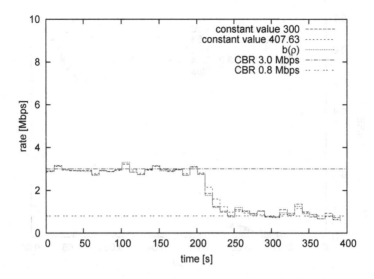

Fig. 6. Comparison of sending rate

to 9.07 ms and 3.23 ms^2 respectively . Then, the amount of decrease derived at the sender node becomes 0.90 Mbps. As a result of drastic rate reduction, the obtained end-to-end delay approaches the target delay again. The instantaneous increase of delay is basically unavoidable, but the duration can be shorten by a shorter control interval, that is, frequent rate control. However, too short control interval decreases the accuracy of variance derivation as a statistic and a sender node cannot precisely capture the fluctuation of a network. We should emphasize here that setting of constant value b does not affect the performance of our proposal very much. It suggests that the prior knowledge or parameter tuning, such that we did in Sect. 3.2, is not necessary.

Figure 7 summarizes results of all simulation experiments conducted 30 times for each settings. Figure 7(a,b) show the relationship between the coefficient of variation and the mean square error, and that between the coefficient of variation and the delay jitter, respectively. In both figures, the closer the point is to the origin, the more stable the end-to-end delay is around the target delay. Figure 7 shows that points of our proposal overlap with each other independently setting of the coefficient b and they are in the lower left region. MSE of our proposal is smaller than that of CBR 0.8 Mbps and much smaller than that of CBR 3.0 Mbps. On the other hand, our proposal results in larger coefficient of variation in some cases and larger delay jitter in all cases than CBR 0.8 Mbps. A reason is that the average delay does not change much before and after the load increase due to the low sending rate with CBR 0.8 Mbps. In contrast, our proposal suffers from the instantaneous increase of delay after the load increase. It makes the delay jitter larger than that of CBR 0.8 Mbps and affects the coefficient of variation as well. From a practical view point, the delay jitter of as much as 1.2 ms on a session of the propagation delay of 15 ms is small enough.

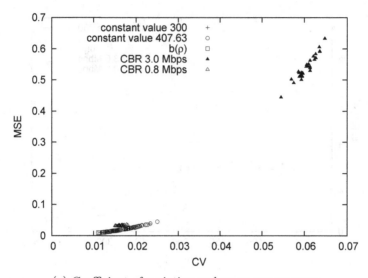

(a) Coefficient of variation and mean square error

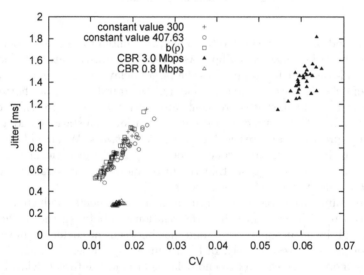

(b) Coefficient of variation and delay jitter

Fig. 7. Performance comparisons

In summary, we can conclude that our proposal can accomplish the stable end-to-end delay facing to the sudden load increase except for the instantaneous growth of delay right after the increase. We further showed that the setting of coefficient b did not influence rate control very much, which supports our motivation not to rely on the prior or detailed knowledge about characteristics of a network. That is, our proposal is insensitive to parameter setting as can be seen in the flexibility and robustness of biological systems.

6   Conclusion and Future Work

In this paper, as an example of application of the attractor perturbation model, we propose a novel rate control mechanism to achieve and maintain the target delay in the dynamically changing environment. We first proved that the attractor perturbation principle held in a packet-based network as well as a general M/D/1 queuing system. Next through simulation experiments, we confirmed that our proposal could achieve the goal and more interestingly the setting of coefficient b did not influence the performance of proposal very much.

As future work, we are going to conduct further evaluation to verify the insensitivity of our proposal to characteristics of a network including the size, topology, and competing sessions and their protocols. Furthermore, we plan the comparison with other non-bio-inspired mechanisms for delay jitter suppression. Some mechanisms such as a playout buffer can be incorporated with our proposal. From a view point of the attractor perturbation principle, the behavior of other incorporated mechanism is another origin of fluctuation of a network system. As such, it only changes the variance of end-to-end delay observed by our rate control mechanism. Then, we can expect that our proposal can work well without any tuning, modification, or extension.

Acknowledgement. This research was supported in part by Grand-in-Aid for Scientific Research (B) 22300023 of the Ministry of Education, Culture, Sports, Science and Technology, Japan.

References

1. Atzori, L., Lobina, M.: Playout buffering in IP telephony: a survey discussing problems and approaches. IEEE Commun. Surv. Tutor. **8**, 36–46 (2006)
2. Fujimoto, K., Ata, S., Murata, M.: Adaptive playout buffer algorithm for enhancing perceived quality of streaming applications. Telecommun. Syst. **25**, 259–271 (2004)
3. Kadur, S., Golshani, F., Millard, B.: Delay-jitter control in multimedia applications. Multimedia Syst. **4**, 30–39 (1996)
4. Verma, D., Zhang, H., Ferrari, D.: Delay jitter control for real-time communication in a packet switching network. In: Proceedings of IEEE TRICOMM'91, pp. 35–43, April 1991
5. Mansour, Y., Patt-Shamir, B.: Jitter control in QoS networks. IEEE/ACM Trans. Network. **9**, 492–502 (2001)
6. Pedrasa, J., Festin, C.: Value-based utility for jitter management. In: Proceedings of IEEE Region 10 Conference, pp. 1–5, November 2006
7. Hay, D., Scalosub, G.: Jitter regulation for multiple streams. IEEE/ACM Trans. Algorithms **6**, 1–19 (2009)
8. Okuyama, T., Yasukawa, K., Yamaoka, K.: Proposal of multipath routing method focusing on reducing delay jitter. In: Proceedings of IEEE Pacific Rim Conference, pp. 296–299, August 2005

9. Busse, I., Deffner, B., Schulzrinne, H.: Dynamic QoS control of multimedia applications based on RTP. Comput. Commun. **19**(1), 49–58 (1996)
10. Sun, Y., Tsou, F., Chen, M.: Predictive flow control for TCP-friendly end-to-end real-time video on the internet. Comput. Commun. **25**, 1230–1242 (2002)
11. Sato, K., Ito, Y., Yomo, T., Kaneko, K.: On the relation between fluctuation and response in biological systems. Nat. Acad. Sci. **100**, 14086–14090 (2003)
12. Leibnitz, K., Murata, M.: Attractor selection and perturbation for robust networks in fluctuating environments. IEEE Netw. **24**, 14–18 (2010)
13. Shanthikumar, J.: On reducing time spent in M/G/1 systems. Eur. J. Oper. Res. **9**, 286–294 (1982)
14. The network simulator ns-2. http://www.isi.edu/nsnam/ns/

The Reactive ASR-FA – An Ant Routing Algorithm That Detects Changes in the Network by Employing Statistical Delay Models

Malgorzata Gadomska-Kudelska[1]([✉]), Andrzej Pacut[1,2], and Michal Kudelski[2]

[1] Institute of Control and Computation Engineering,
Warsaw University of Technology, 00-665 Warsaw, Poland
mgadomsk@elka.pw.edu.pl
[2] NASK Research and Academic Computer Network, 02-796 Warsaw, Poland

Abstract. This paper introduces the Reactive ASR-FA algorithm which is a novel ant routing algorithm that utilizes statistical models of packet delay to detect changes in the network conditions. The algorithm is able to quickly react to various load level changes by temporarily modifying the learning parameter's settings. We show in a set of experiments that using the Reactive ASR-FA significantly speeds up the adaptation process of ant routing algorithms and assures lower values of the mean packet delay. It can be also employed in DoS and DDoS attacks detection.

Keywords: Adaptive routing · Ant algorithms · Computer networks

1 Introduction

Many similarities can be found between swarms of insects and parallel, distributed information systems that are controlled in a decentralized manner. An example of such system is a communication network. The problem of designing an adaptive routing algorithm that works efficiently under various network conditions and traffic patterns is well suited to be solved using the Ant Colony Optimization metaheuristic.

The ant algorithms have already proved to be effective for the purpose of routing, thank to their adaptation abilities to changes in the environment. The results reported in [16] showed that the ant routing algorithms have high adaptation abilities to various changes in the network load distribution, which may occur in time as well as in space (non-homogeneous and non-uniform loads). They easily adapt to periodical load level changes, and they are also able to cope with non-uniform load patterns, such as a Denial of Service (DoS) or a Distributed Denial of Service (DDoS) network attack.

However, the network's adaptation to load level changes may be quite slow. To a large extend, this is a result of the learning parameter's settings [14]. For some settings the adaptation would be fast, but the operation under constant load would be unstable. On the other hand, other settings would result in a very

G.A. Di Caro and G. Theraulaz (Eds.): BIONETICS 2012, LNICST 134, pp. 33–48, 2014.
DOI: 10.1007/978-3-319-06944-9_3, © Institute for Computer Sciences, Social Informatics and Telecommunications Engineering 2014

stable algorithm but lacking of the adaptation abilities. Therefore, the parameter values should be optimized to work well and at the same time to preserve some adaptation abilities.

The above issue is in some way related to insufficient information about the network's state. Typically, an ant routing algorithm adapts to changes of the network traffic by updating the values of simple statistics collected by ants, such as the mean packet delay. There is no explicit mechanism that would effectively alert the routing agent about a sudden change in the network's conditions. We show that the use of more comprehensive information about the network's state can significantly increase the performance of ant routing.

We extend the accessible information by introducing the use of models of the end-to-end packet delay in order to solve the stated problems. In our approach, we build a statistical model that approximates the end-to-end packet delay distribution on the base of information collected by the backward ants. We use statistical models of data packet delay to describe the current traffic conditions on different paths in the network. We believe that thank to these models it will be possible to detect and quickly react on various load level changes.

The paper is organized as follows. In Sect. 2 we shortly introduce the swarm intelligence approach to adaptive routing and describe the ant-routing algorithms. Section 3 describes the modeling scheme of the packet delay distribution for fixed networks. In Sect. 4 we present how to utilize the delay models to develop a modification of the ASR-FA algorithm that is able to detect and quickly react on various load level changes by temporarily modifying the learning parameter's settings. The experimental results are presented in Sect. 5. Section 6 concludes the paper.

2   Swarm Intelligence and Ant Routing

Swarm Intelligence is related to emergent collective intelligence of groups of simple autonomous agents. The need of methods based on observations of natural systems was caused by unsatisfactory solutions provided by classical optimization algorithms (e.g., dynamic programming). Classical methods are often not able to cope with high dimensional problems, complex constraints, or high and difficult to model variability in time. It appears that the use of stochastic algorithms that draw from the natural systems may provide a more successful solution.

Basing on the Swarm Intelligence paradigm and its derivative, the Ant Colony Optimization scheme [11], there were various ant routing algorithms proposed for both fixed and wireless telecommunication networks. The first ant routing algorithm designed for asymmetric packet-switched networks was introduced by Dorigo and di Caro [9]. Their AntNet algorithm implicitly achieves load balancing by employing a probabilistic distribution of packets on multiple paths. The experiments reported in [7–9] proved that AntNet outperforms other competitors, such as Q-routing [4], PQrouting [5], Shortest Path First (SPF) and OSPF [12], in terms of the throughput and the delay. In [16] it was shown that AntNet performs very well under high and time varying load levels, including periodical load level changes, and is also able to cope with non-uniform traffic patterns, such as the Denial of Service (DoS) or a Distributed Denial of Service (DDoS) network attacks.

Various modifications of AntNet have been developed within the following years. Comprehensive reviews of routing protocols inspired by collective behavior of social insects can be found in [3,13]. In this paper, we improve the ASR-FA algorithm which is a modification of the ASR algorithm proposed in [19] and analyzed under various network conditions in [15,16]. We will show that our approach can better detect the changes of load distributions and thus enables quick reaction to DoS or DDoS.

2.1   The ASR Algorithm

The Adaptive Swarm-based Routing (ASR) was proposed in 2004 for packet-switched networks by Yong, Guang-Zhou and Fan-Jun [19] as a modification of the AntNet algorithm [9]. As in AntNet, there are two types of simple agents (ants): the *forward ants* that explore the network in order to find paths, and the *backward ants* that use information collected by the forward ants to improve the routing policies. Every network node k is assigned a routing table T_k that stores the routing policy, and a traffic model M_k including some local traffic statistics. The routing table T_k stores the probabilities $t_k(d, n)$ for each neighbor node n and each destination node d, used to determine the probabilistic routing policy. Both the routing tables and the statistical model are calculated iteratively during the normal operation of the network.

The forward ants $F_{s \to d}$ are launched at regular intervals from randomly selected source nodes s to randomly selected destination nodes d. For each node visited, the forward ant stores its age (i.e., the time elapsed from its launch) and chooses the next node n to be visited according to a probability $t_k(d, n)$. After reaching the destination node, the forward ant $F_{s \to d}$ creates the backward ant $B_{d \to s}$, transfers all the collected knowledge to the backward ant and is then removed from the system. The backward ant travels back the same path as its parental forward ant. In every visited node k it updates the values of the traffic model and the routing probabilities for all the entries corresponding to every node i visited on the path.

2.2   The ASR-FA Algorithm

The ASR-FA algorithm was proposed in [17] as a variant of the ASR algorithm. ASR-FA introduces the same modifications to ASR as those proposed in [9] to improve AntNet. Namely, in the ASR-FA algorithm both forward and backward ants make use of high-priority queues which accelerates the propagation of information about good paths in the network. Moreover, the forward ants do not carry any information about the trip times they experienced. The backward ants update the routing tables in the visited nodes using estimates of forward ants trip times computed at each node k on the base of the dynamics of local links l and the actual queue sizes. On the base of this estimates, the statistical traffic model M_k is built for each node k. The estimated trip time $\widehat{o}_k(n)$ from any node k to its neighbor node n is calculated as:

$$\widehat{o}_k(n) = d_k(n) + (\ell_k(n) + s_a)/B_k(n) \tag{1}$$

where $d_k(n)$ is the links propagation delay, $\ell_k(n)$ is the size of the output queue to neighbor n (in bytes), s_a is the ant's size (in bytes) and $B_k(n)$ is the *link bandwidth* (the measure of available or consumed data communication resources expressed in bits per second). As a result, the estimated trip times are computed later and are more up-to-date than in the case of the basic ASR (where ant's past trip times are used). Thus, the statistical traffic model is also more reliable.

3   Packet Delay Modeling

Packet delay modeling is the first step toward network performance evaluation and optimization. The objective of delay modeling is to find a mathematical model that not only can characterize the data, but also provides tools for performance evaluation and optimization.

Let us consider here a fixed telecommunication network controlled by an ant routing algorithm. In such network, the delay distribution between a source and destination node consists of several peaks. The ant routing algorithms are multipath algorithms, so different packets within a single session can travel along different paths. Thus, the successive peaks of the delay distribution correspond to different paths traveled by a packet.

We model the empirical distribution of packet end-to-end delay in the considered network with a mixture of probability distributions. We intend to build a simple but accurate model that can be easily computed on-line and, at the same time, provides comprehensive information about the network's state. To model the empirical delay distribution of a single path, we introduce the gamma-exponential-delta mixture model, denoted by $\Gamma\mathrm{Ex}\delta$. The model is a mixture of three probability distributions, namely the gamma distribution, the exponential distribution, and the single point distribution, all delayed by a constant time. We use the notation

$$f_\Gamma\left(x|\theta_\Gamma\right) = \begin{cases} \frac{\lambda_\Gamma^{\nu_\Gamma}}{\Gamma(\nu_\Gamma)} x^{\nu_\Gamma-1} e^{-\lambda_\Gamma x} & x \geq 0 \\ 0 & x < 0 \end{cases} \tag{2}$$

for the gamma distribution density with the rate parameter $\lambda_\Gamma > 0$ and the shape parameter $\nu_\Gamma > 0$, where $\Gamma(\nu_\Gamma) = \int_0^\infty z^{\nu_\Gamma-1} e^{-z} dz$ is the gamma function. Further we use

$$f_{\mathrm{Ex}}\left(x|\theta_{\mathrm{Ex}}\right) = \begin{cases} \lambda_{\mathrm{Ex}} e^{-\lambda_{\mathrm{Ex}} x} & x \geq 0 \\ 0 & x < 0 \end{cases} \tag{3}$$

for the exponential distribution density with the rate parameter $\lambda_{\mathrm{Ex}} > 0$ and

$$f_\delta\left(x\right) = \delta(x) \tag{4}$$

for the formal description of the single point distribution. We sometimes group the parameters of each distribution, denoting then $\theta_\Gamma = (\lambda_\Gamma, \nu_\Gamma)$ and $\theta_{\mathrm{Ex}} = (\lambda_{\mathrm{Ex}})$.

For a single hop path, the heuristics for our model follows an approximation of a M/D/1 queue (Poisson arrival process, constant service time, 1 queue) [6]. We use the delta distribution to fit the delay of packets that did not wait in any queue, the gamma distribution approximates the M/D/1 delay for small and medium delays [18], and an exponential distribution approximates the tail of the delay distribution of a M/D/1 queue. According to our experiments, the $\Gamma\mathrm{Ex}\delta$ mixture model can be also used to approximate multi-hop paths with a good accuracy.

The empirical delay distribution from a source to a destination node consists of several peaks that correspond to different possible paths traveled by a packet. Thus, the model of the overall packet delay distribution is a mixture of $\Gamma\mathrm{Ex}\delta$ triplets, $i = 1, \ldots, M$, $t = 1, \ldots, N$, namely

$$f(x_t|\theta) = \sum_{i=1}^{M} (\pi_{\Gamma,i} \, f_\Gamma(x_t - s_i|\theta_{\Gamma,i}) + \pi_{\mathrm{Ex},i} \, f_{\mathrm{Ex}}(x_t - s_i|\theta_{\mathrm{Ex},i}) + \pi_{\delta,i} \, f_\delta(x_t - s_i)) \quad (5)$$

where M is the number of possible paths from the source to the destination, N is the data sample size, $\theta = \{(\pi_{\Gamma,i}, \pi_{\mathrm{Ex},i}, \pi_{\delta,i}, \theta_{\Gamma,i}, \theta_{\mathrm{Ex},i}, s_i), i = 1, \ldots, M\}$ is the parameter vector and s_i reflects the minimum delay for a given path, which depends on the link delay and bandwidth. The mixing parameters satisfy the following constraints $\pi_{\Gamma,i} \geq 0$, $\pi_{\mathrm{Ex},i} \geq 0$, $\pi_{\delta,i} \geq 0$, $\sum_{i=1}^{M}(\pi_{\Gamma,i} + \pi_{\mathrm{Ex},i} + \pi_{\delta,i}) = 1$.

Typically, a mixture model is estimated by using the Expectation Maximization (EM) method. The EM algorithm is used in statistics for finding the maximum likelihood estimates of parameters in probabilistic models. Our models have two particular features:

1. Each path in the network has its minimum delay s, so the $\Gamma\mathrm{Ex}\delta$ model for a given path is delayed by s.
2. The $\Gamma\mathrm{Ex}\delta$ mixture model is a discrete-continuous model, as it contains a single point distribution component (the density has the delta term).

It is easy to notice that the resulting likelihood function is not differentiable with respect to the delay, hence its basic properties are not fulfilled, and the typical estimation procedures may behave erratically. To overcome this problem, we propose a two-stage estimation procedure:

1. **Stage 1.** Elimination of the discrete-type distribution by the estimation of the delay.
2. **Stage 2.** Estimation of the elements of the mixture of continuous distributions using the EM algorithm.

In stage 1, we eliminate the discrete part of the distribution (the delta peaks) together with the delays. In this order, we calculate the empirical cumulative distribution function (ECDF) with a given bin width. We set a $\Gamma\mathrm{Ex}\delta$ model delay at the position of every ECDF jump. In the second stage, we use the Expectation Maximization (EM) algorithm to estimate the parameters of the mixture model. The EM algorithm provides an efficient iterative procedure to compute the Maximum Likelihood (ML) estimates in the presence of missing or unobservable data. It iterates two steps: in the expectation step (E-step) the distribution of the unobservable variable is estimated and in the maximization step (M-step) the parameters which maximize the expected log likelihood found on the E-step are calculated.

The parameters of the gamma components are estimated similarly to the way proposed in [1]. The difference is that we use robust parameter estimation [10]. We weigh each data point in such way that the influence of this observation on the value of the estimated distributions parameters decays with the distance from the distributions mean. Consequently, the EM algorithm is more robust to the noise and small insignificant peaks that we do not want to model, as they may turn out to bias parameter estimates of nearby peaks (details in [10]).

For each possible path $i = 1, 2, \ldots, M$, the conditional probability densities for the gamma components estimated in E-step are calculated as

$$p_{\Gamma,i}(x_t, \theta^k) = \frac{\pi_{\Gamma,i}^k \, f_\Gamma(x_t - s_i | \theta_{\Gamma,i}^k)}{\sum_{j=1}^M \left(\pi_{\Gamma,j}^k \, f_\Gamma(x_t - s_i | \theta_{\Gamma,j}^k) + \pi_{\mathrm{Ex},j}^k \, f_{\mathrm{Ex}}(x_t - s_i | \theta_{\mathrm{Ex},j}^k) \right)} \tag{6}$$

and for the exponential components as

$$p_{\mathrm{Ex},i}(x_t, \theta^k) = \frac{\pi_{\mathrm{Ex},i}^k \, f_{\mathrm{Ex}}(x_t - s_i | \theta_{\mathrm{Ex},i}^k)}{\sum_{j=1}^M \left(\pi_{\Gamma,j}^k \, f_\Gamma(x_t - s_i | \theta_{\Gamma,j}^k) + \pi_{\mathrm{Ex},j}^k \, f_{\mathrm{Ex}}(x_t - s_i | \theta_{\mathrm{Ex},j}^k) \right)} \tag{7}$$

The robust parameter estimates are calculated in the M-step as follows:

$$\pi_{\Gamma,i}^{k+1} = \frac{1}{N} \sum_{t=1}^N p_{\Gamma,i}(x_t, \theta^k) \tag{8}$$

$$\pi_{\mathrm{Ex},i}^{k+1} = \frac{1}{N} \sum_{t=1}^N p_{\mathrm{Ex},i}(x_t, \theta^k) \tag{9}$$

$$\overset{k+1}{\underset{\Gamma,i}{\lambda}} = \frac{\nu_{\Gamma,i}^k \sum_{t=1}^N w_{\Gamma,it} \, p_{\Gamma,i}(x_t, \theta^k)}{\sum_{t=1}^N w_{\Gamma,it} \, (x_t - s_i^k) p_{\Gamma,i}(x_t, \theta^k)} \tag{10}$$

$$\overset{k+1}{\underset{\mathrm{Ex},i}{\lambda}} = \frac{\sum_{t=1}^N w_{\mathrm{Ex},it} \, p_{\mathrm{Ex},i}(x_t, \theta^k)}{\sum_{t=1}^N w_{\mathrm{Ex},it} \, (x_t - s_i^k) p_{\mathrm{Ex},i}(x_t, \theta^k)} \tag{11}$$

$$\nu_{\Gamma,i}^{k+1} = \nu_{\Gamma,i}^k + a_k G_{\nu_{\Gamma,i}}^a (X, \theta^k) \tag{12}$$

The parameters w_{it} weigh the data points using the Mahalanobis distance $d_{it} = |x_t - \mu_i|/\sigma_i$, for $i = 1, \ldots, M$ (see [10] for details).

4   The Reactive ASR-FA Algorithm

In this section we present the Reactive ASR-FA algorithm, which is a modification of the ASR-FA algorithm that utilizes the delay models to speed up the adaptation process. In this solution, every node s in the network maintains a delay model to every possible destination node d. Such the delay model consists of several $\Gamma \mathrm{Ex}\delta$ triplets, corresponding to paths from s to d. We use the models described in Sect. 3 that are calculated on-line in every node during the networks operation, on the base of information gathered by ants.

We use the delay models to detect changes in the network conditions. At given time intervals, we collect a small sample of ants delay values. We use the Anderson–Darling test to assess whether the sample comes from the currently maintained delay model.

The Anderson–Darling test is a modification of the Kolmogorov-Smirnov (K-S) test that gives more weight to the tails than does the K-S test. As a result, it is more sensitive. It makes use of the fact that, assuming the data arise from a hypothetic distribution, it can be transformed into a uniform distribution. The transformed sample data can be then tested for uniformity with a distance test.

The test statistic A verifies whether the sample data come from a distribution with the cumulative distribution function (CDF) F. The data in the sample must be put in ascending order $Y_1 < \ldots < Y_n$. The test statistic A can be computed as

$$A^2 = -n - S \tag{13}$$

where

$$S = \sum_{k=1}^{n} \frac{2k-1}{n} \left(\ln F(Y_k) + \ln(1 - F(Y_{n+1-k})) \right) \tag{14}$$

The statistic A is then averaged over time, namely

$$\overline{A} = \frac{A_m + A_{m-1} + \ldots + A_{m-z}}{z} \tag{15}$$

where z is the number of the averaged values of the statistic A, m is an index, and A_m is the current value of the statistic A.

The Anderson–Darling test is one-sided and the null hypothesis that the distribution is of a specific form is rejected if the test statistic is greater than the critical value A_{thr}. We used $\overline{A}$ as our test statistic.

The main goal of the Reactive ASR-FA is to accelerate the process of adaptation to the changing network conditions. Therefore, when the null hypothesis of the Anderson–Darling test is rejected, one of the following actions is performed:

1. **Action 1. If $\overline{A} \geq A_{thr}$ then $\beta = \beta_m$ else $\beta = \beta_a$**
   The non-linearity parameter β is modified. Parameter $\beta \geq 1$ in the ASR algorithm is a *non-linearity parameter* that adjusts the exploring properties of the routing policy. The higher value of β, the more greedy is the policy [19]. By decreasing the value of β we allow a more exploring policy. This can help the routing algorithm to find new paths in the network that may be a better solution in the changed network conditions.
2. **Action 2. If $\overline{A} \geq A_{thr}$ then $\eta = \eta_m$ else $\eta = \eta_a$**
   The learning rate η is modified. As the average packet delay in the ASR algorithm is calculated according to the exponential moving average (additionally

smoothed), the learning rate η represents the degree of weighting decrease. A higher η discounts older observations faster. Therefore, if a change in the network conditions is detected, the η should be increased in order to limit the history taken into account while calculating the average delay.

3. **Action 3. If $\overline{A} \geq A_{thr}$ then $\eta = \eta_m$ and $\beta = \beta_m$ else $\eta = \eta_a$ and $\beta = \beta_a$**
   Both the non-linearity and learning rate parameters are modified. The approaches described above can be combined in order to increase the speed of reaction on the changing network conditions.

The above improvements are implemented as follows. In every network node and at given time intervals, the $\overline{A}$ is computed. If the statistic value exceeds the given threshold A_{thr}, the parameters of the routing algorithm are modified. The following two approaches have been examined:

1. **Approach 1.** In the first approach, β is decreased and/or η is increased if at least for one destination node the value of the Anderson–Darling statistic exceeds A_{thr}. This means that if β was already decreased, the value of the Anderson–Darling statistic for all destinations must fall below A_{thr} in order to increase β once again. In this approach the routing agent maintains only one value of β and η.
2. **Approach 2.** The second approach assumes maintaining separate values of β and η for each possible destination node. In this approach, if a delay model to a certain destination is found invalid, only the parameters related to this destination are modified.

In our solution, the critical value A_{thr} was experimentally set to 20 by analysis of the obtained delay models. The values of β_a, β_m, η_a, and η_m were set according to our analysis presented in [14], where we studied the influence of various parameter values on the performance of the ASR algorithm. Our experiments showed that for constant load levels the best results can be obtained when setting $\beta = 4$ and $\eta = 0.2$. Lower values of β introduce higher exploration, which results in much longer learning times. Good and stable results can be obtained for $\eta \in [0.2, 0.5]$. Setting $\eta \leq 0.2$ increases the learning time and for $\eta \geq 0.5$ both the learning time and the obtained mean delay have significantly worse values. Thus, we use the following parameter values $\beta_a = 4$, $\eta_a = 0.2$, $\beta_m = 1$ or $\beta_m = 2$, and $\eta_m = 0.35$. Setting $\beta_m = 1$ or $\beta_m = 2$ introduces more exploration in order to allow finding new paths, whereas increasing η_m to 0.35 shortens the history of delays used to compute the average packet delay (see Action 2).

5   Experimental Results

We have performed experiments to test all the approaches described in the previous section. The presented simulations were performed using the NS2 network simulator with additional custom made modules. We used a UDP agent in the transport layer and the load level in the network was generated by CBR traffic sources. At a predefined moment, the load level was increased by decreasing

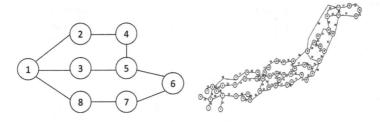

Fig. 1. SimpleNet (left) and the nippon telephone and telegraph (right) networks.

the packet interarrival time (*mpia*) in the CBR traffic sources. As a result, a load level jump was obtained. Next, after a given amount of time, the load level returned to the primary value. We present results for the SimpleNet network proposed in [8], which is a test network structure, and a the Nippon Telephone and Telegraph network (NTT), which is a Japanese backbone network (Fig. 1). We used the parameters of the NTT network as described in [2].All the presented results were averaged over 20 simulations.

Moreover, we present experiments for the NTT network that test the capability of the proposed Reactive ASR-FA algorithm to detect a Distributed Denial of Service (DDoS) network attack. In [16] the authors showed that the ability of adaptive routing algorithms to distribute the traffic over several paths makes them robust to several network attacks, including DoS and DDoS network attacks. Here, we examine whether adding a mechanism for detecting changes in the network conditions can speed up the adaptation process after such attacks.

5.1 Single Values of the β and the η Parameters (Approach 1)

This section examines a performance of the approach in which the routing agent maintains only a single value of β and η for all destination nodes.

First, we present results concerning the SimpleNet network structure for the following simulation scenario. After 500 s the load level in the network is increased by decreasing the *mpia*. Next, after 1500 s from the beginning of the simulation, the load level returns to the primary value. We verify whether the modified ant routing algorithm is able to quickly detect such load level changes.

We compare the results obtained for Actions 1-3, described in Sect. 4, with the original ASR-FA algorithm for different values of the load level jump (Fig. 2). The value of *beta* alternates between 4 and 2 depending on $\overline{A}$ ($\beta_m = 2$).

Let us call the initial process of finding the suitable routing policy - *the learning process*, and the process of modifying the policy after the load jump - *the adaptation process*. In most cases all the modifications speed up both the learning process and the adaptation process (Fig. 2). In case of the learning process, the lowest learning time was obtained when using the combined modification of both β and η (Action 3). The next best result was obtained by modifying solely β (Action 1). In case of the adaptation process, the lowest adaptation time can be obtained when using Action 3 and the second best occurs to be Action 2 (see Sect. 4).

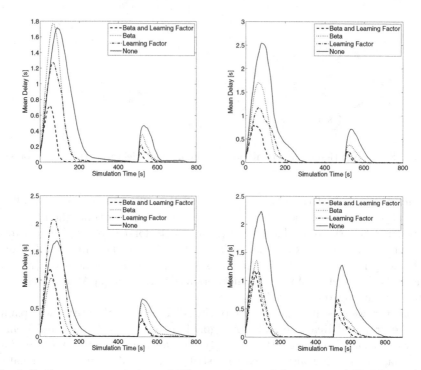

Fig. 2. Performance comparison of different versions of the reactive ASR-FA algorithm under different load level jumps. The jumps were caused after 500 s by *mpia* step decrease = 0.54 (upper-left), 0.6 (upper-right), 0.66 (lower-left) and 0.72 (lower-right). The value of β_m set to 2.

It is worth noticing that we present only the results obtained for the selected set of parameters. We chose the parameters experimentally, taking into account both the adaptation capabilities and the stability of the network's operation. E.g., we were able to achieve better adaptation times when we used the value of $\beta_m = 1$ instead of $\beta_m = 2$. But for the highest tested jump, this caused the algorithm to diverge (it seemed that the level of exploration implied by $\beta = 1$ was too high for the load level of the biggest jump).

We also compared other network performance indicators, namely the resulting mean packet delay after the adaptation process ends and the maximum packet delay during the adaptation process (Fig. 3). In most cases, all actions result in a shorter mean data packet delay (Fig. 3 left). Only for the smallest load level jump, Action 1 results in a bigger delay then the original ASR-FA algorithm. On the other hand, this modification assures the lowest mean data packet delay for the biggest load level jump. Thank to this approach, under high load levels, the algorithm is able to find more efficient routing policies by temporarily allowing for intensive exploration. Under lower load level jumps, the best results can be obtained when modifying solely the learning factor η or both β and η. The maximum data packet delay during the adaptation process shows a similar tendency (Fig. 3 right). In all examined cases all the proposed modifications lower these value in comparison to the original ASR-FA algorithm. The lowest value can be obtained when modifying only the learning factor.

Moreover, we examined whether the Anderson-Darling statistic is a good indicator of the changes in the network's conditions. We present exemplary results concerning the value of the averaged Anderson-Darling statistic for the traffic from Node 1 to Node 5 in the SimpleNet network (Fig. 4). It can be seen that the value of the averaged Anderson-Darling statistic increases rapidly after the increase of the load level, and also after the decrease of the load level. As a result, the value of the β parameter (temporarily) decreases to 2 both after the increase and the decrease of the load level in the network. It is worth noticing that when analyzing the averaged delay between Nodes 1 and 5, no significant peaks can be observed at the moments of the load level changes. This means that in the considered scenario it would not be possible to detect the changes solely

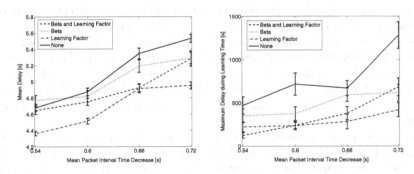

Fig. 3. Performance comparison of different versions of the reactive ASR-FA algorithm in terms of the mean data packet delay (left) and the maximum packet delay during the adaptation process (right)

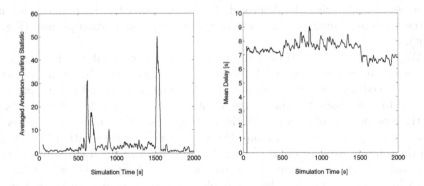

Fig. 4. The value of the averaged Anderson-Darling statistic $\overline{A}$ (left) and the averaged delay (right), for the traffic from node 1 to node 5 in the SimpleNet network

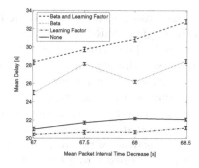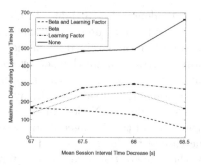

Fig. 5. Performance comparison of different modifications of the ASR-FA algorithm's learning model in terms of the mean data packet delay (left) and the maximum packet delay during the adaptation process (right). The value of β_m set to 1. NTT network

on the base of the simple delay analysis. Thus, employing a more comprehensive information — i.e., the delay models — is well motivated here.

Now, let us validate the results obtained for the SimpleNet network in a different network structure, namely in the NTT network. The load level was increased by decreasing the *mpia* after 500 seconds. Next, after 1800 s from the beginning of the simulation the load level returned to the primary value. Our experiments showed that $\beta_m = 1$ is the adequate value of the decreased β parameter for the NTT network.

As in case of the SimpleNet network structure, we compare the results with the original ASR-FA algorithm for different values of the load level jump. We show the resulting mean packet delay after the adaptation process ends and the maximum packet delay during the adaptation process (Fig. 5). The mean delay is lowest in case of the modification based solely on the learning factor η (Fig. 5 left). The other two modifications result in mean delays worse than in case of the original ASR-FA algorithm (we will see in Sect. 5.2 that much better delays can be achieved when using the Approach 2). On the other hand, all modifications decrease the maximum data packet delay during the adaptation process (Fig. 5 right).

We skipped the detailed plots of delays, as they were similar to Fig. 2. However, we would like to point out the core observed properties. In all cases examined, the modifications speed up both the learning process and the adaptation process. In most cases, the shortest adaptation time can be achieved by combining the modification of β and η. All the routing algorithms managed to converge for every size of the load level jump. However, the combined solution (Action 3) and the solution using solely β (Action 1) result in some small variations of the mean delay after the adaptation process ends. This happens, since lowering of β to 1 introduces a very high level of exploration. If at some moment of time, β is unnecessarily decreased, suboptimal paths may be used and the mean delay may grow until β is increased again. We also noticed that — after the adaptation porcess — the mean delay converges to a higher value in case of Action 1 and Action 3 (modifying β parameter) than in case of Action 2 and the original ASR-FA algorithm (not changing *beta*). Additional experiments showed that we can eliminate both problems by setting $\beta_m = 2$, yet this results in a significant increase of the adaptation times and reduces the advantages of our algorithm.

As for the SimpleNet network structure, we also examined the value of the averaged Anderson–Darling statistic versus the averaged delay measured in selected nodes of the NTT network. Again, the averaged Anderson–Darling statistic increased rapidly after the increase of the load level and also after the decrease of the load level. No such peaks could be seen on the mean delay plot. Thus, our experiments show that it is not always possible to detect the changes in the network conditions based solely on the mean delay, whereas it is possible to successfully detect such changes based on the $\overline{A}$ statistic and the delay models.

5.2 Separate Values of the β and the η Parameters for Each Possible Destination Node (Approach 2)

In this section we present the results for the approach in which the routing agent maintains separate values of β and η for each destination node (Approach 2). We do not present the corresponding results for the SimpleNet network structure, as our experiments showed that this network is small and a change of the conditions near one node implies a change in the whole network's conditions. Therefore, it is reasonable to maintain common β and η parameter values for all destinations of each node. Yet, the NTT network structure is much more complex, so a change of the traffic conditions near one node does not necessarily have to imply a global change. Thus, it is worth to verify the merits in using the separate of β and η for each possible destination of a given node.

In this set of experiments we test the same scenario as in Sect. 5.1. The load level is increased after 500 s and returns to the primary value after 1800 s from the beginning of the simulation.

We compare the results obtained for Actions 1-3 with the original ASR-FA algorithm for different values of the load level jump. It can be seen that using separate values of β and η influences mostly the mean delay of the data packets in the network (Fig. 6). When using one value of β (Sect. 5.1), the mean delays obtained by the ASR-FA were significantly lower than in case of the proposed modifications. Using separate values of these parameters almost in all cases results in lower mean delays than obtained when using the ASR-FA. The learning times are similar for both approaches. The learning time in the modification concerning only the value of β slightly increased, whereas the learning time in the modification using both β and η decreased. However, regardless of the approach (common or separate values of β and η), the proposed modifications in almost all cases decrease the learning time of the ASR-FA algorithm. Best results can be obtained when using the modification of both β and η (Action 3).

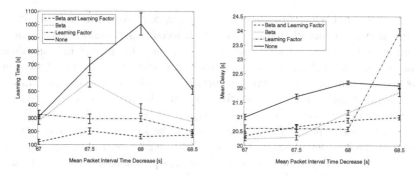

Fig. 6. Performance comparison of different versions of the reactive ASR-FA algorithm in terms of the learning time (left) and the mean data packet delay (right). The value of β_m set to 1, β modified independently for each destination. NTT network

5.3   Using the Reactive ASR-FA Algorithm to Detect a DDoS Network Attack

A Distributed Denial of Service attack (DDoS) occurs when multiple systems flood the bandwidth or resources of a targeted system, usually one or more web servers. In [16] the authors showed that, in comparison to the Shortest Path routing, the ant routing algorithms are able to successfully adapt to non-uniform load patterns, such as the Denial of Service (DoS) or the Distributed Denial of Service (DDoS) network attacks.

We tested the performance of the the Reactive ASR-FA algorithm in presence of such attacks. In this section, we present the most important results for the NTT network structure. In the described simulation scenario, the DDoS attack was aimed at node 20, started after 200 s and lasted for the next 800 s of the simulation. During the attack, all nodes generated extra load towards the node number 20.

We present results concerning the Approach 1, for the Actions 1-3 described in Sect. 4. All results were averaged over 20 simulations.

The experiments show that all the proposed modifications speed up the adaptation process to the DDoS attack at a similar factor (Fig. 7 left). Namely, using the Reactive ASR-FA algorithm shortens the adaptation time by about 50 %. However, the modifications based on β result in a higher mean delay in comparison to the original ASR-FA, whereas the modification based on η maintains a similar mean delay to the ASR-FA. Thus, in this case the best results can be obtained by modifying the learning rate η.

Figure 7 (right) shows how the DDoS attack can be detected by analyzing the averaged value of the Anderson–Darling statistic. There is a high peak on the plot after the attack starts. It can be seen that also the end of the attack is visible as a second peak on the plot.

The DDoS attack can be easily detected when using the Reactive ASR-FA algorithm, as it implies a change in the network conditions. When a DDoS attack occurs, the delay distribution to the attacked destination node changes because

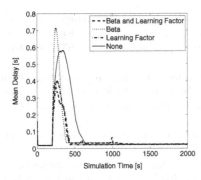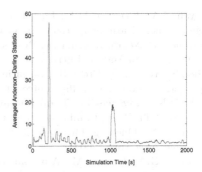

Fig. 7. Performance comparison of different versions of the reactive ASR-FA algorithm in case of a DDoS network attack (left), the averaged Anderson-Darling statistic from node 34 to node 20 (right). The value of β_m set to 1. NTT network

of the extra traffic. As an effect, a different routing policy may be more appropriate. Therefore, it is essential to quickly detect the attack and to modify the learning parameters in such way to enable faster convergence and recovery to the normal network's operation. It is worth noticing that a DoS attack can be as well detected by this mechanism, as it also implies a change in the packet delay distributions.

6   Conclusions

In this paper we showed that using more comprehensive information about the network's state in place of simple statistics can significantly increase the adaptation abilities of ant routing algorithms. We propose to use statistical models of the end-to-end packet delay to describe the current traffic conditions on different paths in the network.

We introduced the Reactive ASR-FA algorithm as a modification of the ASR-FA algorithm. We demonstrated that our algorithm successfully utilizes the proposed delay models to detect changes in the network conditions. In particular, it speeds up the adaptation process and achieves lower values of the mean data packet delay. Moreover, we successfully employed the Reactive ASR-FA to detect the DoS and DDoS network attacks and to adjust the routing policy accordingly.

References

1. Almhana, J., Liu, Z., Choulakian, V., Mcgorman, R.: A recursive algorithm for gamma mixture models. In: IEEE International Conference on Communications 2006, ICC '06, vol. 1, pp. 197–202 (2006)
2. Baran, B., Sosa, R.: Antnet: routing algorithm for data networks based on mobile agents. Intel. Artif. **5**(12), 75–84 (2001)
3. Blum, C., Merkle, D.: Swarm Intelligence: Introduction and Applications. Springer, Heidelberg (2008)

4. Boyan, J.A., Littman, M.L.: Packet routing in dynamically changing networks: a reinforcement learning approach. In: Cowan, J.D., Tesauro, G., Alspector, J. (eds.) Advances in Neural Information Processing Systems, vol. 6, pp. 671–678. Morgan Kaufmann, San Mateo (1994)
5. Choi, S., Yeung, D.: Predictive q-routing: a memory-based reinforcement learning approach to adaptive traffic control. In: Touretzky, D., Mozer, M.C., Hasselmo, M.E. (eds.) Advances in Neural Information Processing Systems 8 (NIPS8), pp. 945–951. MIT Press, Cambridge (1996)
6. Daigle, J.N.: Queuing Theory with Applications to Packet Telecommunication. Springer, New York (2004)
7. Di Caro, G.A., Dorigo, M.: Ant colonies for adaptive routing in packet-switched communications networks. In: Eiben, A.E., Bäck, T., Schoenauer, M., Schwefel, H.-P. (eds.) PPSN 1998. LNCS, vol. 1498, p. 673. Springer, Heidelberg (1998)
8. Di Caro, G., Dorigo, M.: Antnet: distributed stigmergetic control for communications networks. J. Artif. Intell. Res. **9**, 317–365 (1998)
9. Di Caro, G., Dorigo, M.: Two ant colony algorithms for best-effort routing in datagram networks. In: Proceedings of the Tenth IASTED International Conference, PDCS98, pp. 541–546. IASTED/ACTA Press (1998)
10. Dijkstra, M., Roelofsen, H., Vonk, R.J., Jansen, R.C.: Peak quantification in surface-enhanced laser desorption/ionization by using mixture models. Proteomics **6**, 5106–5116 (2006)
11. Dorigo, M., Maniezzo, V., Colorni, A.: Positive feedback as a search strategy. Technical report, Tech. Rep. 91–016, Politecnico di Milano, Dipartimento di Elettronica (1991)
12. Doyle, J.: Routing TCP/IP, vol. I and II. CCIE Professional Development (1998)
13. Ducatelle, F., Di Caro, G., Gambardella, L.: Principles and applications of swarm intelligence for adaptive routing in telecommunications networks. Swarm Intell. **4**, 173–198 (2010)
14. Gadomska, M.: Wykorzystanie algorytmow mrowkowych do problemu adaptacyjnego rutingu w sieciach (in polish). M. Sc thesis, Warsaw University of Technology (2005)
15. Gadomska, M., Pacut, A.: Performance of ant routing algorithms when using TCP. In: Giacobini, M. (ed.) EvoWorkshops 2007. LNCS, vol. 4448, pp. 1–10. Springer, Heidelberg (2007)
16. Gadomska, M., Pacut, A., Igielski, A.: Ant-routing vs. q-routing in telecommunication networks. In: Proceedings of the 20-th ECMS Conference, pp. 67–72 (2006)
17. Kudelska, M.: Ant algorithms for adaptive routing in telecommunication networks. Ph.D. thesis, Warsaw University of Technology (2011)
18. Menth, M., Henjes, R., Zepfel, C., Tran-Gia, P.: Gamma-approximation for the waiting time distribution function of the m/g/1-/spl infin/ queue. In: Proceedings of the 2nd Conference on Next Generation Internet Design and Engineering, NGI '06, pp. 123–130 (2006)
19. Yong, L., Guang-zhou, Z., Fan-jun, S.: Adaptive swarm-based routing in communication networks. J. Zhejiang Univ. Sci. A **5**, 867–872 (2004)

A Mathematical Model for the Analysis of the Johnson-Nyquist Thermal Noise on the Reliability in Nano-Communications

Pietro Santagati[1,2](✉) and Valeriu Beiu[1]

[1] Faculty of Information and Technology,
UAEU, P.O. Box 19551 Al Ain, UAE
`santagati@dmi.unict.it, vbeiu@uaeu.ac.ae`
[2] Department of Mathematics and Computer Science, Citta' Universitaria,
Viale Andrea Doria 6, 95125 Catania, Italy

Abstract. The aim of this paper is to investigate how the length of communication links could affect the reliability during operations, by analysing how the Johnson-Nyquist (thermal) noise on the links affects the probability of failure (defined as the probability of switching) of devices/switches scaled to the limit like, e.g., ion channels but also nanoscale CMOS transistors. To this end, we will consider classical CMOS circuits, and base our analysis on statistical considerations. In particular, our aim is to look for the existence of an optimum wire/link length in this context, which would maximize the reliability of the simplest system formed by a communication link (wire) driving a switch (transistor or ion channel).

1 Introduction

An ideal vacuum tube has not sources of spurious noise. However, in 1918 it was shown that the electrical current is still affected by two types of noise [1]. The first one is due to the thermal motion of the electrons, and is commonly known as thermal noise (or Johnson-Nyquist effect), and occurs in any conductor [2]. The second one is due to the discreteness of the carriers charge, and is called shot noise. This one does not occur in any conductor. The power spectrum, $P(\nu)$, of the noise is given by the Fourier Transform at frequency ν, of the current-current correlation. Thermal noise power spectrum is related to the conductance G [3], as long as $h\nu < KT$, where h is the Plank constant, K the Boltzmann constant and T the temperature. In particular, for all devices whose electronic transfer shows a Poisson statistics, the shot noise has a maximum value

$$P = 2eI, \quad \text{as long as } 1/\nu \ll \tau \tag{1}$$

where I is the time average current and τ the one electron pulse interval. It is expected that these two types of noises would play a key role when investigating the reliability of deeply scaled devices, circuits and systems. Reliability

G.A. Di Caro and G. Theraulaz (Eds.): BIONETICS 2012, LNICST 134, pp. 49–59, 2014.
DOI: 10.1007/978-3-319-06944-9_4, © Institute for Computer Sciences, Social Informatics and Telecommunications Engineering 2014

of nanoelectronic devices is anyhow becoming a very important and limiting aspect, not only for CMOS technologies, but also for other beyond CMOS technologies under investigation, when approaching nanoscale dimensions. At such scales the inputs start having more influence on the circuits reliability, due to internal noises as well as variations and defects. As reported by the International Technology Roadmap for Semiconductor (ITRS) [4], CMOS scaling significantly affects the reliability of nanoscale circuits. Currently, new materials are being developed to implement new channels and source/drain to better silicon. The main goals are to increase the saturation velocity and reduce voltage supply, hence also power dissipation, for 2018 and beyond. Under the light of what has been done during the last few decades, there is a clear need to develop methods to study not only the variability of critical dimensions, but also intrinsic noises such as to identify reliability issues in advance. As the size of CMOS transistors is being shrinked towards nanoscale significant voltage scaling are also required. Hence future nanoscale devices are expected to operate at very low voltage (e.g., for saving power). Reducing voltage levels makes the data signal to approach the thermal noise [5,6]. As a consequence, the signal-to-noise ratio is strongly reduced. In order to design reliable circuits, formed of non-reliable nanoscale transistors, several methods have been proposed. The first methods are due to von Neumann. He introduced the N-tuple Modular Redundancy (NMR) technique as well as the von Neumann Multiplexing. Lately the concept of Stochastic Computation was introduced. This approach includes the error statistics at the architectural and the system level design, and allows a good balance between robustness and energy efficiency. However, other approaches have been presented, where reliability of circuits is not the main objective. The case of Probabilistic CMOS is an example. Still, all the efforts during the last 50 years have focused mostly on gates and devices while the wires have been mostly ignored, with a few cases folding them either in the gates or in the devices.

2   Motivations

Reducing size is the main goal in VLSI design. The integration of high density (very small) devices allowed the growth of memory size and clock frequency, which had a strong impact on the economy. Moreover, the energy consumption played a key role in the design criteria in terms of electronic device autonomy. Several studies [7,8] proved that in CMOS technology wire switching dissipates well over 50 % of the overall dynamic power, a percentage that will only grow with scaling. In this context, wires become extremely important, because of the large power/energy they dissipate. Despite significant improvements of the transistors performances (due to scaling), the wires have hardly enjoyed improvements of performances. Additionally, with scaling line edge roughness (LER) becomes more of an issue [9], leading to the variability in resistance and capacitance, and in turn to variability in delay and power consumption. Therefore, wires have a strong impact on the overall circuit performance, and should be carefully considered besides transistors and their variations. In this context, the switching of a CMOS device should include the interconnecting wires. Figure 1 shows the simplest possible structure: a transistor connected to an interconnecting wire. In such a setting reliability has also been considered with respect to the electromigration [10,11], in order to understand the relation with the current density. This aspect proved to be quite important in terms of reliability, since the current density increases as the transistor and wire sizes are down-scaled. This paper focuses on thermal noise on a wire. In particular, we analyze how the reliability, of the system transistor and interconnect wire, is affected by the thermal noise. In this context we try actually to answer a more fundamental question: Is there a an optimum wire length which could maximize the reliability of transmission over the wire followed by the switching of the transistor? In order to perform our analysis, we consider scaled CMOS transistors [12], while the wire will be considered as a classical connecting element. The analysis is performed in the framework of the statistical modeling, where the voltage fluctuations are described by suitable distribution functions.

3 Analysis

This section aims to present a reliability analysis of a nMOS transistor[1] operation, in order to investigate whether (and eventually how) the thermal (Johnson-Nyquist) noise on the interconnecting wire affects the probability of failure, i.e., the switching probability of the transistor.

3.1 The Probability of Failure: Definition

While preserving the generality, we consider a nMOS transistor and recall shortly the basic concept of *probability of failure*, which will be used here. The threshold voltage v_{Th} is assumed to be distributed as a Normal Gaussian defined by a mean value V_{Th} and standard deviation $\sigma_{V_{Th}}$

$$f_{Th}(v_{Th}) = \frac{1}{\sqrt{2\pi}\sigma_{V_{Th}}} \exp\left[-\frac{(v_{Th} - V_{Th})^2}{2\sigma_{V_{Th}}^2}\right]. \tag{2}$$

To get the nMOS device in conduction, the condition $V_G > V_{Th}$ must be satisfied, where V_G denotes the gate voltage[2]. If there are not external perturbations on the gate voltage V_G, this condition can be represented in Fig. 2, where V_G would denote a specific point on the voltage axis. In our case we assume V_G to be equal the voltage supply V_{DD}. In any case, there will be a non-zero probability for the occurrence of the wrong operation, $V_G < v_{Th}$. As measure of the systems reliability, we define the probability of failure as follows

$$P_f = \int_{V_{Th}+\beta\sigma_{V_{Th}}}^{+\infty} f_{Th}(v_{Th}) dv_{Th}, \tag{3}$$

where β is an arbitrary, positive real parameter[3].

---

[1] The case about pMOS would not change the core of the analysis.
[2] Here, to simplify the notations, $V_G = V_{GS} = V_{Gate} - V_{Source}$.
[3] Usually, and for our analysis, it will be $\beta = 6.0$.

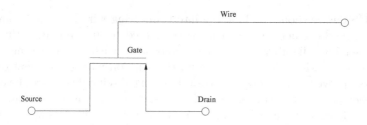

Fig. 1. A nMOS transistor connected to a wire.

3.2 Thermal Noise on Wire Interconnections

If thermal noise disturbances affect the wire (showed in Fig. 1), V_G cannot be considered constant anymore, and the fluctuations on the wire need to be taken into account. To this end, the thermal noise is modelled by a stationary Gaussian stochastic voltage fluctuation process with mean supply voltage V_{DD} and standard deviation σ_{V_G}. We assume that the wire can be treated as a passive device, whose capacitance C_W increases as the wire length increases. In particular, for our computation without losing generality, the wire capacitance is expressed as αC_G, where α is a real positive parameter proportional to the wire length, and C_G is the gate capacitance. Therefore we can write the standard deviation as follows [12]

$$\sigma_{V_G} = \sqrt{\frac{KT}{\alpha C_G}}, \tag{4}$$

where K is the Boltzmann constant and T is the absolute temperature. We assume that the wire voltage distribution is a Normal Gaussian density function, defined by standard deviation σ_{V_G} and mean value V_{DD}, as follows

$$f_G(v_G) = \sqrt{\frac{\alpha C_G}{2\pi KT}} \exp\left[-\frac{\alpha C_G(v_G - V_{DD})^2}{2KT}\right]. \tag{5}$$

3.3 A Wire Driving an nMOS

Once the wire is coupled to the gate of an nMOS transistor (as shown in Fig. 1), the previous definition of probability of failure needs to be extended. Let us suppose that a voltage (of mean value) V_{DD} ($> V_{Th}$) is applied on the gate of a nMOS with mean threshold voltage V_{Th}. The probability of failure depends on two conditions

$$v_G < v_{Th}, \quad v_{Th} > V_{Th} + \beta\sigma_{V_{Th}}. \tag{6}$$

The inequalities (6) have to be regarded as statistical, because v_{Th} and v_G are stochastic variables described by their respective distribution functions (2) and (5). Moreover, these inequalities lead us to observe that the probability of failure must be computed by a joint distribution function, which models the statistical events of failure in terms of the independent Gaussian distributed variables v_G and v_{Th}.

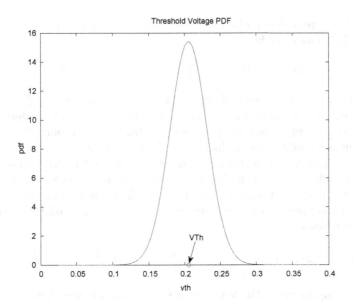

Fig. 2. Threshold voltage PDF: $V_{Th} = 205.6\,\text{mV}$, $\sigma_{V_{Th}} = 25.88\,\text{mV}$.

To carry out our analysis, we define a new variable: $\Delta v = v_G - v_{Th}$. Thus, the probability of failure has to be computed as follows

$$P_f(\alpha, V_{DD}) = \int_{-\infty}^{0} \int_{V_{Th}+\beta\sigma_{V_{Th}}}^{+\infty} f_{\alpha,V_{DD}}(\Delta v, v_{Th}) dv_{Th} d\Delta v, \qquad (7)$$

where α and V_{DD} can be considered real parameters, once the temperature T and the nMOS transistor have been fixed. To construct $f_{\alpha,V_{DD}}$, we start by observing that Δv and v_{Th} are not statistically independent. Furthermore, we remark that the mean value $\langle \Delta v \rangle$ and the standard deviation $\sigma_{\Delta v}$ are[4]

$$\langle \Delta v \rangle = V_{DD} - V_{Th}, \qquad \sigma_{\Delta v} = \sqrt{\sigma_{V_G}^2 + \sigma_{V_{Th}}^2}. \qquad (8)$$

The statistical correlation coefficient $\rho_{\Delta v, v_{Th}}$ can be evaluated as follows

$$\begin{aligned} \rho_{\Delta v, v_{Th}} &= \frac{\text{cov}(\Delta v, v_{Th})}{\sigma_{\Delta v}\sigma_{V_{Th}}} \\ &= \frac{\langle (\Delta v - \langle \Delta v \rangle)(v_{Th} - V_{Th}) \rangle}{\sigma_{\Delta v}\sigma_{V_{Th}}}. \end{aligned} \qquad (9)$$

This simplifies to

$$\begin{aligned} \rho_{\Delta v, v_{Th}} &= -\frac{\sigma_{V_{Th}}}{\sqrt{\sigma_{V_G}^2 + \sigma_{V_{Th}}^2}} \\ &= -\frac{\sigma_{V_{Th}}\sqrt{\alpha C_G}}{\sqrt{KT + \alpha C_G \sigma_{V_{Th}}^2}}. \end{aligned} \qquad (10)$$

---

[4] Here v_G's mean value is denoted by V_{DD} in order to simplify the notation, since it relates to the voltage applied to the gate. As a generic parameter, it should be denoted by V_G.

Two significant remarks can be made here, about the dependence of $\rho_{\Delta v, v_{Th}}$ on α. In the limit cases, we have

$$\lim_{\alpha \to 0^+} \rho_{\Delta v, v_{Th}} = 0, \qquad \lim_{\alpha \to +\infty} \rho_{\Delta v, v_{Th}} = -1, \qquad \forall T > 0. \qquad (11)$$

The first limit in (11) shows that when the wire is not considered ($\alpha = 0$), variables Δv and v_{Th} are fully un-correlated, i.e. statistically independent. This is physically consistent, as in case the wire is not taken into account, the computation of joint probability must reduce to (3). The second limit, on the other hand, shows a complete *anticorrelation*. It is also pretty interesting to observe how the temperature affects the correlation coefficient. This is shown in Fig. 3, and it is related to the nMOS-1 transistor [13] described in Table 1. Once the correlation coefficient is computed, we can construct the joint (multivariate) distribution function as follows

$$f_{\alpha, V_{DD}}(\Delta v, v_{Th}) = \frac{1}{\sqrt{2\pi |\Sigma|}} \exp\left[-\frac{1}{2}(\mathbf{v} - \langle \mathbf{v} \rangle)^T \Sigma^{-1} (\mathbf{v} - \langle \mathbf{v} \rangle) \right], \qquad (12)$$

where the components of the $\mathbf{v}$ vector are the stochastic variables Δv and v_{Th}, while Σ is the covariance matrix[5]

$$\mathbf{v} = \begin{pmatrix} \Delta v \\ v_{Th} \end{pmatrix}, \qquad \Sigma = \begin{bmatrix} \sigma_{\Delta v}^2 & \mathrm{cov}(\Delta v, v_{Th}) \\ \mathrm{cov}(\Delta v, v_{Th}) & \sigma_{V_{Th}}^2 \end{bmatrix}. \qquad (13)$$

Once (12) is used in (7), we can compute the probability of failure, taking into account the thermal noise on the wire, and evaluate how α and the supply voltage V_{DD} affect the overall reliability.

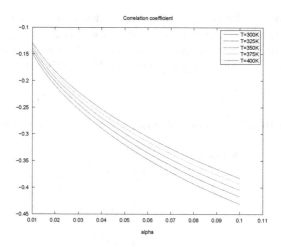

Fig. 3. Correlation coefficient for the device nMOS-1 in Table 1

---

[5] This notation is commonly used in statistical calculations.

4   The Energy Distribution

As we have been considering the supply voltage as a stochastic variable distributed as a Normal Gaussian, it is straightforward also to look for the distribution of energy on the wire. We expect it depends on α, and in general also on V_{DD}. The energy E_w, stored on the wire, once charged under a voltage supply v_G is $E_w = 1/2\alpha C_G v_G^2$. A simple manipulation shows that

$$
\begin{aligned}
E_w &= \tfrac{1}{2}\alpha C_G v_G^2 = \tfrac{1}{2}KT\tfrac{\alpha C_G}{KT}v_G^2 \\
&= \tfrac{1}{2}KT\left(\tfrac{v_G}{\sigma_{V_G}}\right)^2 = \tfrac{1}{2}KT\xi^2.
\end{aligned}
\tag{14}
$$

therefore the energy distribution follow the distribution of the stochastic variable ξ^2. As v_G is a normal Gaussian distributed variable and σ_{V_G} is the related standard deviation, the probability density function f_ξ of ξ^2 is the *non-central chi squared*, with one degree of freedom [14]

$$
f_\xi(v_G; \alpha, V_{DD}) = \frac{1}{2}\exp\left(-\frac{v_G + \lambda_\alpha}{2}\right)\left(\frac{v_G}{\lambda_\alpha}\right)^{-\frac{1}{4}}I_{-\frac{1}{2}}(\sqrt{\lambda_\alpha v_G}),
\tag{15}
$$

where λ_α is the *non-centrality parameter* and $I_{-\frac{1}{2}}(z)$ is the modified Bessel function of the first kind, in the particular case of one degree of freedom[6]

$$
\lambda_\alpha = \left(\frac{V_{DD}}{\sigma_{V_G}}\right)^2, \qquad I_\nu(z) = \left(\frac{z}{2}\right)^\nu \sum_{j=0}^{\infty}\frac{(z^2/4)^j}{j!\,\Gamma(\nu + j + 1)}.
\tag{16}
$$

5   Results and Discussions

The numerical tests that we have performed are based on the data in Table 1 and the results are showed in Figs. 4, 5 and 6. The devices qualitatively present the same behaviour. The probability of failure decreases as V_{DD} increases, although this trend is not preserved as α decreases. This, physically consistent result, is a direct consequence of using the model (4) for the standard deviation of the Gaussian distribution of the voltage on the wire affected by the thermal noise. The probability of failure does not depend significantly on V_{DD} because of its high statistical dispersion (high standard deviation σ_{V_G}). The reliability changes dramatically in a very small α's range, and the results show how reliable the wires are as their capacitance is bigger than the transistor's gate one. Energy stored in the wire follows a non-central chi squared distribution and the values are in agreement with [15]. It is to be noted that as α increases, the local maximum in the probability density function is located at higher energy. In particular, it is possible to have the most likely stored energy value in the wire, once the length is given. This is better showed by the maps in Figs. 7, 8 and 9. All the numerical calculations have been performed at $0.9V_{DD}$, but it does not affect the generality of the method.

---

[6] It is also possible to write the non-central chi distribution in terms of the hypergeometric functions.

Table 1. Parameters of the nMOS transistors, used for the numerical calculations.

Parameters	nMOS-1	nMOS-2	nMOS-3
Technology (nm)	16	22	32
V_{DD}(mV)	700	800	900
V_{Th}(mV)	289.06	321.22	336.23
$\sigma_{V_{Th}}$(mV)	166.22	101.02	615.24
C_G (Faraday)	1.4112E-17	2.7470E-17	5.5134E-17

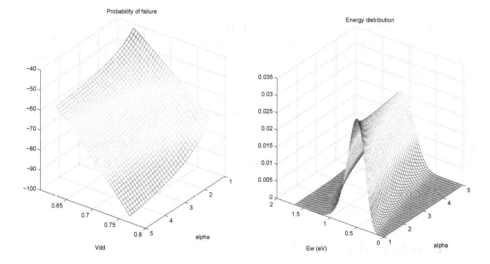

Fig. 4. Probability of failure and Energy distribution: nMOS-1.

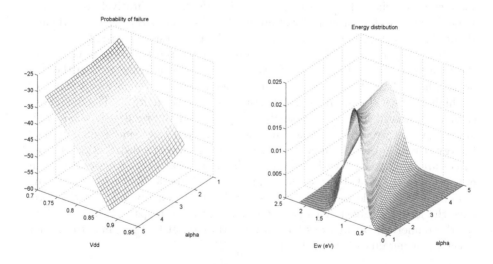

Fig. 5. Probability of failure and Energy distribution: nMOS-2.

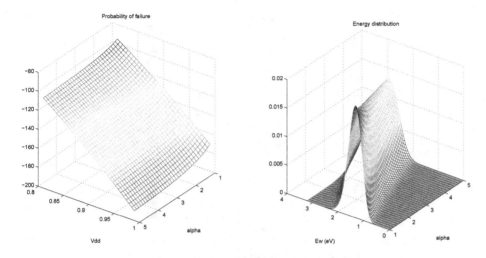

Fig. 6. Probability of failure and Energy distribution: nMOS-3.

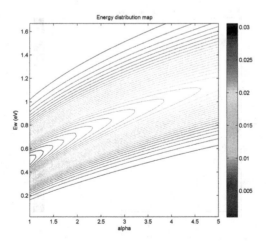

Fig. 7. Energy distribution map: nMOS-1

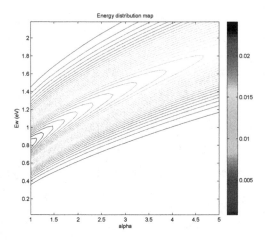

Fig. 8. Energy distribution map: nMOS-2

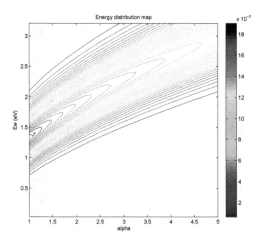

Fig. 9. Energy distribution map: nMOS-3

6 Conclusion

In this paper we developed a statistical mathematical model to analyse the reliability of a nanoscale CMOS when the thermal noise on wires is considered. The results show that, despite an optimum length is not present, the wire length plays an important role in term of reliability, in particular when its capacitance is lower than the CMOS gate one. A map of the stored energy in the wire has been constructed, showing how the energy is distributed, depending on the wire length, for a generic voltage V_{DD}.

Acknowledgements. P. Santagati was supported by UAEU (1108-00451/2010, Brain-inspired Interconnects for Nanoelectronics), while V. Beiu was supported by Intel (2011-05-24G, Ultra Low-Power Application-specific Non-Boolean Architectures).

References

1. Schottky, W.: Über spontane Stromschwankungen in verschiedenen Elektrizitätsleitern. Ann. Phys. (Leipzig) **57**, 541–567 (1918)
2. Johnson, M.B.: Phys. Rev. **29**, 367 (1927)
3. Callen, H.B., Welton, T.W.: Irreversibility and generalized noise. Phys. Rev. **83**, 34–40 (1951)
4. ITRS Report, Emerging Research Device (1951)
5. Sarpeshkar, R., Delbruck, T., Mead, C.: White noise in MOS transistors and resistors. IEEE Circ. Devices Mag. **9**(6), 23–29 (1993)
6. Qian, W., Li, X., Riedel, M., Bazargan, K., Lilja, D.: An architecture for fault-tolerant computation with stochastic logic. IEEE Trans. Comput. **60**(1), 93–105 (2001)
7. Magen, N., Kolodny, A., Weiser, U., Shamir, N.: Interconnect-power dissipation in a microprocessor. In: Proceedings of SLIP'04, pp. 7–13 (2004)
8. Borkar, S.: Design challenges of technology scaling. IEEE Micro **19**, 23–29 (1999)
9. Twaddle, F.J., Cumming, D.R.S., Roy, S., Asenov, A., Trysdale, T.D.: RC variability of short-range interconnects. In: Proceedings of IWCE, Beijing, China, art. 5091143, pp. 1–4 (2009)
10. Ghate, P.B.: Electromigration-induced failures in VLSI interconnects. In: Proceedings of IRPS, San Diego, CA, pp. 292–299 (1982)
11. Lienig, J., Jerke, G.: Current-driven wire planning for electromigration avoidance in analog circuits. In: Proceedings of ASP-DAC, Kitakyushu, Japan, pp. 783–788 (2003)
12. Kish, L.B.: End of Moores law: thermal (noise) death of integration in micro and nanoelectronics. Phys. Lett. A **305**, 144–149 (2002)
13. Li, Y., Cheng, H.W.: Nanosized-metal-Grain-induced characteristic fluctuation in 16 nm complementary metal oxide semiconductor devices and digital circuits. Japan. J. Appl. Phys. **50**, 04DC22 (2011)
14. Siegel, A.F.: The non-central chi-squared distribution with zero degrees of freedom and testing for uniformity. Biometrika **66**, 381–386 (1979)
15. Wawrzyniaka, M., Mackowskia, M., Sniadeckib, Z., Idzikowskib, B., Martinekb, J.: Current voltage characteristics of nanowires formed at the CoGe99.99Ga0.01 interface. Acta Phys. Pol. A **118**(2), 375–378 (2010)

Robotics, Learning, and Neural Networks

Visuomotor Mapping Based on Hering's Law for a Redundant Active Stereo Head and a 3 DOF Arm

Flavio Mutti[✉] and Giuseppina Gini

Dipartimento di Elettronica e Informazione, Politecnico di Milano, Milano, Italy
mutti@elet.polimi.it

Abstract. How humans are able to move the arm so to reach an object in space is far from being completely understood. The problem is often addressed computing a visuomotor mapping between image features and arm joints configurations. We propose a biologically inspired controller able to compute the visuomotor mapping of a simplified humanoid model. The simulated robotic system is composed by a redundant active stereo head and by a 3 degrees of freedom arm with human-like mechanics constraints. The head is driven by a biologically plausible controller based on the Hering's law of equal innervation. The overall system is able to perceive a feature in space through the stereo cameras, compute the head joint angles to foveate it and, without moving the head, to perform the mapping between the foveation angles and the final joints configuration of the arm for reaching that feature. The visuomotor mapping is obtained over a radial basis network, using as input the foveation angles of the head. The network training is performed following the motor babbling schema, using a 10-fold cross-validation technique to validate the robustness of the mapping. Our results show how the visuomotor mapping is able to efficiently cover the whole arm workspace.

1 Introduction

Solving the reaching task problem means to compute the final hand position in space whereas a sensory stimulus -usually vision- has indicated the position of a target. Before computing the trajectory of the arm it is necessary to estimate the target position with respect to the arm frame of reference (FoR). More specifically, the controller has to be able to compute the chain of coordinate transformations from the sensory input to the actuation, considering also that the dimensionality of the input and output spaces often differs.

A way to define these transformations is to compute a sensorimotor map that correlates the sensory input space with the actuators space. According to neuroscience findings and in particular to the theory of neural modelling, the system can learn the sensorimotor mapping over a radial basis framework [19]. In principle, it is possible to compute sensorimotor maps among several sensory

G.A. Di Caro and G. Theraulaz (Eds.): BIONETICS 2012, LNICST 134, pp. 63–74, 2014.
DOI: 10.1007/978-3-319-06944-9_5, © Institute for Computer Sciences, Social Informatics and Telecommunications Engineering 2014

systems and actuators systems but for our purposes we always refer to the visual sensory system; the mapping between the feature space of the visual stimuli and the joints space of the actuator is named visuomotor mapping.

In our study an active stereo head, mimicking the human control strategy, processes the incoming visual information. Given a target, the active stereo head is able to foveate it in a specific joints configuration. If the target is the arm's end-effector the learning strategy correlates the arm joints configuration with the head joints configuration to foveate the target. The learning of visuomotor mapping is then obtained with an active stereo head and an arm. Three degrees of freedom are enough in our study since we only need to reach a position point, regardless of its orientation.

Among the others, this mapping can be learned after foveating several random arm movements; this technique is named *motor babbling* [23]. Motor babbling is a learning schema (or system identification) where the robot autonomously develops an internal model of its body in the environment either randomly or systematically exploring different configurations. This approach could be particularly suitable for those cases in which the robot kinematics is unknown or too difficult to infer.

In this work we present a bioinspired approach for reaching; we show how data from a redundant stereo camera structure, driven by a controller fitting the Hering's law of equal innervation are used to build a visuomotor map in the radial basis framework. The main contributions of this paper are briefly summarized. The first contribution is to successfully exploit the synergy between an active stereo vision system based on the Hering's law and a radial basis network that performs the visuomotor mapping. The second contribution is to show how a redundant stereo camera controller can be effectively used to train a sensorimotor map.

This paper is organized as follows. In Sect. 2 we present several related works, in Sect. 3 we design the system architecture that performs the implicit sensorimotor mapping, in Sect. 4 we present the experimental results and in Sect. 5 we draw our conclusions.

2 Related Works

Even though the problem of sensorimotor representation has been widely treated in literature (see [11] for a review), for the purposes of our work we consider only those papers approaching the visuomotor mapping through motor babbling [23]. We start with a review of recent works on active stereo systems and then we present some approaches to the sensorimotor mapping problem.

The problem of controlling an active stereo head for the foveating task is well known in literature, as surveyed in [3]. The proposed control architectures are either bioinspired or not. Many applications are presented, describing how an active stereo head can be effectively used to solve the arising problems (e.g. localization, object recognition, tracking, and mapping). Among the different surveyed approaches, it is worth noting that the most of bioinspired architectures are based on the disparity energy model [17], directly controlling vergence and version.

The biologically inspired strategies to design a control system for active stereo head are based on autonomous development through reinforcement learning [28,29], on the disparity energy neurons [7,8,21,25,26], or on or on Hering's law of equal innervation [15,24]. Other classical approaches address the problem of controlling the vergence of an active stereo head through algorithms such as fuzzy systems [14], or SIFT [1].

The disparity energy neurons are a model of the neural architecture of the primary visual cortex, dealing with the binocular fusion [17]. They are based on the Gabor-like filters, modelling the receptive fields of the energy neurons. These filters can either be predefined [4,16,20,27] or estimated through statistical learning [12,13,18]. The neural population responses of the disparity energy models are robust features both for the disparity estimation [16,27] and the computation of the vergence angle [26]. The Hering-based models are parametric open-loop controllers, based the law of equal innervation. They receive as input the retinal position of both eyes and produce in output the final joints configuration of the head to foveate the target. The models internally compute the vergence and version angles that are used to compute the movement in the head joints space.

The visuomotor mapping is the correlation between the visual representation of the target and the arm joints configuration to reach it. Chinellato et al. show a bidirectional visuomotor mapping of the environment built on a radial basis function framework that is trained through exploratory actions (gazing and reaching) and implemented on a real humanoid robot [5]. Saegusa et al. propose a method to infer the body schema based on stochastic motor babbling. The babbling is driven by the visuomotor experience [22]. In another work, Saegusa et al. propose a new method to produce a motor behaviour which improves the reliability of the state prediction [23]. Hemion et al. report a competitive learning mechanism to infer the way the robot actuators can influence its sensory input without a preprocessing step of self-detection [10]. Gläser et al. claim the first implementation of a framework that includes a population coding layer for the representation of the *schemata* in a neural map and a basis function representation for the sensorimotor transformations where *schemata* refers to a cognitive structure describing regularities within experiences that is similar to the motor primitives reported for vertebrates [9]. Further references on the radial basis networks used for sensorimotor mapping can be found in [5].

3   System Architecture

The simulated robotic system is composed by a head and an arm. The head is a 4 DOF structure with 2 DOF for the pan command for both eyes, 1 DOF for tilting, and 1 DOF for the neck component; the arm is composed by 2 DOF for the shoulder and 1 DOF for the elbow as in Fig. 1. We define the properties of the head and the arm to be as much as possible compatible with the human characteristics (see Fig. 1 right pane, for the head please refer to [15]).

We evaluate the system for a reaching task: given a target feature in space the system must be able to perceive it with the stereo cameras, compute the joint

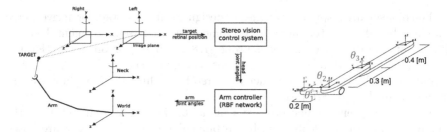

Fig. 1. System architecture. (left pane) The schematic model of the working environment with the active stereo system and the arm initial position. The aim is to detect the target position in space through the stereo cameras, compute the head joint angles to foveate the target and directly compute the final joint configuration of the arm to reach the target location. The sensorimotor map is learned using the end-effector itself as a target for the vision system. (right pane) The schematic of the arm. It has 3 DOF with links lengths compatible with human counterparts. The range of θ_1 is $[-\frac{\pi}{2}, \frac{\pi}{2}]$, the θ_2 range is $[-\frac{\pi}{2}, \frac{\pi}{2}]$ and the θ_3 range is $[0, \frac{3}{4}\pi]$.

angles of the head to foveate that 3D point and, *without* physically foveating it, use these head angles to map the head joints space into the arm joints space for reaching the target position with the end-effector.

The control architecture for reaching can be functionally subdivided in two main modules: the first module is the vergence control system that controls the stereo cameras system, and the second module is the RBF network that computes the final arm joint configuration for reaching, only knowing the joint configuration of the stereo system to foveate the target. Conceptually, the training is equivalent to the motor babbling schema; given a set of random movements of the arm it is possible to correlate the arm joints space and the foveating joint angles of the head. After an initial training phase, where the system explores the environment and gathers data, the radial basis network is trained and the performance is verified on a test set of 3D points (knowing the ideal arm joints configurations to reach them).

3.1   Active Stereo System

The active stereo system is driven by a bioinspired controller based on the Hering's law [15]. The vision system is redundant and it is trained independently, before constructing the sensorimotor map. The system uses a proportional model which needs to be trained to learn the proportional parameters. The fundamental equations are:

$$\dot{\theta}_{version} = K_1(x_L + x_R)$$
$$\dot{\theta}_{vergence} = K_2\delta$$
$$\dot{\theta}_{tilt} = K_3(y_L + y_R)$$
$$\dot{\theta}_{neck} = K_4(x_L + x_R)$$

where x_L and x_R are the feature x-position on the left and right image plane, whereas y_L and y_R are the feature y-position on the left and right image plane. The disparity of the projected feature is represented by δ, and $[K_1, K_2, K_3, K_4]$ are the parameters that must be estimated.

We can compute the pan, tilt and neck angles as following:

$$\dot{\theta}_r = \dot{\theta}_{version} - \dot{\theta}_{vergence}$$
$$\dot{\theta}_l = \dot{\theta}_{version} + \dot{\theta}_{vergence}$$
$$\dot{\theta}_t = -\dot{\theta}_{tilt}$$
$$\dot{\theta}_n = \dot{\theta}_{neck}$$

where $\dot{\theta}_r$ is the pan command for the right camera, $\dot{\theta}_l$ is the pan command for the left camera, $\dot{\theta}_t$ is the tilt command for the head and $\dot{\theta}_n$ is the command for the neck in order to foveate the target.

The parameters K are estimated through gradient descent, based on the minimization of the following cost function:

$$c(X, Y, Z) = e_L^2 + e_R^2 + \sum_j | \dot{\theta}_l |_j + \sum_j | \dot{\theta}_r |_j + \sum_j | \dot{\theta}_t |_j + \sum_j | \dot{\theta}_n |_j$$

$$K = argmin_{K_1, K_2, K_3, K_4} \sum_x \sum_y \sum_z c(x, y, z)$$

where e_L and e_R are the Euclidean distances in the image plane between the image centre and the projection of the target into the left and right image plane, respectively. The sum terms are introduced to avoid trajectory oscillations and to minimize the trajectory duration. After the training, the estimated Ks parameters in [15] are:

$$K1 = 0.0167 \quad K2 = 0.5543 \quad K3 = 0.1584 \quad K4 = 0.3542 \tag{1}$$

The performed experiments show the robustness and the effectiveness of the controller; moreover the trajectories generated from this controller are compatible with the human head trajectories in foveating tasks [2]. Finally, comparing the performance with those reported by [24], the controller solves the redundancy improving both the performance and the robustness of the system.

For the purpose of this work, we take into account only the final joint configuration in foveating tasks regardless the followed trajectory.

3.2 Motor System and Sensorimotor Map

Given a target in 3D space, the motor system should be able to compute the final joint configuration of the arm to reach the target. This computational task is performed by the sensorimotor mapping. The subsystem receives as input only the joint angles of the head to foveate the target and it is able to compute the final configuration of the arm without physically moving the head. The basis functions can be combined linearly to approximate any nonlinear function, e.g. mapping

of the peripersonal space (or the working space of the robot). According to
[19], basis functions are suitable to model the computational framework of the
posterior parietal cortex where the sensorimotor transformations are performed.

The radial basis network is an artificial neural network where the activation
function of each node is a radial basis function. The radial basis network is
typically composed by a nonlinear hidden layer and a linear output layer. Each
nonlinear hidden neuron is represented by a radial basis function. We use a
Gaussian radial function:

$$h(x) = \exp\left(-\frac{(x-c)^2}{r^2}\right) \tag{2}$$

where $h(\cdot)$ is the basis function, x is the head joints vector, c is the basis center
and r is the spread. The output neuron is defined as:

$$f(x) = \sum_{j=1}^{m} w_j\, h_j(x) \tag{3}$$

where m is the number of hidden neurons, w_j is the weight and $f(x)$ is the
estimated arm joints vector.

We will discuss more in the following section the details of the dataset generation. On the dataset we perform a 10-fold cross-validation, selecting with a
uniform distribution samples in the populated dataset [6]. With 9 folds we train
the network and with the last fold we evaluate the performance, repeating this
procedure for each possible combination of folds. This procedure is equivalent
to motor babbling since the folds are populated by randomly choosing samples
from the whole dataset. Moreover, to improve the accuracy of the sensorimotor mapping, we perform a "meta-learning" over all the possible combination of
the folds, varying the spread of the basis to infer the best spread value. Once
we selected the best spread value, we perform the cross-folding validation to
evaluate the radial basis network. The controlled 3 DOF arm has human-like
characteristics (see Fig. 1 right pane). The lengths of the links are compatible
with those of the human counterpart and the range of the joints is similar to
those of the humans.

4   Experiments and Results

In this section we present the obtained results related to the sensorimotor mapping. The learning procedure of the active stereo system controller is not
presented here in detail, since it is discussed in [15].

4.1   Method

Training the active stereo head results in the definition of its parameters [15].
We discretized the arm joint space in fine steps of 12° along each axis, for a total

number of 1062 samples. For each position in the joint space the direct kinematics is computed. Knowing the position of the end-effector and its projection onto the image planes of the stereo cameras, we compute the vergence-version angles to foveate the end-effector itself. With the calibration parameters of the stereo cameras we compute the foveating point in 3D space in order to compute the Euclidean error between the 3D position of the end-effector and the foveated point in space. This is an intrinsic error of the active visual controller and it does not depend on the arm controller. The trained network should be able to manage it, estimating the end-effector position regardless the foveation error. Figure 2 (left pane) shows the generated dataset; each point represents a valid end-effector position. Moving selectively the arm in space we build a dataset where each sample is composed by:

- 3D position of the end-effector
- arm joints configuration
- head joints configuration, foveating the 3D position of the end-effector (projected into the image planes of the cameras)
- Euclidean error in 3D space between the foveated point and the position of the arm (due to the vision system intrinsic error).

We split the dataset in 10 folds to perform the cross-validation in the training phase of the radial basis network where each testing fold is composed by 118 samples. Cross-validation is widely used to evaluate the performance of the radial basis network, using the mean square error (MSE) as evaluation criterion to control the stopping of the training phase. In order to improve as much as possible the accuracy of the network, we implemented an optimization loop to detect the best spread value for the basis that minimizes the mean Euclidean error between the estimated arm joint configuration and the real arm configuration. The range of the evaluated spread values is $[0.5, 1.3]$ rad.

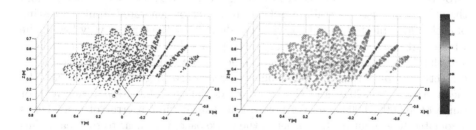

Fig. 2. (left pane) Complete dataset. The points in space represent the end-effector positions used as targets for the active stereo head. In the dataset, each end-effector position is associated with the corresponding arm joint configuration, the foveating joint angles of the head, and the Euclidean error between the foveation point and the end-effector position. This dataset is used for the cross-folding validation. (right pane) The Euclidean error between the real end-effector position and the one estimated by the radial basis network; the error is quite low except for that 3D points that are very near to the head and to the shoulder.

4.2   Results

In this paper we extend previous results adding a 3 DOF arm, to quantitatively evaluate the capability of learning a visuomotor map. We briefly summarize how to train the head control system, according to [15]. We had trained the head controller on the whole space with the gradient descent technique; the obtained results pointed out the robustness of the system. It is worth noting that the head is redundant with 4 DOF. According to previous results, we confirm that the best performances are obtained with a decoupled control eye-neck. The trajectories generated by the head controller are compatible with humans [2]. The experiments show the capability of the head controller first to foveate targets in a wide region of the 3D space starting from a initial head position and, second to foveate a specific 3D target starting from a generic head joints configuration.

On the other hand, we implemented the radial basis network to encode the visuomotor mapping between the head joints space and the arm joints space. The evaluation of the radial basis network is mainly composed by two phases:

1. choose the best spread value
2. evaluate the network performance with the cross-folding technique (using the optimal spread value).

In the first phase, we execute an optimization loop over the 10-fold cross-validation, varying the spread value of the basis functions. In the optimization phase, we found that the best spread value is equal to $1.25\,[rad]$. After the estimation of the optimal value, we use it to perform the 10-fold cross-validation over the dataset, evaluating the overall performance related to the reaching task.

Figure 2 (right pane) shows the scatter of the Euclidean error between the real position of the end-effector and the estimated one for the test set. As previously said, we use the 10-fold cross-validation so the dataset is split in 10 folds and the error is given for each fold used in testing (each fold is used as testing set and the other 9 are used as training set). Here the error generated from the testing of each fold is plotted. The figure clearly shows that the error is very low in the whole workspace with the exception of the points that are very near to the head and shoulder. Knowing the foveation angles of the head, the network is able to correctly compute the arm joint angles to reach the targets that are seen. The mean error is $0.0320\,m$ and the standard deviation is $0.0591\,m$ inside a error range of $[0.0001, 0.9603]\,m$. We notice that the maximum error of $0.9603\,m$ is associated to a point in the border of the workspace, very near to the head. This error is due to the constraints that we have imposed to the arm joints range, when the forearm is at $132°$ over $135°$.

Figure 3 shows the Euclidean distance between the estimated arm position and the desired target. The blue dots represent the targets and the red lines are the distance in space between the target and the arm position computed by the network. In the left pane is shown the projection on the X-Z plane of the world frame of reference; the error is distributed in the whole space but decreases when the Z value increases. This is due to the constraints imposed on the arm

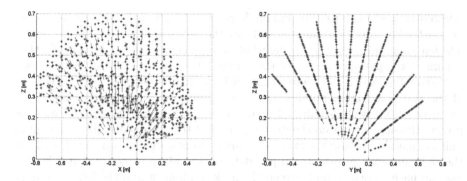

Fig. 3. Error directions projected on different planes of the world frame of reference. The blue dots represent the targets and the red lines are the distance in space between the target and the arm position computed by the network. For visualization reasons, we do not plot the estimated end-effector position. (left pane) Error projection into the plane X-Z. (right pane) Error projection into the plane Y-Z (Color figure online).

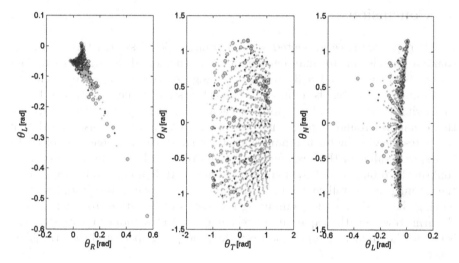

Fig. 4. Radial basis centers distribution in the input space. The red circle represents a basis center, the blue dots are the testing values and the cyan dots are the values used for the training phase (Color figure online).

range. The right pane shows the same projection on the Y-Z plane and clearly indicates that the error is not due to the shoulder component in the tilt direction, since the red lines are along the same shoulder position. This is due to the fact that the network in input receives only one tilting component of the head respect to the horizontal head movements that are generated by the left, right and neck components.

Figure 4 shows the radial basis centers distribution in the input space and is referred to a single training of the cross-folding validation. The training phase has selected $m = 82$ neurons. The red circle represents a basis center, the blue dots are the testing values and the cyan dots are the values used for the training phase. The left pane shows the centers distribution projected on the plane of θ_L-θ_R that command the pan movements of the camera. The centers are distributed along a straight line and it is in agreement with the highly cooperation-competition of the two pan angles. Furthermore, the θ_L and θ_R values used as input for the network are well-distributed along the same line. In the middle pane we project the same center on the θ_N-θ_T plane and the centers distribution is quite uniform mainly because the two angles are conceptually independent. Finally in the right pane we project the input data on the plane θ_N-θ_L and also in this case we have a correlation between the neck component and the left camera. It is clearly shown that the neck performs the raw movements and the camera only executes small movements to correct the vergence. Also in this case the centers are mainly distributed in the region containing the common movements but for the outliers, there are ad-hoc centers created during the training phase.

5   Conclusions

In this paper we have presented a novel computational system that is able to compute the visuomotor map between a target (perceived through a redundant active stereo head) and a 3 DOF arm, designed with human-like mechanics constraints. The system is composed by an active stereo camera controller, fitting the Hering's law, that is able to foveate a generic point in space and a 3 DOF arm that can move around its peri-personal space through a radial basis network that computes the visuomotor mapping. Through an optimization phase we select the best spread value for the basis functions that are part of the neural network. We confirm the robustness of the mapping through a 10-fold cross-validation. After the training the overall controller is able to detect in space a target and without moving the head, but just computing the angles of foveation, to reach it with the arm. However, the capability of reaching a target without moving the head is due to the implicit mapping between the head joints space and the arm joints space.

The results confirm the robustness and the accuracy of the system to reach target in peri-personal space. Moreover, the Hering-based stereo controller generates the joint angles that are robust features to drive the network for the visuomotor mapping. The next step is to validate the model with a real robot with the same characteristics described above and adding new degrees of freedoms to the arm to make it redundant and more humanoid.

Acknowledgments. Partial support to this work has been provided by the 2010-2012 Italian-Korean bilateral project (ICT-CNR and KAIST). The authors gratefully acknowledge the contribution of the NVIDIA Academic Partnership for providing GPU computing devices.

This work was partially developed at Artificial Intelligence Laboratory, Department of Informatics, University of Zurich.

References

1. Aragon-Camarasa, G., Fattah, H., Siebert, J.P.: Towards a unified visual framework in a binocular active robot vision system. Robot. Auton. Syst. **58**(3), 276–286 (2010)
2. Chen, L.L.: Head movements evoked by electrical stimulation in the frontal eye field of the monkey: evidence for independent eye and head control. J. Neurophysiol. **95**, 3528–3542 (2006)
3. Chen, S., Li, Y., Kwok, N.M.: Active vision in robotic systems: a survey of recent developments. Int. J. Robot. Res. (2011). doi:10.1177/0278364911410755. http://ijr.sagepub.com/content/early/2011/08/16/0278364911410755.abstract
4. Chen, Y., Qian, N.: Coarse-to-fine disparity energy model with both phase-shift and position-shift receptive field mechanisms. Neural Comput. **16**, 1545–1577 (2004)
5. Chinellato, E., Antonelli, M., Grzyb, B.J., del Pobil, A.P.: Implicit sensorimotor mapping of the peripersonal space by gazing and reaching. IEEE Trans. Auton. Ment. Dev. **3**(1), 43–53 (2011). doi:10.1109/TAMD.2011.2106781
6. Duda, R.O., Hart, P.E., Stork, D.G.: Pattern Classification. Wiley, New York (2001)
7. Gibaldi, A., Canessa, A., Chessa, M., Sabatini, S.P., Solari, F.: A neuromorphic control module for real-time vergence eye movements on the iCub robot head. In: Proceedings of 11th IEEE-RAS International Conference on Humanoid Robots (Humanoids), pp. 543–550 (2011). doi:10.1109/Humanoids.2011.6100861
8. Gibaldi, A., Chessa, M., Canessa, A., Sabatini, S.P., Solari, F.: A cortical model for binocular vergence control without explicit calculation of disparity. Neurocomputing **73**, 1065–1073 (2010)
9. Glaser, C., Joublin, F., Goerick, C.: Learning and use of sensorimotor schemata maps. In: Proceedings of IEEE 8th International Conference on Development and Learning ICDL 2009, pp. 1–8 (2009). doi:10.1109/DEVLRN.2009.5175509
10. Hemion, N.J., Joublin, F., Rohlfing, K.J.: A competitive mechanism for self-organized learning of sensorimotor mappings. In: Proceedings of IEEE International Conference on Development and Learning (ICDL), vol. 2, pp. 1–6 (2011). doi:10.1109/DEVLRN.2011.6037364
11. Hoffmann, M., Marques, H., Arieta, A., Sumioka, H., Lungarella, M., Pfeifer, R.: Body schema in robotics: a review. IEEE Trans. Auton. Ment. Dev. **2**(4), 304–324 (2010). doi:10.1109/TAMD.2010.2086454
12. Hoyer, P.O., Hyvarinen, A.: Independent component analysis applied to feature extraction from colour and stereo images. Netw. Comput. Neural Syst. **11**, 191–210 (2000)
13. Hyvarinen, A., Hoyer, P.O.: A two-layer sparse coding model learns simple and complex cell receptive fields and topography from natural images. Vis. Res. **41**, 24132423 (2001)
14. Kyriakoulis, N., Gasteratos, A., Mouroutsos, S.G.: Fuzzy vergence control for an active binocular vision system. In: Proceedings of 7th IEEE International Conference on Cybernetic Intelligent Systems CIS 2008, pp. 1–5 (2008). doi:10.1109/UKRICIS.2008.4798931
15. Mutti, F., Alessandro, C., Angioletti, M., Bianchi, A., Gini, G.: Learning and evaluation of a vergence control system inspired by hering law. In: IEEE International Conference on Biomedical Robotics and Biomechatronics (BIOROB) (2012)
16. Mutti, F., Gini, G.: Bio-inspired disparity estimation system from energy neurons. In: ICABB (2010)

17. Ohzawa, I., De Angelis, G.C., Freeman, R.D.: Stereoscopic depth discrimination in the visual cortex: neurons ideally suited as disparity detectors. Science **249**, 1037–1041 (1990)
18. Olshausen, B.A., Field, D.J.: Emergence of simple cell receptive field properties by learning a sparse code for natural images. Nature **381**, 607–609 (1996)
19. Pouget, A., Sejnowski, T.J.: Spatial transformations in parietal cortex using basis functions. J. Cogn. Neurosci. **9**(2), 222–237 (1997)
20. Qian, N.: Computing stereo disparity and motion with known binocular cell properties. Neural Comput. **6**, 390–404 (1994)
21. Qu, C., Shi, B.E.: The role of orientation diversity in binocular vergence control. In: Proceedings of International Joint Conference on Neural Networks (IJCNN), pp. 2266–2272 (2011). doi:10.1109/IJCNN.2011.6033511
22. Saegusa, R., Metta, G., Sandini, G.: Own body perception based on visuomotor correlation. In: Proceedings of IEEE/RSJ International Conference on Intelligent Robots and Systems (IROS), pp. 1044–1051 (2010). doi:10.1109/IROS.2010.5650974
23. Saegusa, R., Metta, G., Sandini, G., Sakka, S.: Active motor babbling for sensorimotor learning. In: Proceedings of IEEE International Conference on Robotics and Biomimetics ROBIO 2008, pp. 794–799 (2009). doi:10.1109/ROBIO.2009.4913101
24. Samarawickrama, J.G., Sabatini, S.P.: Version and vergence control of a stereo camera head by fitting the movement into the Hering's law. In: Proceedings of Fourth Canadian Conference on Computer and Robot Vision CRV '07, pp. 363–370 (2007). doi:10.1109/CRV.2007.69
25. Shimonomura, K., Yagi, T.: Neuromorphic vergence eye movement control of binocular robot vision. In: Proceedings of IEEE International Conference on Robotics and Biomimetics (ROBIO), pp. 1774–1779 (2010). doi:10.1109/ROBIO.2010.5723600
26. Tsang, E.K.C., Lam, S.Y.M., Meng, Y., Shi, B.E.: Neuromorphic implementation of active gaze and vergence control. In: Proceedings of the IEEE International Symposium on Circuits and Systems, pp. 1076–1079 (2008)
27. Tsang, E.K.C., Shi, B.E.: Disparity estimation by pooling evidence from energy neurons. IEEE Trans. Neural Netw. **20**(11), 1772–1782 (2009)
28. Wang, Y., Shi, B.E.: Autonomous development of vergence control driven by disparity energy neuron populations. Neural Comput. **22**, 730–751 (2010)
29. Wang, Y., Shi, B.E.: Improved binocular vergence control via a neural network that maximizes an internally defined reward. IEEE Trans. Auton. Ment. Dev. **3**(3), 247–256 (2011). doi:10.1109/TAMD.2011.2128318

A Distributed Multi-level PSO Control Algorithm for Autonomous Underwater Vehicles

Raffaele Grandi$^{(\boxtimes)}$ and Claudio Melchiorri

DEI, Department of Electrical, Electronic and Information Engineering,
University of Bologna, Bologna, Italy
{raffaele.grandi,claudio.melchiorri}@unibo.it

Abstract. This paper presents a distributed control technique based on the Particle Swarm Optimization algorithm and able to drive in unknown environments a group of autonomous robots to a common target point. In this paper, we consider in particular the case of underwater vehicles. The algorithm is able to deal with complex scenarios, frequently found in benthic exploration as e.g. in presence of obstacles, caves and tunnels, and to consider the case of a mobile target. Moreover, asynchronous data exchange and dynamic communication topologies are considered. Simulations results are provided to show the features of the proposed approach.

Keywords: AUVs · Particle swarm optimization · Multi-robot systems · Distributed control · Obstacle avoidance

1 Introduction

In this article, a navigation control technique able to drive a group of autonomous underwater vehicles (AUVs) is presented. Main features of this technique are its distributed structure and intrinsic robustness with respect to the presence of unknown obstacles, tunnels and dead ends, and also with respect to losses of one or more units. Moreover, the case of mobile final target has also been considered.

This technique is based on the fairly recent theory of *Swarm Intelligence*, [1] which takes into account the study of self-organizing systems. This means that the action expressed by the whole group results from the combination of coordinated actions by individual entities. Initially, simple motion rules have been defined from the study of these actions [2]. More recently, more complex techniques have been introduced, such as the optimization algorithm initially considered for the development of our navigation technique. The Particle Swarm Optimization algorithm (PSO) [3] is a meta-heuristic algorithm biologically inspired by flocks of birds. It is a non-gradient and direct-search based optimization strategy, in which a set (a population) of $\mathcal{N}$ possible solutions (a swarm of particles) is iterated in parallel. They search for the best solution in a multidimensional space (or domain). As a matter of fact, this technique has some interesting features that can be exploited in the guidance of a swarm of robots,

G.A. Di Caro and G. Theraulaz (Eds.): BIONETICS 2012, LNICST 134, pp. 75–90, 2014.
DOI: 10.1007/978-3-319-06944-9_6, © Institute for Computer Sciences, Social Informatics and Telecommunications Engineering 2014

such as its reliability and flexibility, its intrinsic simplicity, the robustness to failures and the relatively small amount of information needed to create desired emergent behaviors. PSO approach was originally developed in 1995 to study social interactions, [4]. Later, many modifications to original formulation of the algorithm have been proposed in order to improve its efficiency [5] and, more recently, PSO algorithms have been applied to robot navigation ([6,7]) and path planning ([8]). PSO has also been used in a hybrid fashion with other bio-inspired techniques: Genetic Algorithm (GA) [9] and Artificial Neural Network (ANN) [10]. Concerning control strategies for autonomous underwater vehicles, many papers have been presented in the literature, see e.g. [11,12]. Some applications of the PSO algorithm on AUVs can even be found e.g. in [13,14], although probably the most interesting modification to original PSO algorithm, useful to guide a swarm of robots, has been only recently introduced: the asynchronous and completely decentralized features described e.g. in [15], where each particle evaluates PSO rules autonomously and a dynamic communication topology is considered [16]. Recent applications of these features are presented in e.g. [7,17,18].

In this article, a novel improvement of the algorithm is presented, starting from its asynchronous version in order to use it in the navigation of underwater vehicles. The purpose is to make the group of robots suitable for seabed and cave exploration, being simultaneously able to avoid collisions with obstacles and team-mates. Navigation algorithm is completely decentralized and structured in three levels. The two upper levels, which are directly based on PSO, provide navigation way-points and obstacle avoidance respectively. The third one is devoted to AUV control with standard techniques. Each agent is autonomous and it exploits not only its own knowledge of the environment, but also the data received from reachable neighbors. A particular feature of this technique is the addition of obstacle avoidance. Unknown environment barriers, perceived by vehicle sensors, are displayed by the algorithm as constraints into the search space. During the navigation, agents exchange data about best locations and encountered obstacles. The sharing of these information creates a kind of collective memory that is used in the selection of future way-points towards final destination. This procedure runs asynchronously and in an intermittent fashion, depending on a free communication topology which is a function of inter-agent distances.

This paper is organized as follows: In Sect. 2, background material on PSO and related algorithm is provided. The main result of the paper is reported in Sect. 3, where our PSO technique is described. The simulations used to validate our approach are presented and discussed in Sect. 4 and, finally, conclusions and future work are reported in Sect. 5.

2   Background on PSO

In literature, PSO is usually considered as an optimization algorithm, i.e. it is typically used to solve optimization problems defined in a m-dimensional space by a fitness function $f(\cdot)$, that is subject to predefined constraints and whose value has to be minimized (or maximized). In the original PSO formulation, optimal solution is computed by simulating a group of n particles that explore the search space of the problem in order to find the best fitness value. Each agent (or particle) moves in the solution search space with known position and velocity, and with the ability to communicate with other agents. In particular, PSO is based on a population $\mathcal{P}$ of n particles that represents a set of possible solutions of the given m-dimensional problem, i.e. $\|\mathcal{P}\| = n$ where the operator $\|\cdot\|$ computes the cardinality of a given set. Position and velocity of the i-th particle at the k-th iteration of the algorithm are identified by the m-dimensional vectors

$$\boldsymbol{p}_i(k) = [p_{i,1}(k) \ldots p_{i,m}(k)]^T \\ \boldsymbol{v}_i(k) = [v_{i,1}(k) \ldots v_{i,m}(k)]^T \qquad i = 1 \ldots n$$

When the algorithm is initialized, a random position $\boldsymbol{p}_i(0)$ and a starting random velocity $\boldsymbol{v}_i(0)$ are assigned to each particle. At each iteration, a set of candidate solutions is optimized by the particles which are moving through the search space toward better values of the fitness function $f(\cdot)$. Each particle knows the value of the fitness function corresponding to its current position in the m-dimensional search space and it is able to *remember* data from previous iterations. In particular, each particle is able to remember the position where it has achieved the best value of fitness function, namely $\boldsymbol{p}_i^*$ (called *local best*). This value can be exploited by neighbors to change their behavior. In fact, assuming the possibility of a global communication between the particles, each of them can gather the positions $\boldsymbol{p}_j^*$ of other team-mates where they have detected their best fitness value. Therefore, the best value $\boldsymbol{p}_i^+$ of all the particles can be defined (*global best*). The propagation of $\boldsymbol{p}_i^+$ through the swarm depends on communication topology, thus it may happen that non-communicating particles have temporarily different values. Moreover, each agent can use the best current fitness value of its neighbors $\boldsymbol{p}_i^\times$ (called *neighborhood best*) as another term to define its behavior. Figure 1 clarifies this concept in a 2D space.

In conclusion, at each iteration of the algorithm, the behavior of each particle is defined by a proper function made of three factors: $\boldsymbol{p}_i^*$ (the local best), $\boldsymbol{p}_i^+$ (the global best), and $\boldsymbol{p}_i^\times$ (the neighborhood best). From these considerations, it follows that the communication topology chosen to route information among the particles is an important feature able to drastically change the behavior of the whole swarm. Therefore, it has to be carefully chosen. There are many works on the importance of swarm topological configurations, see e.g. [16,19]. As an example, in case of a *static* communication topology, the contribution of $\boldsymbol{p}_i^+$ generates a quite *static behavior*. Therefore it is not guaranteed that the search space is fully and properly explored. As a matter of fact, in this case the movements of each particle must be somehow limited in order to remain connected to the neighbors. The consequence is that the search possibilities also become limited. It follows that this approach could drive the system to a local minimum and thus to a local optimal solution. Furthermore, if the chosen static communication topology has not enough connections, each agent could even have

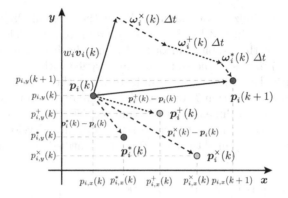

Fig. 1. Original PSO algorithm applied to particles moving in a 2D environment.

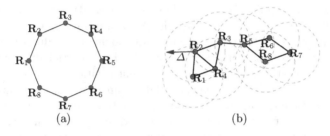

(a)   (b)

Fig. 2. Examples of a 2D communication graph in case of $\|\mathcal{P}\| = 8$ with $\Delta_i = \Delta$, $i = 1 \ldots 8$: a *ring* graph, 2(a), and a Δ-disc graph, 2(b).

a different value for p_i^+ since the data could not be updated properly. In this case, the swarm could eventually reach dispersed configurations. The ring graph depicted in Fig. 2(a) is an example of a static communication topology where each agent chooses a well defined neighbor, always keeping the same connections.

In order to avoid these problems, we have assumed a dynamic communication topology that depends on inter-particle distances. Each particle can move without constraints in the search space and can communicate with any other neighbor in its communication range. In this way we have obtained a better exploration of the search space and a faster update of data related to the environment. If a particle has better data on the environment, its neighbors automatically tend to aggregate around it to reach the best available position.

Communication graphs, whose topology depends on inter-agents distance, are usually addressed as Δ-disc graphs, see Fig. 2(b). This kind of topology is typically applied to 2-D networks. With abuse of notation we have addressed distance-based communication topology as Δ-disc graphs also in case of agents defined in $m > 2$ dimensions. To this purpose, we have defined the neighbor set of the i-th particle as the set of all the particles whose distance from p_i is smaller than a predefined threshold Δ_i

$$\mathcal{N}_i = \{j \in \mathcal{P} : \|p_i - p_j\| \leq \Delta_i\} \tag{1}$$

where $|| \cdot ||$ computes the length of a vector. At the k-th iteration of PSO algorithm, the state of each particle is updated as follows

$$v_i(k+1) = \xi_0 \left(w_i v_i(k) + \omega_i^*(k) + \omega_i^+(k) + \omega_i^\times(k) \right) \qquad (2)$$

$$p_i(k+1) = p_i(k) + v_i(k+1)\, \Delta t \qquad (3)$$

where w_i is a scalar constant that represents the *inertia* of the particle, ξ_0 is a parameter called *constriction factor* [20] (introduced to avoid the dispersion of the particles), the term $w_i v_i(k)$ is addressed in literature as *persistence* and represents the tendency of a particle to preserve its motion direction, Δt is the simulation time step. The terms $\omega_i^*(k)$, $\omega_i^\times(k)$ and $\omega_i^+(k)$ represent *historical* and *social* contributions to control action and are defined as

$$\omega_i^*(k) = \phi \, r_i^*(k) \;\; (p_i^*(k) - p_i(k))/\Delta t$$

$$\omega_i^+(k) = \phi \, r_i^+(k) \;\; (p_i^+(k) - p_i(k))/\Delta t$$

$$\omega_i^\times(k) = \phi \, r_i^\times(k) \;\; (p_i^\times(k) - p_i(k))/\Delta t$$

The first term represents the contributions to the velocity given by the *best individual* value of the self particle. The second and third elements represent the *swarm* and the *neighborhood* contribution respectively. Each term is a vector attracting each particle towards the corresponding point into the search space. The parameters $r_i^*(k)$, $r_i^+(k)$ and $r_i^\times(k)$ are usually selected as uniform random numbers in $[0, 1]$ and they are computed at each iteration of the algorithm to give a range of randomness to the particle's behavior. The parameter ϕ is used to modulate the maximum influence of this random behavior. Because of its importance in tuning PSO parameters, the value of ϕ has been determined in literature by using many methods, see e.g. [21], often with empirical approaches as in our case. As a matter of fact, after many simulations and considering in particular the capability of the swarm to overcome obstacles without loosing any particles, we have set $\phi = 3.5$, see Sect. 4.

3   The Algorithm

The algorithm presented in this paper has been developed with the aim to drive a swarm of underwater vehicles $\mathcal{V}_i$, $i = 1, \ldots, n$ moving in unknown 3D environments. Each vehicle is autonomous: this means in particular that a local version of PSO algorithm is executed individually by each agent, and that all the decisions are made on the basis of the data locally available at the k-th step. In order to apply PSO algorithm to AUV 3D navigation, there are some problems to be solved: the matching of each particle to a single vehicle; the matching of the PSO search space with the environment surrounding the swarm and how to embed obstacle avoidance capabilities; local minima avoidance; the transformation of exchanged information, collected by each vehicle, into a unique collective map.

3.1   Matching Particles with Physical Vehicles

In this Section, we describe the model of the vehicles considered in the PSO algorithm (see also Sect. 4), the general control structure, and the matching between the mobile robots and the particles of the algorithm.

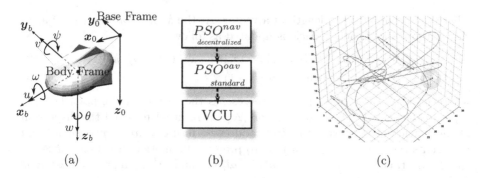

(a)   (b)   (c)

Fig. 3. Figure 3(a) shows the AUV with the reference frames and the state variables; the global control scheme is depicted in Fig. 3(b), while Fig. 3(c) shows a way-point navigation test.

Table 1. Kinematic variables

Name	Description	Unit	Name	Description	Unit
x	Position with respect to x_0	m	u	Linear velocity along x_b	m/s
y	Position with respect to y_0	m	v	Linear velocity along y_b	m/s
z	Position with respect to z_0	m	ω	Linear velocity along z_b	m/s
Ω	Roll angle with respect to x_0	rad	w	Angular velocity about x_b	rad/s
Ψ	Pitch angle with respect to y_0	rad	ψ	Angular velocity about y_b	rad/s
Θ	Yaw angle with respect to z_0	rad	θ	Angular velocity about z_b	rad/s

Model of the AUV. Since our interest is mainly focused on the distributed control algorithm, we have taken into account only the kinematic model of the vehicle, neglecting its dynamics as also done in many other papers. We define a body reference frame $^bF_i = \{x_{b,i}, y_{b,i}, z_{b,i}\}$ for each vehicle, and a common inertial (base) frame $^0F = \{x_0, y_0, z_0\}$, as shown in Fig. 3(a) (for the sake of simplicity, in the following the subscript i will be neglected). Moreover, we assume that each vehicle is equipped with an engine generating a linear velocity u along the x_b direction, and with four fins that can change the orientation of the vehicle itself with respect to ^0F. We have used RPY angles to describe the orientation of the vehicle with respect to ^0F, as shown in Fig. 3(a). In this way, the state of the vehicle can be defined by: the body-fixed linear velocity vector $^b\dot{\mathbf{x}} = [u, v, w]^T$, the body-fixed angular velocity vector $^b\boldsymbol{\omega} = [\omega, \psi, \theta]^T$ and the three RPY angles (see Table 1). The correspondence between velocity in body and inertial frame is expressed by (see also [11])

$$\dot{\mathbf{x}}_v = \begin{bmatrix} \dot{x} \\ \dot{y} \\ \dot{z} \end{bmatrix} = \mathbf{R}_{zyx}(\Theta, \Psi, \Omega) \, ^b\dot{\mathbf{x}} \qquad \dot{\boldsymbol{\Gamma}}_v = \begin{bmatrix} \dot{\Omega} \\ \dot{\Psi} \\ \dot{\Theta} \end{bmatrix} = \frac{1}{c_\Psi} \begin{bmatrix} c_\Psi & s_\Omega s_\Psi & c_\Omega c_\Psi \\ 0 & c_\Omega c_\Psi & -s_\Omega c_\Psi \\ 0 & s_\Omega & c_\Omega \end{bmatrix} \, ^b\boldsymbol{\omega} \quad (4)$$

where

$$\mathbf{R}_{zyx}(\Theta,\Psi,\Omega)=\mathbf{R}_z(\Theta)\mathbf{R}_y(\Psi)\mathbf{R}_x(\Omega)=\begin{bmatrix} c_\Theta c_\Psi & c_\Theta s_\Psi s_\Omega - s_\Theta c_\Omega & c_\Theta s_\Psi c_\Omega + s_\Theta s_\Omega \\ s_\Theta c_\Psi & s_\Theta s_\Psi s_\Omega + c_\Theta s_\Omega & s_\Theta s_\Psi c_\Omega + c_\Theta s_\Omega \\ -s_\Psi & c_\Psi s_\Omega & c_\Psi c_\Omega \end{bmatrix}$$

is the matrix that represents the relative orientation of the two frames, and $c_a = \cos(a)$, $s_a = \sin(a)$. Notice that this model presents a singular configuration if $\Psi = \pm\frac{\pi}{2}$ rad: this is acceptable since the vehicles typically do not operate close to this configuration. However a different representation, free of this kinematic singularity, can be defined as in [22], where quaternions have been used. Notice also that, due to the non-holonomic characteristics of the vehicle, velocities v and w are always null.

Finally, the inverse homogeneous transformation matrix between ^bF and ^oF is:

$$^b\mathbf{T}_o = \begin{bmatrix} ^o\mathbf{R}_b^T & -^o\mathbf{R}_b^T \boldsymbol{x}_v \\ \mathbf{0}^T & 1 \end{bmatrix} \tag{5}$$

where $^o\mathbf{R}_b^T = \mathbf{R}_{zyx}(\Theta,\Psi,\Omega)$. Control inputs available to drive the vehicle are engine thrust, that in our model corresponds to the velocity u along x_b, and the position of the four fins. In particular (see Fig. 4), if 'fin up' ($\mathcal{F}_u$), 'fin down' ($\mathcal{F}_d$), 'fin left' ($\mathcal{F}_l$), and 'fin right' ($\mathcal{F}_r$), represent the four fins, three fin configurations can be defined:

- α: $\mathcal{F}_u$ and $\mathcal{F}_d$ both rotated of an α angle on the right of vertical axis. This produces a *Yaw* angle rotation of the vehicle.
- β: $\mathcal{F}_l$ and $\mathcal{F}_r$ both rotated of a β angle over the top of horizontal axis. This produces a *Pitch* angle rotation of the vehicle.
- γ: all the fins are rotated on the left of the relative fin's axis. This produces a *Roll* angle rotation of the vehicle.

We have assumed that the range of motion of each fin is bounded, i.e. α, β, $\gamma \in [-0.26, 0.26]$ rad. Furthermore, when the vehicle is moving and fin positions are not null, angular velocities are not null as well. We have modelled the relation between fin positions $[\alpha, \beta, \gamma]^T$ and angular velocities $[\omega, \psi, \theta]^T$ in ^bF with a (non-linear) function

$$^b\boldsymbol{\omega} = \boldsymbol{K}_\omega\, u\, \sin\left([\gamma, \beta, \alpha]^T\right) \tag{6}$$

where $\boldsymbol{K}_\omega$ is a proper diagonal gain matrix. In conclusion, taking into account that control input consists in the vector $\boldsymbol{\tau} = [u, \alpha, \beta, \gamma]^T$ and that velocities v and w are null, the kinematic model of each AUV is given by (4) and (6).

General Structure of the Control System. On the basis of the results presented in [11], we have designed the vehicle's low-level control (the *Vehicle Controller Unit* (VCU)) as an autopilot that provides a point-to-point navigation mode (see Fig. 3(b)). To this purpose, we have assumed that the AUV state variables (see Table 1) are available, that a desired way-point $^b\boldsymbol{x}_d^{VCU} = [x_d, y_d, z_d]^T$

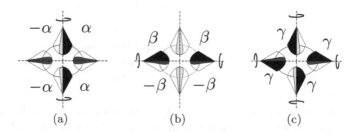

Fig. 4. Rear view of AUV with directional fins in α, β, γ configurations.

and a desired velocity vector ${}^b\dot{\boldsymbol{x}}_d^{VCU}$ have been assigned in the body-frame (provided by PSO algorithm, see below). Then, VCU has to generate both the linear speed u and the fin angular positions α, β, γ. Linear speed is computed by using the following simple control law

$$
{}^b\dot{\boldsymbol{x}}_v = {}^b\dot{\boldsymbol{x}}_d^{VCU} + \boldsymbol{K}_p ({}^b\boldsymbol{x}_d^{VCU} - {}^b\boldsymbol{x}_v) \tag{7}
$$

from which $u = ||{}^b\dot{\boldsymbol{x}}_v||$. $\boldsymbol{K}_p$ is a proper constant (diagonal) matrix used to tune the proportional term. The steering variables $[\alpha, \beta, \gamma]^T$ are also computed on the basis of the desired position ${}^b\boldsymbol{x}_d^{VCU}$. In fact, with a simple transformation from Cartesian to spherical coordinates given by

$$
\begin{cases}
\Psi_d = atan2(y_d, x_d) \\
\Theta_d = atan2(z_d, \sqrt{x_d^2 + y_d^2}) \\
\epsilon_d = \sqrt{x_d^2 + y_d^2 + z_d^2}
\end{cases} \tag{8}
$$

azimut (Ψ_d) and elevation (Θ_d) values are obtained and used to compute the desired rotational velocities in ${}^bF$ as

$$
\psi = k_\Psi (\Psi_d - \Psi), \quad \theta = k_\Theta (\Theta_d - \Theta) \tag{9}
$$

from which, by exploiting (6) and assuming $\omega = 0$, it is finally possible to obtain fin positions α and β useful to reach the given way-point as

$$
[0, \beta, \alpha]^T = \arcsin \left(\frac{1}{u} \boldsymbol{K}_\omega^{-1} [0, \psi, \theta]^T \right) \tag{10}
$$

Notice that it is not necessary to reach a specific final AUV orientation, but only to reach the target. For this reason, the γ fin configuration, for the control of the Ω rotation, has not been considered. The parameter ϵ_d in (8) is used by the algorithm as an estimation of remaining distance between the robot and the way-point target. A way-point navigation test is depicted in Fig. 3(c).

As described in Sect. 2, PSO algorithm considers each particle/unit as a single integrator in x, y, z coordinates, i.e. it calculates a velocity vector $\boldsymbol{v}_i(k) \in \mathbb{R}^3$ without taking into account vehicle's kinematic constraints deriving from its non-holonomic characteristic. In order to solve this problem, we have developed an algorithm structured into two levels, called PSO^{nav} and PSO^{oav}. They provide navigation and obstacle avoidance respectively, see Fig. 3(b). Each level is managed by a proper software agent ($\mathcal{SA}$).

Both levels' agents run on-line in a decentralized fashion on each AUV and compute their respective targets as detailed below. PSO^{nav} has a distributed structure and each $\mathcal{SA}^{nav}$ considers itself as a particle and the neighbor vehicles as other particles inside its work-space. By exploiting the information given by sensors and team-mates, the $\mathcal{SA}^{nav}$ computes its future position according to the PSO basic rules, see Sect. 2. This position is taken as a temporary way-point x_{wp}^{nav} of the global path in order to reach the final target. In the lower level, the PSO^{oav} agent adopts as optimization target the way-point x_{wp}^{nav} provided by $\mathcal{SA}^{nav}$, and it tries to find the best way to reach this point while avoiding obstacles (see next Section on obstacle avoidance). $\mathcal{SA}^{nav}$ runs only when the previous way-point is reached by the vehicle. Instead, $\mathcal{SA}^{oav}$ runs continuously, in a real-time fashion, providing a series of better *local* positions over the time, namely x_{wp}^{oav}. Notice that the unit computes both the PSO^{nav} and the PSO^{oav} in an asynchronous and completely decentralized fashion. Finally, the way-point x_{wp}^{oav} is transformed from the base frame ^oF to the body frame ^bF by using (5) and it is exploited by the unit's VCU as the desired $^bx_d^{VCU}$ in (7) and (8). When the vehicle reaches x_{wp}^{nav}, the loop starts again until final target is reached.

3.2 Matching the Environment with the Search Space

The solution of this problem is not easy because in real space there are physical obstacles that need to be mapped, creating therefore a *constrained* search space. We will show how the search space on which PSO algorithms work can be modified to embed the detected obstacles, and how $\mathcal{SA}^{oav}$ exploits these modifications for obstacle avoidance. As first step, it is necessary to define the fitness function used to solve optimization problems.

The fitness function used in both PSO levels is a simple distance function between two points in a 3D space, i.e.

$$f(p_i^j, p_{goal}^j, k) = \gamma_i^j \, \|p_{goal}^j(k) - p_i^j(k)\|, \qquad i = 1, \ldots, n_j \quad j = nav, oav \quad (11)$$

where the parameter γ_i can be tuned to modify the relevance given by each particle/agent to the target point. Considering PSO^{nav} algorithm, the parameter $p_i = [x_i, y_i, z_i]^T$ is the position of the ith–AUV ($\mathcal{V}_i$), while p_{goal} is the global target that the *swarm* must reach. Notice that from VCU's point of view $p_i = x_v$. The parameter n_{nav} is equal to i^{th}–AUV's neighbors. On the other hand, in case of PSO^{oav}, the parameter $p_i = [x_i, y_i, z_i]^T$ identifies the position of the ith *scout* particle used by the algorithm to perform obstacle avoidance, p_{goal} is the way-point given by PSO^{nav} and n_{oav} is the number of *scout* particles.

Obstacle Detection. In the simulated environment, we have used for simplicity spherical obstacles $\mathcal{O}_j$, where $j \in [1 \cdots n_o]$. Moreover we have assumed that each vehicle is equipped with a spherical proximity sensor device that detects

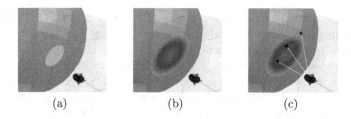

(a)   (b)   (c)

Fig. 5. Fig. 5(a) shows a collision area. Figure 5(b) shows its corresponding virtual patch with a 2D-Gaussian probability distribution created in the search space as a constraint. Figure 5(c) shows $\mathcal{SA}^{oav}$ that computes PSO^{oav} for obstacle avoidance.

obstacles in surrounding space. When a vehicle $\mathcal{V}_i$ detects an obstacle $\mathcal{O}_j$, it appends that the sensor sphere $\mathcal{S}_i$ intersects it in a circular section $\mathcal{C}_{i,h} = \mathcal{S}_i \cap \mathcal{O}_j$ (Fig. 5(a)). This *section*, centred in (x_0^p, y_0^p), is used to create a *virtual patch* $\mathcal{P}_{i,h}^{\mathcal{C}}$ (Fig. 5(b)) in the search space of $\mathcal{SA}^{oav}$ with the same positions of $\mathcal{C}_{i,h}$. Indexes i, h represent respectively the vehicle which detects the collision and a time-marker that identifies each patch during the *exchanging information phase* later explained. The patch data-set stored in unit's internal database consists of several elements: the point (x_0^p, y_0^p), the set of angles $[Pitch, Yaw]$ of the section's orientation (respect to ^oF) and the collision's probability data. Indeed each patch is characterized by a collision probability function, described with a Gaussian distribution (see Fig. 5(b)) as

$$f(x^p, y^p) = A \, \exp\left(-\left((x^p - x_0^p)^2/2\sigma_{x^p}^2 + (y^p - y_0^p)^2/2\sigma_{y^p}^2\right)\right) \tag{12}$$

x^p, y^p are the coordinates of the points belonging to the intersection plane which incorporates $\mathcal{C}_{i,h}$. Considering a probability threshold $c_{th} = 0.5$, we have tuned Gaussian function's parameters $\sigma_{x^p}, \sigma_{y^p}$ and A, in order to adapt the function to the $\mathcal{P}_{i,h}^{\mathcal{C}}$ border, in correspondence of c_{th}. This way it has been created a plateau of values whose collision is certain and thus we have considered the points with coordinates x^p, y^p, belonging to the patch $\mathcal{P}_{i,h}^{\mathcal{C}}$ whose value is $f(x^p, y^p) \in [0.5 \ldots 1]$, as valid collision points. In this fashion $\mathcal{P}_{i,h}^{\mathcal{C}}$ becomes a *virtual wall* in the search space. The procedure is repeated for all the obstacles encountered. Roughly speaking we can consider this procedure like as throwing a balloon full of fluorescent paint against a wall in the dark. After several launches (i.e. collision detections) a virtual wall made of fluorescent patches appears. So obstacles become visible and it is possible to plan the right path.

Obstacle Avoidance. As we have explained before, obstacle avoidance is directly performed by the lower PSO control level. Each patch is created inside the $\mathcal{SA}^{oav}$ search space and this modifies AUV's movements as follows. $\mathcal{SA}^{oav}$ computes PSO^{oav} algorithm by using n_{oav} particles called *scout*-particles. When it generates next particle positions $p_i^{oav}(k+1)$, $i = 1 \ldots n_{oav}$ exploiting (3), a proper *validation procedure* checks if the segments that connect current with future particle positions cross an existent patch. If so, the algorithm recovers collision values of all the cross-points from the related patch data-sets. New positions are considered valid if collision probability values of related cross-points are below c_{th}. Only valid particles are considered by $\mathcal{SA}^{oav}$ that performs PSO^{oav} computation over $n_k = 10$ iteration steps. Among all the positions reached by particles during this session, the one with best fitness value is selected as global "low-level best" (see Fig. 5(c)), that is useful to VCU's way-point navigation. The n_k value has been empirically chosen as best trade-off between computational speed and good results. During AUV's flocking, many different collisions can be detected, therefore the sum of all the patches $\sum_{i,h} \mathcal{P}_{i,h}^{\mathcal{C}}$ creates a *distributed wall structure* that limits particle movements in search space and AUV's movements in real space. Moreover, during the flocking, team-mates are perceived like obstacles but AUVs' positions are shared in communication's data-flows, so that a simple check on inter-vehicle distances, performed by connected neighbors, allows them to discard team-mates from static obstacles. Other units are only considered by $\mathcal{SA}^{oav}$ to perform the on-line obstacle avoidance but they are not stored in internal database. PSO^{nav} also helps to avoid obstacles: the same *validation procedure* that acts on next positions of the particles belonging to PSO^{oav} is exploited to select a valid high-level free-of-collision way-point. Finally notice that $\mathcal{SA}^{oav}$, basing its computation on scout particle's always available data, exploits a standard structure version of PSO algorithm, that is different from the asynchronous and decentralized one of PSO^{nav}.

Local Minima Avoidance. The presence of local minima is always a problem because a single AUV or all the swarm could remain stuck in the same position for a long time, loosing the possibility to achieve final target. When a robot detects a possible local minimum, suggested by the long time elapsed in the same position far from the target, a *virtual spherical patch* $\mathcal{P}_{i,h}^{\mathcal{V}}$ is produced by the stuck AUV on its position. This patch affects the search space, creating a virtual obstacle with the same benefit of the 2D patch previously described. This way the dangerous zone is avoided by remaining team-mates and the good result given by this position, in term of fitness value, is cancelled. So the agent possibly entrapped in a local minimum can follow its better fitness values outside the virtual patch area or it can be dragged out by the team-mates' data contribution.

3.3   Exchanging Information

As already mentioned, from a global point of view, $\mathcal{SA}^{nav}$ runs by using a completely decentralized version of PSO algorithm. This way each vehicle is completely autonomous. In order to permit the algorithm's correct behavior, information exchanged between vehicles is very important. As described in Sect. 2, the communication topology adopted in this work allows agents to exchange data with any other close agents, meanwhile distributing information on environment. Any information exchanged have a time-stamp to identify them in temporal flow. Briefly we can sum up the set of exchanged information between two rendez-vous agents as: (1) current positions of the known AUVs; (2) individual global best value; (3) collision patch information. Taking into account for example a two team-mate rendez-vous, respective current position recorded at each n_k time step is exchanged and stored in a proper table with time-stamp and vehicle's owner number. Moreover the two vehicles do not only exchange their two data-sets but also past data-sets, deriving from previous exchanging with other vehicles. This way a vehicle's data can be viewed as broadcasted on vehicles' network established during every rendez-vous. As previously described, individual *global* best value is provided by PSO^{nav} algorithm of each vehicle during the flocking and it is shared in the same fashion of position's data. $\mathcal{SA}^{nav}$ that receives this value, it uses this p_i^* data as a neighbor value useful to compute PSO^{nav}. Data belonging to other vehicles are updated whenever two or more vehicles have a match. The data managing background routine computes the same sharing procedure on collision patches' data. Each patch individually collected by each AUV is shared with all the connected team-mates. Update cycle is continuous and information run on established network. Finally, the last procedure executed by control algorithm, is the merging of all the received data, in which each unit collects most recent data among the ones received. To conclude, each unit updates environmental information with "best zones" and team-mate positions. Ideally, the updates of environmental data are broadcasted to the entire swarm and so it is possible to consider a sort of *shared collective memory* used by all the units. Notice that a long-lived connectionless situation can produce a warp perception of the environment by unconnected units and a delay on task accomplishment. In our simulations we have detected that in order to reach the target point at least two agents must be connected.

4   Simulations

The presented PSO algorithm has been intensively tested in a Matlab 3D simulated environment. In particular, we have considered an environment volume whose dimensions range in $[30 \ldots 300]\,[m^3]$ and three different types of scenarios: a map with obstacles randomly disposed (Fig. 6(a)), different simulated tunnels (Fig. 6(b)) and caves (Fig. 6(c)). In order to maintain the simulated physical parameters closer to real devices ([23]), each unit has been modelled as an ellipsoid whose dimensions are $[0.7 \times 0.4 \times 0.3]\,[m]$, maximum velocity $u_{max} = 5\,[m/s]$, four fins with equal opening speed $w_{max} = 0.07\,[rad/s]$, max fin's opening angle $\delta_a = 0.26[rad]$, sensor range $r_S \in [5 - 25]\,[m]$ and communication range $\Delta \in [10 - 50]\,[m]$. A sample time of $\Delta t = 0.02\,[s]$ has been selected for the controller. Notice that the *virtual* time step Δt_v perceived by the high-level agent $\mathcal{SA}^{nav}$ is equal to 0.2 [s] due to the fact that the computation session of PSO^{oav} runs over $n_k = 10$ time steps. The number of scout particles $n_{oav} = 5$. The first scenario (e.g. Fig. 6(a)) has been created by using obstacles modelled as spheres of arbitrary radius $r_O \in [0.5 \cdots 5]\,[m]$ randomly deployed in the middle of the testing volume. On the contrary, in the second and in the third ones the sphere's radius is fixed at $r_O = 2.5\,[m]$ but it has been changed the dimension of the built structures. In the second scenario (e.g. Fig. 6(b)) it has been created a simulated tunnel by composing a different number of spherical obstacles

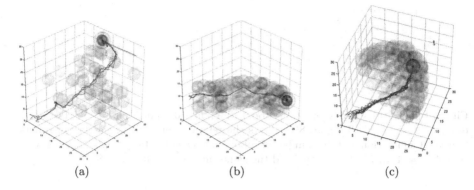

(a)   (b)   (c)

Fig. 6. Figures depict some simulation snapshots. Scenario in Fig. 6(a) presents obstacle avoidance technique with mobile target performed in a random map. In Fig. 6(b) a navigation test through a tunnel is shown and in Fig. 6(c) instead is shown a navigation inside a cave.

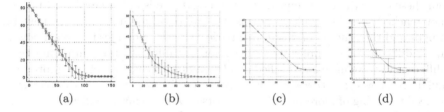

(a)   (b)   (c)   (d)

Fig. 7. Statistical results for scenario 1 are depicted with respect to a fixed (Fig. 7(a)) and a mobile target (Fig. 7(b)). In Fig. 7(c)–(d) for scenarios 2 and 3 respectively. All the figures represent the mean of the units' distances between respective start and goal areas. In the simulation depicted in Fig. 7(c) the number of units is 20

disposed along a spline. The aperture radius of the tunnel r_a^t in the different simulation varies in $[4\ldots 8]\,[m]$. The last type of scenario (Fig. 6(c)) tests the algorithm in a simulated cave whose dimensions are WxHxD = $[14,\ 10,\ 15]\,[m]$.

Statistical Results and Parameter Selection

In order to gather statistical results about the performance of the proposed PSO algorithm, we have performed hundreds of simulations for each scenario. In each simulation of the first scenario the obstacles positions and dimensions have been changed. Despite of this, it is possible to notice in Fig. 7(a) that AUVs are able to overcome obstacles and achieve the target even in case the target is moving, as reported in Fig. 7(b) mobile target velocity $\dot{\boldsymbol{p}}_{goal}^{nav} = [-4, 0, 0]\,[m/s]$. The enlargement of the standard deviation in the middle of the statistic graphs describes the various path followed by vehicles to overcome obstacles. Figure 7(a) shows data related to $\varphi = 3.5$ in scenario 1 that produces the best algorithm's performances. This choice has been done on the basis of the simulations made by changing $\varphi = 1, 1.5, 2$, depicted in Fig. 8(a). From top to bottom it is possible

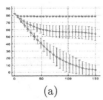
(a)

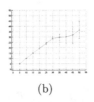
(b)

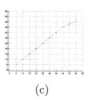
(c)

Fig. 8. Figure 8(a) shows simulations about different values of φ parameter. From top to bottom $\varphi = 1, 1.5, 2$. Figure 8(b) shows the relation between the number of robots that start simulation and the number of robot that reach the target on scenario 2 with tunnel $r_a^t = 4$. In Fig. 8(c) is depicted the same simulation with $r_a^t = 8$

to notice how performances increase. When $\varphi > 3.5$ randomness is too high, standard deviation increases and the swarm's behavior is not controllable any more, so units do not converge. This φ value has been adopted for all three scenarios. We have considered swarms with a number of units ranging from 2 to 50. From these case-studies, it appears that in case there is only one vehicle, target is not reached even in simple scenarios, the information on the environment is too poor and the random part of the particle's velocity provides too much noise, making difficult to accomplish the task. Even in tuning the random parameters in different way, we have not reached good results with a single vehicle. Moreover simulations shows that there are some problems with robots number in scenario 2, indeed the using of more than 25–30 robots create too many collisions and the algorithm fails to maintain the group compact. In this case, the swarm breaks up into smaller groups that reach the target autonomously. Enlarging tunnels aperture this problem tends to disappear but it reoccurs increasing the number of units.

5   Conclusions and Future Work

In this paper, a novel version of the Particle Swarm Optimization algorithm has been proposed in order to drive a group of AUVs in unknown underwater environments from a starting point to a final one, that can be possibly mobile. In particular, we have used a double-level and a completely asynchronous, decentralized version of PSO to manage swarm's coordination. Moreover, PSO algorithm has been modified in order to consider robots' physical constraints, i.e. their dimensions and their interaction with the environment. Vehicles know their own position, and are able to broadcast data to neighbors and to sense the surrounding environment by using spherical sensors. During the navigation, units collect data on the environment not only to reach the target but also to map the environment. This technique could be useful for example in benthic exploration or search and rescue. The performances of the algorithm have been validated using data collected in several simulations, where different groups of AUVs have been simulated in many complex environments. Future work will include deeper analysis of the relation between environmental features, parameters optimization and performances. Also we will study the possibility of controlling team-mates distances according to environmental conditions. Moreover, the possibility of using *dynamic patches* to map both moving and evanescent obstacles, exploiting proper predictive algorithms will be considered. Finally, applications of this algorithm to more complex systems, like groups of heterogeneous robots (e.g. groups of robots including both naval and aerial vehicles) and to more complex environments will be considered.

References

1. Kennedy, J., Eberhart, R.C.: Swarm Intelligence. M. Kaufmann Pub., San Francisco (2001)
2. Reynolds, C.W.: Flocks, herds and schools: a distributed behavioral model. In: ACM SIGGRAPH Computer Graphics, vol. 21, pp. 25–34. ACM (1987)
3. Eberhart, R., Kennedy, J.: A new optimizer using particle swarm theory. In: Proceedings of the 6th International Symposium on Micro Machine and Human Science, MHS'95 (1995)
4. Kennedy, J., Eberhart, R.: Particle swarm optimization. In: Proceedings of the IEEE International Conference on Neural Networks, vol. 4, pp. 1942–1948, November–December 1995
5. Poli, R.: Analysis of the publications on the applications of particle swarm optimisation. J. Artif. Evol. Appl. **2008**(4), 1–10 (2008)
6. Doctor, S., Vanayagamoorthy, G.K.: Unmanned vehicle navigation using swarm intelligence. In: Proceedings of International Conference on Intelligent Sensing and Information Processing, 2004, pp. 249–253 (2004)
7. Grandi, R., Falconi, R., Melchiorri, C.: A navigation strategy for multi-robot systems based on particle swarm optimization techniques. In: Proceedings of the 10th IFAC Symposium on Robot Control, (SYROCO), September 2012
8. Masehian, E., Sedighizadeh, D.: A multi-objective PSO-based algorithm for robot path planning. In: 2010 IEEE International Conference on Industrial Technology, pp. 465–470 (2010)
9. Pugh, J., Martinoli, A.: Particle swarm optimization for unsupervised robotic learning. In: Swarm Intelligence Symposium (2005)
10. Pugh, J., Martinoli, A.: Parallel learning in heterogeneous multi-robot swarms. In: 2007 IEEE Congress on Evolutionary Computation, pp. 3839–3846 (2007)
11. Girard, A.R., Sousa, J.B., Silva, J.E.: Autopilots for underwater vehicles: dynamics, configurations, and control. In: OCEANS 2007, pp. 1–6, June 2007
12. Joordens, M.A.: Underwater swarm robotics consensus control. In: Systems, Man and Cybernetic, pp. 3163–3168 (2009)
13. Shen, J., Zhang, J.: Route planning for underwater terrain matching trial based on particle swarm optimization. In: 2010 Second International Conference on Computational Intelligence and Natural Computing, pp. 226–229 (2010)
14. Tang, X., Yu, F.: Path planning of underwater vehicle based on particle swarm optimization. In: Intelligent Control and Information, pp. 123–126 (2010)
15. Akat, S.B., Gazi, V.: Asynchronous particle swarm optimization based search with a multi-robot system: simulation and implementation on real robotic system. Turk. J. Electr. Eng. Comput. Sci. **18**(5), 749–764 (2010)

16. Burak Akat, S., Gazi, V.: Particle swarm optimization with dynamic neighborhood topology: three neighborhood strategies and preliminary results. In: 2008 IEEE Swarm Intelligence Symposium, pp. 1–8, September 2008
17. Pugh, J., Martinoli, A.: Inspiring and modeling multi-robot search with particle swarm optimization. In: 2007 IEEE Swarm Intelligence Symposium, (Sis), pp. 332–339, April 2007
18. Hereford, J.M.: A distributed particle swarm optimization algorithm for swarm robotic applications. In: IEEE Congress on Evolutionary Computation, 2006, CEC 2006, pp. 1678–1685 (2006)
19. Kennedy, J., Mendes, R.: Population structure and particle swarm performance. In: Proceedings of the 2002 Congress on Evolutionary Computation, CEC'02 (2002)
20. Eberhart, R.C., Shi, Y.: Comparing inertia weights and constriction factors in particle swarm optimization. In: Proceedings of 2000 Congress on Evolutionary Computation (2000)
21. Trelea, I.C.: The particle swarm optimization algorithm: convergence analysis and parameter selection. Inf. Process. Lett. **85**, 317–325 (2003)
22. Chaturvedi, N.A., Sanyal, A.K., McClamroch, N.H.: Rigid-body attitude control. IEEE Control Syst. **31**(3), 30–51 (2011)
23. Burrowes, G., Khan, J.Y.: Short-range underwater acoustic communication networks, chapter 8. In: Autonomous Underwater Vehicles. InTech, Rijeka (2011)

A Novel Communication Technique for Nanobots Based on Acoustic Signals

Valeria Loscrí[1](✉), Enrico Natalizio[2], Valentina Mannara[1], and Gianluca Aloi[1]

[1] University of Calabria, Unical, 87036 Arcavacata di Rende, Italy
vloscri@deis.unical.it
[2] Lab. Heudiasyc - UMR CNRS 7253, Universite de Technologie Compiegne, Compiegne, France

Abstract. In this work we present the simulation of a swarm of *nanobots* that behave in a distributed fashion and communicate through vibrations, permitting a decentralized control to treat endogenous diseases of the brain. Each *nanobot* is able to recognize a cancer cell, eliminate it and announces through a communication based on acoustic signals the presence of the cancer to the other *nanobots*. We assume that our nano-devices vibrate and these vibrations cause acoustic waves that propagate into the brain with some intensity that we evaluated by taking into account the specific physical factors of the context, the nano-metric nature of the vibrant devices and the characteristic of the fluid where the devices are immersed. An important aspect of our approach is related to the communication based on vibrations. This choice is related to the application context where is not advisable either to use indiscriminate chemical substances or electromagnetic waves. Whereas, ultrasonic waves are used in the most frequent diagnostic techniques and the use of this kind of techniques should not have negative collateral effects. Specifically, we propose an approach based on bees' behavior in order to allow our devices to communicate, coordinate and reach the common objective to destroy the cancerous tissues. In order to evaluate the effectiveness of our technique, we compared it with other techniques known in literature and simulation results showed the effectiveness of our technique both in terms of achievement of the objective, that is the destruction of the cancerous cells, and velocity of destruction.

Keywords: Acoustic communication · Nanobots · In-vivo applications

1 Introduction

Nanotechnologies are a new approach based on comprehension and deep knowledge of the properties of the matter at the nanoscale level: a nanometer corresponds to the length of a small molecule. At nanoscale, behaviors and characteristics of the matter drastically change, making it necessary a synergy among

G.A. Di Caro and G. Theraulaz (Eds.): BIONETICS 2012, LNICST 134, pp. 91–104, 2014.
DOI: 10.1007/978-3-319-06944-9_7, © Institute for Computer Sciences, Social Informatics and Telecommunications Engineering 2014

many different disciplines. A great percentage of the nanosytems world is represented by technologies derived from micro-electronics, pushed up to nanoscale, in order to obtain electronic, optical, fluidic, integrated mechanic functions to be applied to different fields ranging from microelectronic, telecommunications and sensor, to environmental and bio-medical. The latter field could be literally revolutionized from the potential applications of nano-devices, both in the diagnostic and pharmaceutic fields [9]. One of the most interesting applications is the controlled release of drugs over time and exactly localized in cells or organs that need it, drastically reducing the side effects. Nanotechnology is already used in the field of diagnostics through the use of synthetic tracer molecules for investigating biological processes in a non-invasive fashion. In the cancer fight, as for the treatment of some diseases related to the cardiovascular system more and more often you hear speak about nano-therapy. Among the first applications to be postulated in the early '90s and also among the most fascinating of nanomedicine is the idea of using nanorobots [8]. A device of a few nanometers (nanorobots will typically be 0.5 to 3 microns large with 1–100 nm parts), could be introduced into the body without causing injury and, if equipped with sensors that transmit precise images could facilitate the early diagnosis of cancer and carry drug to the target or to perform other tasks that would otherwise require invasive surgery [1,3]. In order to prevent attacks from immune system a nanorobot *in vivo* should be characterized with a smooth and flawless diamond exterior, because this prevents Ieukocytes activities since the exterior is chemically inert and have low bioactivity [2].

A *nanorobot* is a system able to modify the surrounding in a controlled and predictable fashion, with size at the molecular or even atomic scale. In practice, a *nanorobot* is either a passive or active structure able to detect, signal and elaborate information. The very limited size of the devices implies limited capabilities and reduced computation resources, therefore it is necessary to make devices collaborate by applying design techniques such as *Swarm Intelligence* in order to realize complex systems through the interaction and the cooperation of very simple agents [11].

This work is inspired by the paper of Lewis Anthony and Bekey George developed in 1992 at Southern California University and entitled "The Behavioral Self-Organization of Nanorobots Using Local Rules" [4], where the objective was the evaluation of a swarm of nanobots, organized without any centralized unit, only trough simple local rules, to destroy cancer cells inside the brain. Thanks to the technological development, the assumptions made in [4] are plausible and it is possible to design molecular nanobots constituted from two polymers able to deactivate the production of a protein that causes the death of the cells attached [7]. Lewis's work has an early work for the assumption of using robots for *in vivo* applications through a swarm technique. In [4] and in other next works [5,6], the communications are performed by chemical mechanisms, that is, the report of either some events or object is realized through the release of chemical substances that diffuse by attracting other nanorobots through the gradient associated to the signal intensity. Specifically, Lewis's strategy is based on three chemical substances that attract the swarm of nanobots in a gradual fashion to achieve the total elimination of the diseased cells. An early part of the work is devoted to the tuning of the characteristic parameters of the algorithm where objective difficulties emerged, first of all the extremely reduced dimensions of the devices, their limited computational capabilities and the structural space needed to install sensors and interfaces to capture all the chemical substances involved. For these reasons, we considered a different approach inspired from bees' behaviors that use vibration to communicate the distance and the position of food sources to the rest of the colony. Our choice for this specific kind of communication has been lead by the simplicity of the components required from the devices, and the fact that many diagnostic instruments are already based on microwaves, which are considered not dangerous for the human. A useful summary of communication paradigms to interconnect nanobots in a body area network is given in [13]. Preliminary results of Bee's algorithm are presented in [15]. The rest of the paper is structured as follows. In Sect. 2 we give some details about Lewis's approach. In Sect. 3 we describe our bees's approach. In Sect. 4 we give the details about the simulation environment and the results. Finally, we conclude the paper in Sect. 5 and we give some directions for future works in Sect. 6.

2   Lewis Algorithm

When Lewis and Bekey presented their work in 1992, the *Swarm Intelligence* concept had not yet been fully defined. In fact, authors refer to a $\mu - colony$ as a set of robots that coordinate activities autonomously to perform some specific tasks. The colony could be constituted by hundreds or thousands of $\mu - robots$, and each of them has simple computational capabilities like a colony in nature. As we already outlined, the technique they proposed was based on chemical communication. Each $\mu - robot$ is able to mark its surrounding through chemical substances, recognizes the different chemical signals and follows the different gradients until it reaches the cancer cells. The colony is injected close to the cancer. Once injected into the body, the $\mu - robots$ move randomly until they reach cancer cells. After the first contact the $\mu - robot$ emits a substance in its surrounding called CHEM-1. This substance is absorbed by the body after a certain time. A certain percentage of $\mu - robots$ differentiate in *guidepost*, stop and start to secrete substances that permit the transmission of the signal over long distances. These substances are CHEM-2 and CHEM-3 and are used as repeaters. The number of $\mu - robots$ that differentiate in *guideposts* determines the efficiency, which is the convergence of $\mu - robots$ to defeat the tumor. The number of *guideposts* is very important. In fact, if too many nanobots differentiate it is difficult to reach and destroy the tumor. On the other hand, if the number of *guideposts* is too low, the colony is not able to complete the task. For this reason, authors considered a differentiation probability $p = 0.01$ and related the total number of *guideposts* to the total number of $\mu - robots$ n and the current time t, *pnt*. In what follows we show the pseudo-code of the Lewis-Bekey approach:

Algorithm 1. Lewis and Bekey Algorithm

IF There is no chemical Markers and No Tumor
THEN Do a random Walk

IF A Tumor is detected
THEN Destroy the cell, Differentiate and Broadcast CHEM-1; w/prob 1

IF The Mag. of CHEM-1 is greater than Θ
THEN Do a random Walk

IF CHEM-1 is detected
THEN Move up the gradient of CHEM-1; w/prob p
or differentiate and Broadcast CHEM-2; w/prob(1-p)

IF CHEM-2 is detected
THEN Move up the gradient of CHEM-2; w/prob p
or Differentiate and broadcast CHEM-3; w/prob(1-p)

IF CHEM-3 is detected
THEN Move up the gradient of CHEM-2; w/prob 1

The algorithm's efficiency is strictly related to some additional parameters, like the threshold value of CHEM-1, Θ. The environment considered is the cerebral cortex that is represented in the simulator NSl [10] as a cells array. The modeled chemical communication takes into account the generation, the diffusion of the substances immersed in a fluid and the absorption factor. A single cell is the measure unit of the space and only one nanobot at a time can occupy a single cell. The cancer and the colony are positioned inside the grid. Each robot is able to move step by step in one of the 8 adjacent cells. Once a $\mu-robot$ reaches a cancer cell it is able to perceive the tumor and to signal its presence to other robots. In our simulation, we adopted two different convergence criteria. The first one is represented by the total elimination of the cancer cells and the other is a maximum number of iterations (1000), that is not an actual convergence criterion but we need to stop the simulation when the algorithm fails to eliminate all the cells. Even if the algorithm is strictly related to the setting of different parameters, the aim of the authors was mainly to show how behavioral algorithm is able to make the colony reach a common objective. It is worth to note that, the use of three chemical signals implies at least three different interfaces to capture the different substances. Moreover, a single nanobot needs to physically transport three different substances. By considering the reduced dimensions it seems very difficult to imagine how this algorithm could be realized.

3 Bees' Approach

As we already outlined above, the acoustic communication we propose in this work is inspired from bees exploration technique for food searching. In practice, our nanobots borrow from the bees the capabilities to communicate through vibration (waggle dance for the bees). The self-organization of the bees is based on very simple rules related to the behavior of each individual. Moreover, the concept of swarm applied in the context of *in vivo* application has several advantages compared to isolated nanobot. For example, acoustical nanobots could form *in vivo* communication networks that could transfer data across much larger distance than possible with direct transmission by considering the attenuation at high frequencies [16]. Generally, nanorobots can improve their performance and they are able to accomplish complex tasks, by coordinating their actions in a decentralized fashion. A similar approach is considered in [17]. We refer to this bees' inspired technique as NanoBee and is supported by the possibility of using acoustic waves as transmission means in communications *in vivo* without specific risks associated to. We exploited the analysis made in [9] to characterize our simulation model and tune the parameters. The vibrations associated to the devices generate acoustic waves that propagate in an elastic medium and cause pressure variations and movement of the particles that compose the medium and that can be perceived and detected from an acoustic detector. This perturbation, that carries both the information and the energy, propagates while every particle, also in the case of a fluid, remains nearby its original position. In practice, there are local vibrations (compression and rarefaction) of the particles and in the case of fluid, where there can not be cutting efforts, the vibrations are parallel to the propagation direction of the wave, i.e. longitudinal waves. The dimension or magnitude of sound can be indifferently expressed as sound power, sound intensity or sound pressure. Sound power is the total amount of acoustic energy emitted from a sound source and is measured in watt. Sound intensity is the ratio of the power of a sound wave and the crossed superficial area, it is usually measured in $watt/meter^2$. Sound pressure is the value of the pressure variation of a corps in a generic point inside the sound field and is measured in $newton/meter^2$. It is worth to note that for a given sound, sound power is constant while both intensity and pressure depend on the specific measure conditions. The simplest case is sound or noise that propagate freely without any obstacle. With this assumption and when the medium considered is non-dissipative, intensity (I), power (P) and pressure (p) are correlated as it follows:

$$I = \frac{W}{2\pi r^2} = \frac{p^2}{\rho c} \tag{1}$$

where ρ is the density of the fluid and c is the light speed. It is worth to note that both intensity and pressure decrease with the square of the distance (r). Based on these considerations we assume the possibility to design a technique based on acoustic communication by modeling the propagation of signals with sound intensity. NanoBee technique considers nanobots as point sources, since their size is small compared to the distance from the receiver. Each nanobot is able to signal the presence of a cancer cell by stopping its movement and starting vibrating, mimicking the bees behavior, in order to transmit a sound signal to alert other agents (nanobots). Every nanobot that receives the signal moves towards the gradient of the intensity received. Vibrating time is limited since it depends on the force associated to each device to accomplish the movement. In order to simulate the spatial variation of intensity sound, we assume three different probabilistic intervals, that simulate both the spatial propagation and the attenuation of the sound. Specifically, we choose higher probabilities values close to the nanobot that is dancing and probabilities values smaller when distances are greater by simulating in this way the attenuation with the square of the distance. A more realistic version of the algorithm must take into account the temporal attenuation of a sound signal which introduces a decrease of both the intensity and the distance reached by the signal. The first version, where we do not take into account the temporal attenuation of the signal is referred as NanoBee ON-OFF, where ON and OFF indicate whether the device is "ON" (it vibrates with the maximum power) otherwise it is "OFF" (and does not vibrate at all). The second version of our algorithm is referred as NanoBeeEvan and is related with a temporal attenuation of the sound signal. From a computational point of view, our devices are very simple since the algorithm only requires that each device has capabilities of:

- recognition of a cancer cell;
- destruction of a cancer cell;
- emission of vibrations to signal the position of the cancer;
- detection of acoustic waves.

In what follows we give the pseudo-code of the NanoBeeAlgorithm:

Algorithm 2. NanoBee

Repeat
for each "active" nanobot i:
pick up any sound signal in its surrounding;
IF there are not any signal/cancer cells
search randomly;
IF discovered a cancer cell
THEN eliminate the cell and starts to dancing;
IF a signal has been received
THEN moves towards the higher intensity of the signal;
Until there is an "active" nanobot

A nanobot is active when it has enough energy, above a certain predetermined threshold and it has not discovered any cancer cell yet.

4   Simulation Results

We simulate the systems previously described with Neural Simulation Language (NSL), [10]. NSL possesses many features that facilitated the simulation development, including graphics capabilities, and language constructs for handling layers of grid object. In order to evaluate our techniques in different situations we consider the possibility to have different initial configurations of the cancer. In fact, based on the specific geometry of the processes, different cancer forms characterize different cancer types. Specifically, we consider the possibility to have metastases detached from the original tissue. Usually, metastases are a group of cells and only rarely they are isolated, but for completeness we considered the three different scenarios: "aggregated", "scattered groups" and "isolated".

4.1   Different Initial Configuration

The space where nanobots move and act is a grid of cells modeling the vertebral cortex. In Fig. 1 we show the initial configuration when the "aggregated" case is considered, in Fig. 2 the initial configuration of the cancer is with "groups" of cells and in Fig. 3 we show the initial configuration of the cancer with "isolated" cells. Specifically, the number of cancer cells is 42 in every configuration and the cancer cells are represented in blue, while the nanobots, that are 289 in every configuration, (as in [4]), are represented in black.

4.2   Convergence Time

The input parameters we consider in order to evaluate the performance of the various techniques are summarized in Table 1. The algorithms considered in this work are based on the concept of nanobots swarm, which allows us to consider a

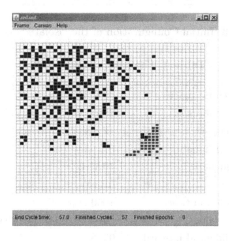

Fig. 1. Initial Configuration of the "aggregated" case.

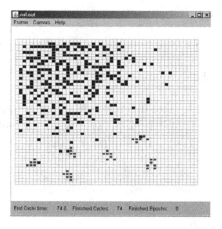

Fig. 2. Initial Configuration of the "scattered groups" case.

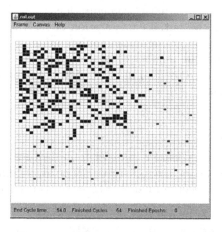

Fig. 3. Initial Configuration of the "isolated" case.

Table 1. Simulation parameters

Searching space	50 × 50 cells
N Nanobots	289
Cancer size	42 cells
Size of nanobots	0.5 μm
Frequency	10 MHz
Power	0.5 pW
Simulation time	1000 runs
Prob values	[1-0.7],[0.69-0.3],[0.29,0]
Confidential interval	95 %

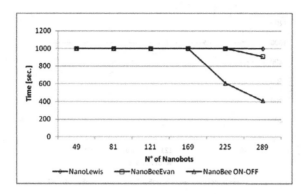

Fig. 4. Convergence Time in the case of "aggregated" configuration.

reduced dimension of devices with available power smaller than that necessary if only a device is considered as observed in [12]. Frequencies ranging from 1 to 5 MHz are typically used for the head [12], thus, we choose the value of frequency equal to 1 MHz. Moreover, in our bees' approach, communication activity is only related to the achievement of the target. In practice, nanobots start to vibrate when they discover a tumor and not when they move. We apply the method of Independent Replications with a confidence interval of 95 %.

In Fig. 4 we show the convergence time of each considered algorithm for the "aggregated" case. The first important observation is related to Lewis approach (referred in the diagram as NanoLewis) that is not able to totally destruct the cancer cells even if we consider the maximum number of nanobots. Concerning Lewis'approach we had to set many parameters and after conducting a sensitivity analysis we chose the set of parameters that guarantee the best performance in terms of convergence. During the simulation we observed as the combination of the three chemical substances creates a kind of "barrier effect". In practice, after the first nanobots enters in contact with cancer cells and a part of the devices differentiate and start working as a kind of relay by sending CHEM-1. At this point, it is very difficult for the other nanobots to attack the most internal cancer cells. When the stations differentiate like "*guidepost*" and start send CHEM-2 if they already entered in contact with CHEM-1 or CHEM-3 if they were touched from CHEM-2, the situation worsens. Bees' approaches behave similarly, with the difference that NanoBee ON-OFF is able to eliminate all the cancer cells with less nanobots than NanoBeeEvan. NanoBee ON-OFF is more effective in this case cause of the signal propagation that reaches farther during the time and is able to attract more nanobots. NanoBeeEvan consideres the attenuation of the signal power during the time. Hence, after a certain period the signal only reaches a shorter distance.

In Fig. 5 we can observe how the NanoLewis is not yet able to destruct all the cancer cells, in the "scattered group" algorithm. In this case we observe a little improvement of the NanoBeeEvan. This latter, through the "evanescence" of the signal in the time, implies more randomness of movement of the nanobots.

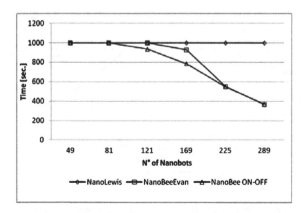

Fig. 5. Convergence Time in the case of "scattered groups" configuration.

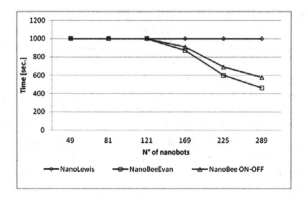

Fig. 6. Convergence Time in the case of "isolated" configuration.

Hence, when tumor cells are scattered in the region of interest, this randomness is useful to reach the various points of interest more quickly.

This reasoning is confirmed in Fig. 6 where NanoBeeEvan overcomes the NanoBee ON-OFF, for the case of "isolated" cells.

It is worth to note how both the nano-bee approaches require less nanobots to reach the goal to eliminate all the cancer cells in respect of those needed by Lewis algorithm. In our opinion the analysis of the performance in terms of both effectiveness, that is the capability to destruct all the cancer cells, and the rapidity to do it by varying the number of nanobots, is very important because it is very important to use the minimum number of nanobots in a similar context. Furthermore, it is very important to try to reduce both the chemical and the acoustic messages as much as possible.

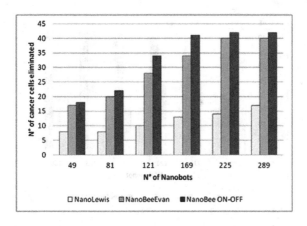

Fig. 7. Number of eliminated cancer cells in the case of "aggregated" configuration.

4.3 Amount of Eliminated Cancer Cells

In this section we present another type of analysis related to the variation of the number of nanobots but also related to the capability to eliminate the cancer cells. The simulation parameters are always those summarized in the Table 1.

In Fig. 7 we can observe how the number of cancer cells eliminated by the NanoLewis technique is always smaller than the other two techniques. As already outlined before, we observe that the algorithm is effective when the first nanobot attaches the tumor, but the barrier created through the chemical substances does not allow to penetrate inside the body of the cancer. In fact, even if the number of nanobots increases, above all, in the first configuration (where we have only a single tumor), the effect does not change. This shows that the scarce efficiency is not related to the small number of nanobots. On the contrary, with our bees techniques we have that an increasing number of nanobots corresponds to an increased capability to attack the cancer. In fact, in the "aggregated" case, we pass from around 18 cells to 42 cells.

In Fig. 8 we can observe how the barrier effect of the NanoLewis approach is mitigated in the "scattered groups" configuration. Moreover, for this specific configuration the bees approaches behave very similarly.

In Fig. 9 we can notice how the effects of the randomness improves the behavior of the NanoBeeEvan that is very close in terms of performance to the NanoBee ON-OFF.

From the analysis conducted is clear that the algorithms considered with the associated parameters in terms of size of devices, available power, etc. are feasible for nanobots operating in the specific environment as we defined (i.e. the brain). When a different environment is considered, the characteristic parameters have to be re-defined and the effectiveness of the technique needs to be proven. *In-vivo* applications need a very accurate choice of available parameters, that can change based on the specific application they are considered for.

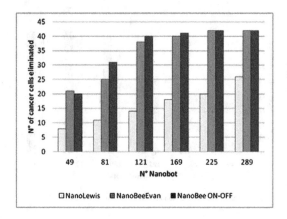

Fig. 8. Number of eliminated cancer cells in the case of "scattered groups" configuration.

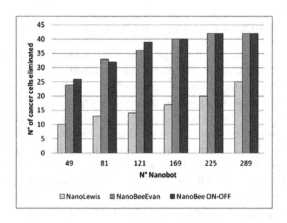

Fig. 9. Number of eliminated cancer cells in the case of "isolated" configuration

5   Conclusions

In this work we considered and evaluated the possibility of using and exploit a swarm of nanobots in medical applications for the treatment of endogenous diseases and through this preliminary study we showed how this is possible. Specifically, we focused on the communication and organization aspects, necessary for the control of a single nanobot through very simple rules and based only on local information exchange. We took inspiration from a previous work where the behavior of the swarms of nanobots was defined through chemical signals able to attract the devices. After an accurate evaluation of the types of communication physically feasible for in vivo communications and based on the bees behavior in the supply phase, we proposed an algorithm based on acoustic signals, NanoBee. This choice is supported from the common use of sound waves for medical applications. By assuming the possibility for a nanobot to generate a vibration when it picks up a cancer cellular tissue, we evaluated this approach and compared the algorithms via simulation by a neural simulation tool, NSL. We showed that nano-bee techniques are more effective to face cancer with different initial shape than Lewis'approach, by eliminating all the malignant cells. Communications are more efficient and nodes only need to perceive the "sound" by moving a step at each time. We showed how the proposed techniques based on the bees behaviors are effective also when the number of nanobots is smaller than that considered in the Lewis and Bekey technique.

6 Future Works

Concerning the future works there are many aspects that are worth to be considered in a deep way. The first one is concerning the chemical communication approach considered from Lewis and Bekey. In fact, during the simulation we were able to observe that the "barrier effect" worsens when CHEM-2 and CHEM-3 are activated. Maybe, an approach with only a single chemical substance would be more effective, and in terms of nanobot design it would be more feasible. An other important aspect is the validation of the probabilistic approach we adopted to simulate the attenuation of the signal with the square of the distance. It would be interesting to evaluate a real model of the brain and compare it with our probabilistic approach. Another important aspect to be considered is related to a more realistic simulation environment, where "obstacles" (i.e. others cells) are explicitly taken into account and the interaction of the signals is analyzed.

References

1. Cavalcanti, A., Shirinzadeh, B., Zhang, M., Kretly, L.: Nanorobot hardware architecture for medical defense. ACM Trans. Program. Lang. Syst. 8(5), 2932–2958 (2008)
2. Davis, M., Zuckerman, J., Choi, C., Seligson, D., Tolcher, A., Alabi, C., Heidel, J., Ribas, A.: Nanorobots in brain tumor. Int. Res. J. Pharm. 2(2), 53–63 (2011)
3. Amato, P., Masserini, M., Mauri, G., Cerofolini, G.: Early-stage diagnosis of endogenous diseases by swarms of nanobots: an applicative scenario. In: Dorigo, M., et al. (eds.) ANTS 2010. LNCS, vol. 6234, pp. 408–415. Springer, Heidelberg (2010)
4. Lewis, A., Bekey, G.: The behavioral self-organization of nanorobots using local rules. In: IEEE/RSJ IROS Conference Proceedings, pp. 1333–1338, July (1992)
5. Cavalcanti, A., Hogg, T., Shrinzadeth, B., Liaw, H.C.: Nanorobot communication techniques: a comprehensive tutorial. In: IEEE International Conference on Control, Automation and Vision, ICARCV, pp. 1–6, December (2006)
6. Hla, K.H.S., Choi, Y.S., Park, J.S.: Mobility enhancement in nanorobots by using particle swarm optimization. In: IEEE International Conference on Computational Intelligence and Security, CIS, pp. 35–40, December (2008)

7. Davis, M., Zuckerman, J., Choi, C., Seligson, D., Tolcher, A., Alabi, C., Heidel, J., Ribas, A.: MoEvidence of RNAi in humans from systematically administered siRNA via targeted nanoparticle. Nature **464**(1064), 1067–1070 (2010)

8. Atakan, B., Akan, O.B.: Body area nanonetworks with molecular communications in nanomedicine. IEEE Commun. Mag. **50**(1), 28–34 (2012)

9. Freitas, R.: Nanomedicine, Volume I: Basic Capabilities. Landes Bioscience, Georgetown (1999)

10. Weitzenfeld, A., Arbib, M., Alexander, A.: The Neural Simulation Language. The MIT Press, Cambridge (2004)

11. Dorigo, M.: Optimization, learning and natural algorithms. Ph.D. thesis, Italy (1992)

12. Hog, T., Freitas, Jr. R.A.: Acoustic communication for medical nanorobots. Nano Commun. Netw. (2)1–38 (2012). doi:10.1016/j.nancom.2012.02.002

13. Akyildiz, I.F., Brunetti, F., Blazquez, C.: Nanoneworks: a new communication paradigm. Comput. Netw. **52**(12), 2260–2279 (2011)

14. National Center for Biotechnology Information. http://www.ncbi.nlm.nih.gov

15. Loscrí, V., Mannara, V., Natalizio, E., Aloi, G.: Efficient acoustic communication techniques for nanobots. In: 7th International Conference on Body Area Networks, Oslo, Norway, 24–26 September 2012

16. Stojanovic, M.: On the relationship between capacity and distance in an underwater acoustic communication channel. In Proceedings of the 1st ACM International Workshop on Underwater Networks (WUWNet06), pp. 41–47 (2006)

17. Hogg, T., Sretavan, D.W.: Controlling tiny multi-scale robots for nerve repair. In: Veloso, M., Kambhampati, S. (eds.) Proceedings of the 20th National Conference on Articial Intelligence (AAAI2005), pp. 1286–1291. AAAI Press (2005)

What Could Assistance Robots Learn
from Assistance Dogs?

Márta Gácsi[1(✉)], Sára Szakadát[2], and Ádám Miklósi[2]

[1] MTA-ELTE Comparative Ethology Research Group,
Pázmány Péter sétány 1/c, Budapest 1117, Hungary
marta.gacsi@gmail.com
[2] Department of Ethology, Eötvös Loránd University, Budapest, Hungary
{szurtrikk,amiklosi62}@gmail.com

Abstract. These studies are part of our broader project that aims at revealing relevant aspects of human-dog interactions, which could help to develop and test robot social behaviour. We suggest that the cooperation between assistance dogs and their disabled owners could serve as a model to design successful assistance robot–human interactions.

In Study 1, we analysed the behaviour of 32 assistance dog–owner dyads performing a fetch and carry task. In addition to important typical behaviours (attracting attention, eye-contact, comprehending pointing gestures), we found differences depending on how experienced the dyad was and whether the owner used a wheel chair or not.

In Study 2 we investigated the reactions of a subsample of dogs to unforeseen difficulties during a retrieving task. We revealed different types of communicative and displacement behaviours, and importantly, dogs showed a strong commitment to execute the insoluble task or at least their behaviours lent a "busy" appearance to them, which can attenuate the owners' disappointment. We suggest that assistant robots should communicate their inability to solve a problem using simple behaviours (non-verbal vocalisation, orientation alternation), and/or could show displacement behaviours rather than simply not performing the task.

In sum, we propose that assistant dogs' communicative behaviours and problem solving strategies could inspire the development of the relevant functions and social behaviours of assistance robots.

Keywords: Assistance robot · Dog behaviour model · Ethological approach · Social interaction

1 Introduction

Considering the aging western industrialized societies, in the near future it will be an absolute necessity that the elderly and disabled people could successfully communicate and cooperate with home assistance robots. There are many different endeavours to design technological aids helping the rehabilitation of the physically disabled, ranging from intelligent wheelchairs to different types of assistant robots [1]. Most of these robots are designed for some specific roles and functions. Assistance systems are

G.A. Di Caro and G. Theraulaz (Eds.): BIONETICS 2012, LNICST 134, pp. 105–119, 2014.
DOI: 10.1007/978-3-319-06944-9_8, © Institute for Computer Sciences, Social Informatics
and Telecommunications Engineering 2014

also developed, for example the small home robot, Mamoru-Kun is able to inform his owner where an object is located communicating verbally or by pointing at the object, and it can also cooperate with his humanoid buddy and ask it to get the object for the user. This bigger robot is able to clean up rooms, manipulate dishes, open and close doors, do the laundry, and it even learns from its mistakes [2]. Another example can be the mobile robotic assistant of the Nursebot project, Pearl. She has two functions: reminding the elderly about their routine activities and guiding them through their environments [3].

However, the commercialization of either type of assistance robots is planned only in ten years, partly because people usually find them disturbing and/or complicated to operate or communicate with. To make humans feel the interactions with the robot more natural, in addition to the technical aid it is essential for the robot to act in accordance with the given social context and to show relevant social abilities [4].

Based on the general assumption that people find a similar companion easier to deal with, the most common strategy both in scientific and commercial environment is to design humanoid robots that have (seemingly) humanlike features and capacities. One of the potential problems with the use of anthropomorphic behaviours lies in the controversy that humanoid robots are likely to raise the expectations of the user in terms of its capabilities and interactional affordances, but present-day robots are simply not able to fulfil these expectations [5]. Recently an alternative suggestion has come up arguing for the use of non-human social animals as models for robot social behaviour: emphasizing the fact that human–animal interaction provides a rich source of knowledge for designing social robots that are able to interact with humans under a wide range of conditions [6]. The idea that in addition to its technical help a robotic assistant could be a suitable companion directed the attention of researchers to use pets as potential behavioural models [7].

In case of assistant robots, the type of function and the limited abilities (compared to that of human assistants) suggest an asymmetrical social relationship between the human user and the robot, which is similar in many ways to the dog-owner relationship [8]. Recently, scientific evidence has supported the idea that dog-human interactions provide a promising model system to study the emergence of social competence [9]. Assistant dogs can successfully communicate and cooperate even with a disabled owner, and show social behaviours, such as attachment, which humans can easily understand without massive prior learning. This way, beside technical assistance, these trained dogs provide social support for their owners [10].

It has already been suggested that using service dog models to design subtle motor behaviours for the manipulation skills of assistance robots is a highly beneficial and cost effective method [11]. We would go further and give dogs a more ambitious role: based on the arguments above we suggest that the social behaviour of dogs and specifically the owner–assistant dog interactions should be used as a model in developing robot companions.

As there were efforts to implement the ethological model of the dog to Sony's developmental robot, AIBO [12], we want to stress that our aim is not to develop a doggy robot. We do not believe that a robot that resembles a dog, for example AIBO, would be an ideal assistance robot. Rather, we propose the use of an embodiment optimally fitted for the specific function of the robot and applying specific behavioural models for different technical/social functions, such as service dogs for the disabled owners or hearing dogs for deaf people.

In order to do so we plan to create plenty of rich, relevant and realistic contexts for the interactions, in which we can observe and evaluate the joint activities of dogs and humans. After extracting the relevant set of behaviour elements based on the desired function, we can adapt the applicable ones to robots of different embodiments and capacities.

Our first attempt to support the dog-model idea applied one of the typical scenarios between service dogs and their disabled owners: when the dog helps the transportation of objects. In this study an appearance-constrained Pioneer robot used dog-inspired affective cues to communicate with its owner and a guest in a fetch and carry scenario. The findings suggested that even limited modalities for non-verbal expression (proxemics, body movement and orientation, camera orientation) offered by the robot were effective for developing/helping the communication [13]. In the present paper we wanted to provide deeper insight into such interactions by detailed analyses of the behaviour of a large and diverse sample.

To develop a life-like and flexible behavioural set for assistance robots it is inevitable to prepare assistant robots for problem situations because it does not seem too pessimistic to assume that they will face several insoluble problems and cannot meet their owners' requirements while performing their tasks. When a 'machine' cannot fulfil our requirements, we are disappointed and annoyed. In contrast when a trained assistant dog cannot obey the owner's commands the owners seem to be rather indulgent and forgiving. We assume that dogs may show different types of communicative behaviours, such as vocalisation or gaze alternation to signal the problem to the owner. Moreover, they may try to perform some alternative activities that are connected to the original task to some extent. In addition, we expected the dogs to perform displacement behaviours, which can influence or even inhibit the negative feelings of the owners, similarly to "guilty" behaviour after transgression [14]. Displacement behaviour emerges both in humans [15] and non-human animals [16] in conflicting situations. For instance, when dogs find themselves in a situation when they are unable to solve a task or a problem, they show typical out-of-context behaviours (e.g. mouth-licking, yawning, sniffing the ground, scratching the ground, pace up and down between the owner and the showed place), which reflect their confusion.

In this paper we observed the interaction of assistance dog–owner dyads in two studies. In Study 1, we investigated the types of behaviours and interactions that can be observed in dog-owner dyads when they perform a fetch and carry task. Moreover, we examined whether there are any difference in the above behaviours depending on how experienced the dyad was and whether the owner used a wheel chair or not. In Study 2, we observed a subsample of dogs in two different types of inhibited trials and described both their typical responses and individual differences when they encountered an unsolvable task commanded by the naïve owners.

2 Study 1 – Fetch and Carry

The fetch and carry task may include picking up, carrying and placing objects at home or outside the house. This ability of the assistance dogs increases the owners' independence and allows a greater range of activities, because while assistive, still requires movement and actions on the part of the owner, thus increasing the amount of physical activity available to him rather than decreasing it.

In this study the dog was supposed to fetch and carry a basket from a start point to a given target place. An assistant capable of performing the task properly need to be able to: go to the person on command, understand human communicative cues to identify the target object, hold and carry the basket, follow or go with (beside) owner or go ahead (with continuous feedback for owner's orientation/instructions), put down the basket on command.

The only instruction for the owners was that they cannot touch the dog or the basket enabling us to reveal spontaneous cues and behaviours of the dyads during the interactions.

2.1 Method

Subjects

We observed 32 dog–owner dyads. All dogs were trained assistance dogs, either therapy dogs or assistance dogs for the disabled. They were tested with their owner (O). Half of the dyads were *novice*, with dogs having the same training as the other group, but not as much experience in various contexts. The other 16 dyads were *experienced*; these dogs had been working with their Os for years. In case of 8 dyads in both the novice and experienced group Os were wheel-chaired. All dogs were more than one year old. (Table 1 contains the data of the dogs.) The training of these dogs were based on the principle that they should be eager to please their owner, that is, they must do their best to find out the task the owner communicates and to cooperate in its execution, even if the human's communication is not completely clear.

Table 1. Independent data of the dog sample

N = 32	Novice	Experienced
Owner	4 males, 4 females	3 males, 5 females
	Mean age: 4 yrs.	Mean age: 3.8 yrs.
	Breeds: Belgian sheepdog, sheltie, golden retriever, border collie, Airedale, 2 lab mixes	Breeds: 3 Belgian sheepdogs, vizsla, lab mix, 3 mongrels
Wheel-chaired owner	5 males, 3 females	5 males, 3 females
	Mean age: 2.6 yrs.	Mean age: 5.6 yrs.
	Breeds: Belgian sheepdog, golden retriever, cavalier, cocker, poodle, lab mix, leonberger mix, mongrel	Breeds: 3 Belgian sheepdogs, 3 golden retrievers, collie mix, lab mix

Procedure

The tests were conducted at a visually separated location in a park that was familiar to the dogs. The dog and O were positioned at the start point. The experimenter (E) placed the basket in front of them. The target place was positioned 10 m from the start point and marked by three 80 cm long sticks forming an equilateral triangle. The behaviour of the dog was recorded from the side by E from a distance of 5–6 m. The records were analysed later (Fig. 1).

Fig. 1. Behavioural sequence of the test: joint attention, pointing, picking up basket, carrying basket ahead of owner, placing basket at target area

Data Analysis

During the analysis our main focus was on the communication between dog and O (paying and getting attention, communicating target object, target location and expected actions) and the dynamics of their movements (relative position of the dog while carrying the object). Since the duration of the whole task depended greatly on O's motor ability, the three main parts of the test; (1) picking up, (2) carrying, and (3) placing the basket, were analysed separately. The following variables were coded: joint attention (yes-no), number of verbal instructions, pointing gestures (yes-no), position of the dog (relative to the owner) when performing the carrying task (ahead – beside – behind), duration of the picking up and placing tasks (s).

Multivariate analysis of variance was applied to compare durations and number of commands. Kruskal-Wallis test and Mann-Whitney test were used to compare the position of dogs in different groups.

2.2 Results

Types of Verbal Communications
The owners' verbal instructions could be categorised as commands (verbs), name of dog, name of object ("basket"), praises ("well done"), and inhibitions ("no", "don't") in all three parts of the test. The proportion of the categories during the entire test duration was as follows: commands (verbs) 63 %, name of dog 14 %, name of object 8 %, praises 13 %, and inhibition 2 %.

Picking Up (Interaction Initiation)
We observed joint attention in case of all dog-owner dyads at the very beginning of the task. This means that the dog and O looked at each other before the dog got hold of the basket. In most cases eye-contact was spontaneous, that is, the dog oriented to O right after having been seated at the start point. If the dog oriented somewhere else (N = 10), O tried to attract its attention by calling its name before giving any instructions. After this, all but four Os pointed at the basket when communicating the task to the dog. Three of the exceptions were from the advanced/wheel-chaired group, where Os either needed both hands for driving the mechanical wheel chair, or could not move their arms. All four Os "pointed" toward the basket with their head.

The number of verbal instructions and the duration of the task showed significant correlation, and both varied according to the experience, because novice dogs needed more time and more verbal instructions to execute the task.

Carrying
Compared to the novice group, experienced dogs tended to carry the basket more frequently by going ahead of O than going behind. Using a wheel chair in itself did not make a general difference in this respect, however, post hoc tests revealed that in the experienced wheel-chaired group dogs had the tendency to carry the basket going ahead of their Os (Fig. 2).

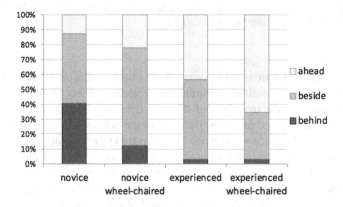

Fig. 2. Position of dogs during carrying: proportion of dogs that walked ahead, beside or behind the owner in the four groups

Novice dogs needed more verbal instructions to fulfil the task. Although the carrying part of the test took more time for the wheel-chaired Os, they did not talk more to their dog.

Placing

The number of verbal instructions and the duration of the task showed significant correlation. We found significant effect of both the experience and if the owner was wheel-chaired or not. The significant interaction of these effects revealed that the novice wheel-chaired group needed much more time and instructions for successfully placing the basket than the other three groups (Table 2).

Table 2. Number of dyads where joint attention and/or pointing gesture could be observed during the given task

	Pick up		Carry		Place	
	joint att	point	joint att	point	joint att	point
Novice	8	8	1	0	4	2
Novice – wheel-chaired	8	8	7	4	7	7
Advanced	8	8	4	3	8	7
Advanced – wheel-chaired	8	5	6	2	4	5

3 Study 2 – Inhibited Tasks

In this study after the dogs successfully performed a well-known and easy task several times, we presented them with an insoluble version of the same task. We assumed that dogs mainly would show two types of response to the situation. They would either react actively by trying to communicate with the owner (e.g. vocalizing, gaze alternation) [17], or passively, by showing displacement behaviour (e.g. lip licking, yawning, stretching). These behaviours may help the owner to realize that his/her request is ambiguous or not executable. Analysing dogs' responses in such scenarios can help to reveal how a robotic assistant should behave in such situations and/or how they should communicate a problem to the human.

3.1 Method

Subjects
A subsample of dyads was tested in this study: novice group (N = 4), novice–wheel-chaired group (N = 4), experienced group (N = 5), experienced–wheel-chaired group (N = 6).

Procedure
The two tests took place in an unfamiliar laboratory room at the Department of Ethology, ELTE, Budapest. The dogs entered the room together with O and were allowed to explore the test room for a few minutes before the test.

The two tests were both simple fetch and carry tasks and always followed each other in a fixed order. In test 1, the task was insoluble because the basket was not at the place where it was supposed to be. In test 2, the dog encountered a non-cooperative experimenter (E) who inhibited the fulfilment of the task by not handing over the basket. E who manipulated the basket was always a trained woman. The tests were video recorded and analysed later.

Test 1

There were two chairs inside: one in the middle for O and one in the corner for the dog's leash. The dog was sitting next to its O off leash. In the other end of the room there was a small cupboard and a barrier attached to it making one of the corners invisible for the dog and the owner. Opposite to the barrier there was an open door.

Prior to the test E explained the scenario shortly to O and laid down two rules: (1) to use only the following 3 commands: "bring", "basket" or "come" (to avoid the use of commands such as "seek" or "find") and (2) after sending the dog to fetch the basket, they are not allowed to give further commands till the dog initiates eye contact with them.

Warm up phase

The test began with a basic basket retrieving task, which is a basic task for these dogs given that this is a cornerstone of their training (and an everyday activity for the experienced ones). First E showed the basket to the dog, put it down to the floor in the middle of the room about 2 m from the dyad, closed the door and settled at the wall. The dog was commanded by O to retrieve the basket. Then in the two following trials after showing the basket to the dog it was hidden behind the folding, and the dog had to retrieve it from there when asked. After the completion of the task the dog always returned to its original position at the side of O. Next, the scenario was completed with a new element to prepare for the test trial: after hiding the basket O had to cover the dog's eyes and also close his/her own eyes. Non-visible to them E closed the door, stepped back to her place, and instructed the O to open her or his eyes, uncover the dog's eyes and execute the basket collecting task in the same way as before. This procedure was repeated two times.

Testing phase

The fifth trial was the inhibited one. When the eyes of the dog and O were closed, E removed the basket to the neighbouring room (far enough from the door, so the dog was not able to smell it from the experimental room), closed the door, returned to her usual position and instructed the owner to execute the same procedure as usual. The duration of the trial was 2 min measured by E from the moment that the dog was faced with the lack of the basket behind the cover. E stood still at her usual place. After 2 min E opened the door and gave the basket to the dog that returned it to O and was praised.

Test 2

E showed the basket to the dog, sat down onto a chair placed about 2 m from the dyad in the middle of the room, and instructed O to command the dog to bring back the basket. However, when the dog got hold of the basket E did not hand it over, but clung on it. The trial lasted for 1.5 min measured by E from the moment the dog discovered that E did not allow it to take away the basket. The same rules were applied for giving commands to the dog as in Test 1. After 1.5 min E released the basket, the dog could return it to O and was praised.

Data Analyses

The analyses of these tests focused mainly on the behaviour description of the dogs in the specific situations.

The following variables were coded in Test 1: latency of looking at owner and experimenter (s), approaching owner (yes-no), vocalization (yes-no), displacement behaviours (yawn, stretch, paddle, scratch itself, lick its lip, shake) (yes-no), fetching other object (yes-no), duration of looking for basket in 1 m around cupboard (s) and at other places in the room (s).

The following variables were coded in Test 2: latency of looking at owner (s), looking at experimenter when not manipulating the basket (yes-no), approaching owner (yes-no), vocalization (yes-no), displacement behaviours (yawn, stretch, paddle, lick its lip, shake) (yes-no), fetching other object (yes-no), duration of pulling (s) and chewing (s) the basket.

Due to the relatively small number of dyads from the specific groups the data were mostly analysed as one sample. Multivariate analysis of variance was applied to compare durations of searching in different places across the groups.

3.2 Results

Test 1

All but one dogs looked at O at least once during the inhibited trial, the mean number of gazes were 4.2, and the mean latency of the first gazing was 32 s. A large proportion of dogs (63 %) also approached O, and 74 % of them even gazed at E. The latency for gazing E was bigger than that of gazing at O. Several dogs showed even more explicit forms of communication: 26 % vocalised while confronting the problem.

Passive forms of expressing confusion could be also observed: 32 % of the dogs exhibited displacement behaviours, and 26 % fetched or manipulated some other object, mainly the leash or tiny pieces of the test set up (tape).

Moreover, dogs kept on looking for the basket on the average for 35 s around the cupboard and for 58 s at other places in the empty room. Dogs of wheel-chaired Os looked for the basket less at irrelevant places (Fig. 3).

Test 2

When E did not allow them to take the basket, 63 % of dogs looked at O, the mean latency for gazing was 33 s. Moreover, 32 % of the dogs approached O, and 67 % of them gazed at E. Many dogs (32 %) vocalised while facing the non-cooperative E, and a relatively large proportion of dogs (42 %) exhibited displacement behaviours.

Dogs tried to manipulate the basket in different ways. All dogs pulled the basket strongly; mean duration for pulling was 28 s. All but 4 dogs also chewed the basket; the mean duration for chewing was 27 s. The durations of pulling and chewing did not correlate with each other. In this test only two dogs retrieved another object to O (Fig. 4).

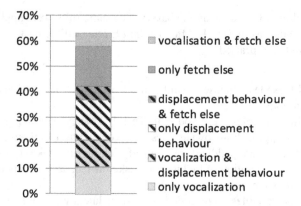

Fig. 3. Proportion of dogs that vocalised, exhibited displacement behaviours, fetched other objects or exhibited some combination of these behaviours when they could not find the object the owner asked for.

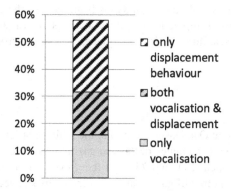

Fig. 4. Proportion of dogs that vocalised, exhibited displacement behaviours or did both when the experimenter did not hand over the object

4 Discussion

These studies range from finding the plausible channels of communication in human-robot-dog swarms applied for specific rescue purposes, through adapting the behavioural characteristics of dog-human attachment bond to robot-human dyads [18], to modelling exploratory behavior of assistance robots designed for home-care applications [19]. Although nowadays mainly assistance dogs help the elderly and disabled humans in their every-day life, in ten years assistance robotics will certainly make a significant breakthrough in this field. In order to enhance the effectiveness and quality of the interaction between humans and robots, users should be able to interact with assistance robotic systems in a natural way. Presently most assistance robots try to mimic human capacities, embodiment and behaviour, even though neither their function nor their abilities make them comparable to a human assistant yet. We argued that the adaptation of some relevant behaviour patterns observed in assistance dog–owner interactions could improve the ability of assistance robots of different embodiments to provide successful social and physical assistance of elderly people in a home environment. In order to design such a robot we have to examine owner-dog interactions with a special focus on the situations the assistant robot will face during its duty.

Our observations of a fetch and carry task showed that even the crucial elements of the interaction, such as attention getting behaviours and the human partner's verbal communication, can show both context specific and dyad specific features (see for similar results [20].). The contribution of the simple verbal instructions to the successful execution of the different tasks (fetching, carrying, placing) needs further analyses, but importantly, during the interactions owners' verbal activity was complemented by non-verbal communication. This draws attention to the significance of the ability for parallel perception and processing of human verbal and non-verbal communication, for instance the need to be able to interpret gestures in assistance robots (see e.g. [21]). We revealed that joint attention was required for the initiation of task execution. As a rule, owners draw their dog's attention before giving any instruction, and they look at each other before performing further actions. This implies that even very simple non-humanoid robots should show their "attentional state" by at least the orientation of their body and/or head to show "attention" and facilitate the initiation of the interaction.

Pointing at an object to be manipulated by the assistant also proved to be a typical action, and dogs were very successful in relying on this gesture. Convincing evidence proves that for this no special training is required because dogs are successful in spontaneous use of human pointing gesture [22]. Dogs' ability to select the requested object from several objects placed close to each other need further investigation, but we suggest that the interpretation of simple pointing gestures (using extended vectors of the outstretched arm/pointing finger) could be a very useful skill for assistance robots and ongoing research (e.g. [23]) should proceed forward in this area applying results from dog cognitive studies (e.g. [24]).

During the carrying task the dogs seemed to apply different strategies depending on their experience. Novice dogs followed the owner toward the target area or walked beside them [25], however, experienced ones could use more subtle information to find out the correct direction, using the owner's pointing gesture or verbal instructions, or by extrapolating the direction of the initial movement of the owner to predict the location of the target. The more advanced performance of the wheel-chaired group is important to note, because in real life situations the ability to navigate the assistant in front of them and not being forced to look back regularly could be especially useful for owners in wheel-chair. Although this particular activity (carrying) is mainly driven by the user's capabilities, because the movement speed of the user regulates the time needed for dog, wheel-chaired owners did not use more verbal instructions for this task. This implies that the number of commands is not simply linked to the duration of the task, but rather to the quality of the cooperation.

In addition to group level differences, we could observe marked within group (individual) differences during the interactions. Placing an object to a predefined place could be difficult for the novice wheel-chaired group because the usual end of the basket carrying task is to hand over the basket to the owner and some inexperienced dogs were not used to flexibly combine the learnt behaviour elements on command (that is, placing the object at the target point).

In Study 2 dogs faced two situations they might encounter in everyday activities when they are not able to perform the tasks they had been asked for. Both scenarios were based on the fetch and carry abilities of dogs, but the situations differed in a crucial element. In Test 1, there was no object at the location indicated by the naïve owner. In Test 2, the object was there but the execution of the required action was actively inhibited by the experimenter. As expected, most dogs tried to express their confusion and negative inner state in at least one way. In contrast to Study 1, during the inhibited trials dogs displayed some specific vocalisation, that is, they emitted high pitched voices that could reflect their inner state in an explicit manner, which needs no training to interpret by humans [26].

There was also a tendency to use more passive forms of expressive behaviours by showing displacement behaviours. We must admit that exhibiting such behaviours cannot be regarded as communication of the ambiguity of the situation, and might only be effective if the owner is sensitive or experienced enough to read these unintentional behavioural signals. However, one of the approaches of social psychology [27, 28] sees the roots of human embarrassment in the appeasement behaviours of animals. These unintentional actions have appeasement-related functions, pacify partners in case of social transgressions by reducing aggression and evoking social approach in others, thus restoring the social interaction and relation. So the calming value of these behaviours – processed by the owners either consciously or unconsciously – can account for the difference in the reactions of humans to the failures of the dogs and robotic agents.

The major message of the results of the inhibited trials is that dogs do not give up easily if they face a "seemingly" insoluble task. Almost each dog showed a strong commitment to execute the task or at least the behavioural elements discussed above lent a "busy" appearance to them. This strong endeavour to fulfil the task can be an attractive characteristic in the eyes of the owners and can attenuate their disappointment. These behaviours (e.g. hesitantly turning around, moving around close to the place of the task to be executed, looking/moving back and forth between the user and the aimed object, emitting some high pitched sound) should be adapted to assistance robots to enhance their similarity to a living helper. Thus we could count on the empathy, understanding and forgiveness in humans that living beings are able to evoke in such situations. Obviously, we do not suggest that an assistance robot should whine, yawn or scratch itself when it is in an ambiguous situation. We propose that the problem solving strategies of assistant dogs could inspire the development of the relevant functions and social behaviours of assistance robots. We suggest that it would make HRI more fluent and less stressful if assistant robots could communicate their inability to solve a problem using simple behaviours like non-verbal vocalisation or orientation alternation, and/or could show displacement behaviours rather than simply not performing the task.

In sum, it is important to stress once again that the dog model does not imply that the robot needs to have a dog-like appearance (as a matter of fact we suggest that it should not look like a dog); instead we try to identify simple basic behaviours available to even a mechanical-looking embodiment, such as orientation, proxemic behaviour, gross body movements, comprehending some gestural and verbal communication of the human partner, and being able to show complex and variable repertoire of simple behaviour elements in relevant social situations.

In one of the first application of the above theory we investigated the dynamics of feedback processes during the interaction between a robot used as a hearing aid and human subjects who played the role of the deaf owner [29]. The behaviour patterns of the robot, which had no arms and verbal capacities, were designed based on the interactions between hearing dogs and deaf owners. Dog-inspired behaviour sequences and decision making strategies were used to program and control the robot during the trials. Findings indicate that untrained and uninformed participants could correctly interpret the robot's actions, and that head movements and gaze directions during signalling, leading, and feed-back processes were important and effective for communicating the robot's intentions.

Acknowledgement. The research was supported by the European Union (LIREC-215554) and MTA 01 031. A.M. also received support from the Swiss National Science Foundation (SWARMIX: Synergistic Interactions in Swarms of Heterogeneous Agents).

We would like to thank Barbara Gáspár for her help in conducting the tests. We are grateful to the Hungarian Dogs for Humans charity and the owners for their participation.

References

1. Amat, J.: Intelligent wheelchairs and assistant robots. In: de Almeida, A.T., Khatib, O. (eds.) Autonomous Robotic Systems. LNCIS, vol. 236, pp. 211–221. Springer, Heidelberg (1998). doi:10.1007/BFb0030807
2. Yamazaki, K., Ueda, R., Nozawa, S., Mori, Y., Maki, T., Hatao, N., Okada, K., Inaba, M.: Tidying and cleaning rooms by a daily assistive robot – an integrated system for doing chores in real world. J. Behav. Robot. **1**(4), 231–239 (2011)
3. Pollack, M.E., Brown, L., Colbry, D., Orosz, C., Peintner, B., Ramkrishnan, S., Engberg, S., Matthews, J., Dunbar-Jacob, J., McCarthy, C., Thrun, S., Montemerlo, M., Pineau, J., Roy, N.: Pearl: a mobile robotic assistant for the elderly. In AAAI Workshop on Automation as Caregiver (2002)
4. Bartneck, C., Reichenbach, J., Breemen, A.V.: In your face, robot! The influence of a character's embodiment on how users perceive its emotional expressions. In: Proceedings of Design and Emotion, Ankara (2004)
5. Miklósi, Á., Gácsi, M.: On the utilization of social animals as a model for social robotics. Front. Psychol. **3**, 75 (2012)
6. Dautenhahn K.: Robots we like to live with?! – a developmental perspective on a personalized, life-long robot companion. In: Proceedings of the IEEE International Workshop on Robot and Human Interactive Communication, Kurashiki, Okayama, Japan, 20–22 September 2004
7. Jones, T., Lawson, S., Mills, D.: Interaction with a zoomorphic robot that exhibits canid mechanisms of behaviour. In: Proceedings of IEEE International Conference on Robotics and Automation (ICRA 2008), Pasadena (2008)
8. Topál, J., Kubinyi, E., Gácsi, M., Miklósi, Á.: Obeying social rules: a comparative study on dogs and humans. J. Cult. Evol. Psychol. **3**, 213–239 (2005)

9. Topál, J., Miklósi, Á., Gácsi, M., Dóka, A., Pongrácz, P., Kubinyi, E., Virányi, Z., Csányi, V.: The dog as a model for understanding human social behavior. Adv. Study Anim. Behav. **39**, 71–116 (2009)

10. Fallani, G., Prato Previde, E., Valsecchi, P.: Do disrupted early attachments affect the relationship between guide dogs and blind owners? Appl. Anim. Behav. Sci. **100**, 241–257 (2006)

11. Nguyen, H., Kemp, C.C.: Bio-inspired assistive robotics: service dogs as a model for human-robot interaction and mobile manipulation (2008)

12. Arkin, R.C., Fujita, M., Takagi, T., Hasegawa, R.: Ethological modeling and architecture for an entertainment robot. In: 2001 IEEE International Conference on Robotics and Automation, Seoul, Korea, pp. 453–458 (2001)

13. Syrdal, D.S., Koay, K.L., Gácsi, M., Walters, M.L., Dautenhahn, K.: Video prototyping of dog-inspired non-verbal affective communication for an appearance constrained robot. In: 19th IEEE International Symposium on Robot and Human Interactive Communication, (RO-MAN 2010), Viareggio, Italy, pp. 632–637 (2010)

14. Hecht, J., Miklósi, Á., Gácsi, M.: Behavioral assessment and owner perceptions of behaviors associated with guilt in dogs. Appl. Anim. Behav. Sci. **139**, 134–142 (2012)

15. Barrett, K.C.: The origins of social emotions and self-regulation in toddlerhood: new evidence. Cogn. Emot. **19**(7), 953–979 (2005)

16. Maestripieri, D., Schino, G., Aureli, F., Troisi, A.: A modest proposal: displacement activities as an indicator of emotions in primates. Anim. Behav. **44**, 967–979 (1992)

17. Miklósi, Á., Pongrácz, P., Lakatos, G., Topál, J., Csányi, V.: A comparative study of the use of visual communicative signals in interactions between dogs (Canis familiaris) and humans and cats (Felis catus) and humans. J. Comp. Psychol. **119**, 179–186 (2005)

18. Kovács, S., Vincze, D., Gácsi, M., Miklósi, Á., Korondi, P.: Ethologically inspired robot behavior implementation. In: 4th International Conference on Human System Interactions Yokohama, Japan, pp. 64–69 (2011). doi:10.1109/HSI.2011.5937344. ISSN 2158–2246. ISBN 978-1-4244-9638-9

19. Numakunai R., Ichikawa T., Gácsi M., Korondi P., Hashimoto H., Niitsuma M.: Exploratory behavior in ethologically inspired robot behavioral model. In: RO-MAN 2012, Paris, France, pp. 577–582 (2012)

20. Faragó, T., Miklósi, Á., Korcsok, B., Száraz, J., Gácsi, M.: Social behaviours in dog-owner interactions can serve as a model for designing social robots. IS submitted (2012)

21. Nandy, A., Mondal, S., Prasad, J.S., Chakraborty, P., Nandi, G.C.: Recognizing & interpreting indian sign language gesture for human robot interaction. In: 2010 ICCCT, pp. 712–717 (2010)

22. Miklósi, Á., Kubinyi, E., Topál, J., Gácsi, M., Virányi, Z., Csányi, V.: A simple reason for a big difference: wolves do not look back at humans, but dogs do. Curr. Biol. **13**, 763–766 (2003)

23. Nickel, K., Stiefelhagen, R.: Visual recognition of pointing gestures for human–robot interaction. Image Vis. Comput. **25**(12), 1875–1884 (2007)

24. Lakatos, G., Gácsi, M., Topál, J., Miklósi, Á.: Comprehension and utilisation of pointing gestures and gazing in dog–human communication in relatively complex situations. Anim. Cogn. **15**, 201–213 (2012)

25. Young, J.E., Kamiyama, Y., Reichenbach, J., Igarashi, T., Sharlin, E.: How to walk a robot: a dog-leash human-robot interface. In: RO-MAN, pp. 376 – 382. IEEE (2011)

26. Molnár, C., Pongrácz, P., Miklósi, Á.: Seeing with ears: Sightless humans' perception of dog bark provides a test for structural rules in vocal communication. Q. J. Exp. Psychol. **63**, 1004–1013 (2010)

27. Keltner, D.: The signs of appeasement: evidence for the distinct displays of embarrassment, amusement, and shame. J. Pers. Soc. Psychol. **68**, 441–454 (1995)
28. Miller, R.S., Leary, M.R.: Social sources and interactive functions of embarrassment. In: Clark, M. (ed.) Emotion and Social Behavior, pp. 322–339. Russell Sage Foundation, New York (1992)
29. Koay, K.L., Lakatos, G., Syrdal, D.S., Gácsi, M., Bereczky, B., Dautenhahn, K., Miklósi, Á., Walters, M.L.: Hey! There is someone at your door. A Hearing Robot using Visual Communication Signals of Hearing Dogs to Communicate Intent. submitted SSCI 2013 Alife

Bioinspired Obstacle Avoidance Algorithms for Robot Swarms

Jérôme Guzzi[✉], Alessandro Giusti, Luca M. Gambardella,
and Gianni A. Di Caro

Dalle Molle Institute for Artificial Intelligence (IDSIA), Manno-Lugano, Switzerland
{jerome,alessandrog,luca,gianni}@idsia.ch
http://www.idsia.ch/~gianni/SwarmRobotics

Abstract. Recent work in socio-biological sciences have introduced simple heuristics that accurately explain the behavior of pedestrians navigating in an environment while avoiding mutual collisions. We have adapted and implemented such heuristics for distributed obstacle avoidance in robot swarms, with the goal of obtaining human-like navigation behaviors which would be perceived as friendly by humans sharing the same spaces. In this context, we study the effects of using different sensing modalities and robot types, and introduce robot's emotional states, which allows us to modulate system's group behavior. Experimental results are provided for both real and simulated robots. The extensive quantitative simulations show the macroscopic behavior of the system in various scenarios, where we observe emergent collective behaviors – some of which are similar to those observed in human crowds.

Keywords: Dynamic obstacle avoidance · Human-friendly navigation · Mobile robots · Emerging macroscopic behaviors · Emotions

1 Introduction

Which mechanisms do pedestrians use in order to navigate around moving obstacles in a dynamic environment? Social anthropologists and researchers concerned with simulating crowd behavior have proposed several different models for explaining human obstacle avoidance behaviors. Among these, Moussaïd et al. [1] have recently proposed a simple heuristic (inspired by specific functions of the human eye and the brain) which is able to accurately predict observed trajectories, as well as macroscopic behaviors, observed in crowds – such as the spontaneous formation of ordered lines of opposite flow in corridors.

We propose to adopt the same obstacle avoidance heuristic in robot swarms, and present a working implementation on real robots and simulations. If robots follow the same mobility criteria as pedestrians, their trajectories will be *predictable* and *legible* to humans sharing the same spaces. The resulting behaviors would then be perceived as *friendly* and *acceptable*, ultimately enabling efficient sharing of spaces among humans and robot swarms, which is the long-term goal of our research.

G.A. Di Caro and G. Theraulaz (Eds.): BIONETICS 2012, LNICST 134, pp. 120–134, 2014.
DOI: 10.1007/978-3-319-06944-9_9, © Institute for Computer Sciences, Social Informatics and Telecommunications Engineering 2014

Unlike humans, which can be modeled as agents with practically homogeneous dynamic and sensing abilities, robots are characterized by a much larger variability in terms of operational capabilities. For example, different kinds of robots may move at quite different speeds, or may be able to sense other agents based on different fields of view and accuracy. Given these core differences, in which conditions can we still observe human-like macroscopic behaviors? Can we introduce simple variations to the core heuristic to promote the emergence of novel collective behaviors useful in typical swarm robotic scenarios? This paper presents our first results in answering such questions by means of quantitative simulations of large robot swarms. In turn, simulation results are validated with an implementation of the same algorithms on real robots, the *foot-bots* [2].

The rest of the paper is organized as follows. Section 2 relates our work to other obstacle-avoidance algorithms in robotics and to recent research in crowd modeling. In Sect. 3 we briefly describe the core obstacle-avoidance heuristic, first proposed in [1], then detail our implementation in robotics and describe its parameters. Moreover, we discuss simple variations aimed at promoting new meaningful behaviors in robot swarms, such as the autonomous, emotion-driven formation of homogeneous groups of robots with similar attitudes and characteristics. Section 4 shows the details of our implementation for real robots and our simulation environment, which in used in Sect. 5 to run a number of quantitative experiments considering different scenarios.

2   Related Work

The problem of dynamic obstacle avoidance is widely studied both in robotics and social sciences. Our works builds upon results in both fields.

In robotics, the most common approach is based on the concept of *velocity obstacle* [3], also known as *collision cone* or *forbidden velocity map*, meaning the sets of velocities that will lead a robot to collision: choosing a velocity outside such set ensures that no collision will occur. Different variations of these ideas have been presented to improve the prediction accuracy of other agents' trajectories, to add recursion and account for sensing errors in a probabilistic framework [4], and to ensure smooth trajectories by sharing the responsibility to avoid a collision with other agents (*reciprocal velocity obstacle* [5]). Applications to very crowded scenarios also introduce asymmetries in the obstacle velocity construction [6], requiring to enforce conventions to allow smooth and deterministic interactions between agents. The velocity-obstacle model was also successfully applied to explain certain characteristics of pedestrian behavior [7].

All the mentioned works basically build on a mechanicistic and artificial approach to navigation, which is designed to ensure safety, and is adapted to produce smooth trajectories. On the contrary, our work stems from a cognitive heuristic [1] modeling human behavior – which produces paths with good efficiency, smoothness, and legibility – to which we add some modifications to also ensure safety (a primary objective in robotics). This paper represents the first implementation of this heuristic to robotics. Implementation-wise, such a heuristic allows us to decouple the computation of speed and desired heading for the navigating agent. This leads to a simpler implementation than velocity-obstacle approaches, which requires a search over the two-dimensional velocity space.

Animals are able to visually control locomotion through optical flow [8], which provides a direct estimation of the current *time to contact* τ with a moving obstacle. For instance, humans use optical flow to control the speed of walking [9] and to perceive upcoming collisions [10], both from internal and external point of views, drivers use it to control braking ($\dot{\tau}$ strategy) and to adjust for a safe distance from the preceding car [11]. This sensing information is fed into a spatial-temporal integration and *temporal prediction* cognitive layer [12] for collision prediction and, in general, is incorporated into higher level cognitive processes. Through neuromodulation [13] internal brain states, like those related to emotions, modulate the sensing and behavioral process. In robotics, this provides a framework to improve learning, flexibility, robustness and control the emergence of cooperative behaviors [14,15]. Following these observations, we also include the use of emotional states as modulators of navigational behaviors.

Mutual avoidance and sharing of space in human groups has been extensively studied in sociological research for the prediction of the behavior of crowds, among other topics that have been considered. The original models are based on the study of *proxemics* [16], which formalizes the concept of *personal* and *social* space, where pedestrian behavior based on *social forces* [17] enforces people to keep a minimum distance from neighbors whenever possible. Such modeling approach was successfully used for crowd simulation and also inspired several human tracking and avoidance models in robotics. The density dependence of the speed of a flow of people along a street can be explained by the *net-time headway* mechanism [18], according to which each pedestrian keeps a constant τ time away from the surrounding pedestrians to avoid collisions and stops walking if this would imply a low speed. Moussaïd et al. [1,19] recently incorporated this rule in a new model of pedestrian navigation based on a simple heuristic, which we build upon and extend it for use in robot swarms. Among other topics the authors also addressed the bidirectional pedestrian flow scenario that is relevant for virtual crowds simulation and for the study of emerging behaviors like lines formation [20]. We also consider similar situations.

3　The Models for Human and Robot Navigation

We first discuss the heuristics explaining human behavior for obstacle-avoidance as introduced by [1] (Sect. 3.1), then present our adaptation to robotics (Sect. 3.2) and an extension for including artificial emotional states, which depend on the presence of close by robots with similar attitudes and characteristics, and is used to modulate group behavior (Sect. 3.3).

3.1　Human Behavioral Model

Given a 2D reference frame F, a moving agent directed to a target point $\vec{O}$ is characterized by an optimal (open space) moving speed v_{opt} and a horizontal

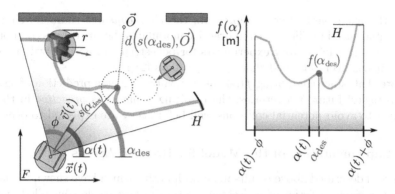

Fig. 1. Illustration of variables and functions for the human navigation model.

field of view 2ϕ (in radians). In the reference frame F, $\vec{x}(t)$ and $\vec{v}(t)$ are agent's position and velocity vectors at time t, and $\alpha(t)$ is agent's heading, (i.e., the direction it is facing with respect to F's horizontal axis). We assume that agent's ground occupancy is approximated by a circle of known radius r.

To direct its movements, the agent makes use of a cognitive function $f(\alpha)$, $\alpha \in [\alpha(t)-\phi, \alpha(t)+\phi]$, based on visual information, that maps each heading α within its field of view to the distance that the agent can travel in $\alpha's$ direction before colliding with any visible obstacle when moving at speed v_{opt}. The distance is bounded by a maximum horizon H. With $s(\alpha)$ we denote the 2D segment connecting $\vec{x}$ with the point at distance $f(\alpha)$ along the direction α (i.e., the point of first collision for heading α). When computing $f(\alpha)$, all obstacles are assumed to keep their current heading and speed, thus moving according to a uniform linear motion (Fig. 1).

Given the above notations, the agent walking behavior can be explained by the following simple heuristic rules.

First, the agent determines its desired heading $\alpha_{\text{des}}(t)$ as the direction allowing the most direct path to destination point $\vec{O}$, taking into account the presence of obstacles:

$$\alpha_{\text{des}}(t) = \text{argmin}_\alpha \, d\left(s\left(\alpha\right), \vec{O}\right) , \qquad (1)$$

where $d(\cdots)$ denotes the minimal distance between a line segment and a point.

The desired velocity vector $\vec{v}_{\text{des}}(t)$ is then determined. Its direction is defined by the heading $\alpha_{\text{des}}(t)$, and its modulus $v_{\text{des}}(t)$ is set to allow stopping in a fixed time τ_1 within the free distance $D(\alpha_{\text{des}}) \in [0, H]$, *currently* perceived along direction α_{des}:

$$v_{\text{des}}(t) = \min\left(v_{\text{opt}}, \frac{D\left(\alpha_{\text{des}}\right)}{\tau_1}\right) . \qquad (2)$$

The actual velocity vector $\vec{v}(t)$ is continuously adjusted depending on $\vec{v}_{\text{des}}(t)$:

$$\frac{d\vec{v}}{dt} = \frac{\vec{v}_{\text{des}}(t) - \vec{v}(t)}{\tau_2}, \qquad (3)$$

where the fixed parameter τ_2 represents the *time constant* characterizing the exponential speed profile, which in practice modulates the smoothness of motion. From controlled laboratory experiments for pedestrians in normal walking conditions it has been observed that $\tau_1 = \tau_2 = 0.5\,\mathrm{s}$ [21].

Note that, since computing $f(\alpha)$ values involves a rough prediction of agent's and obstacles' future trajectories, the resulting behavior is *proactive* in that it attempts to avoid potential collisions well before they are expected to occur.

3.2 Implementation of the Model for Robotic Agents

The model described above results in smooth paths, which have shown to closely match the characteristics of pedestrian motion in large-scale controlled experiments, both for single trajectories and macroscopic crowd motion patterns [19]. Robots following the same rules, and with human-like sensing and locomotion ability, are therefore expected to exhibit the same large-scale, macroscopic behaviors observed in human crowds. They would also exhibit behaviors which are predictable (and thus acceptable) by humans sharing the same environment.

Unfortunately, the presented heuristic does not ensure collision-free behavior: in fact, small collisions (e.g., shoulder to shoulder) among humans happen frequently and contribute to shape the behavior of tightly-packed crowds. However, collisions are not really acceptable in robotics. The problem is further aggravated by inaccuracies in localizing other agents (especially when using low resolution cameras, as typical in swarm systems). Therefore, we introduce a system parameter *safety margin* m_s, defined by augmenting the radius of all obstacles, to be accounted for during the computation of the $f(\alpha)$ function (in the experiments we set $m_s = 0.1\,\mathrm{m}$).

Through an experimental study we investigate the small-scale (Sect. 5.2) and macroscopic (Sects. 5.3 and 5.4) properties of the robot navigation algorithm as described above. In the next section we discuss a specific enhancement of the model which promotes the emergence of interesting macroscopic behaviors in robot swarms, which is experimentally tested in Sect. 5.5.

3.3 Enhancements for Socially Active Robots

We consider the setting in which the robots in the swarm belong to *different classes*. Such classes may represent different types of robots, possibly with different locomotion characteristics and navigation attitudes. However, even in a swarm composed by physically homogeneous robots, different classes may represent other types of relations, such as different roles or responsibilities in the swarm, or different levels of behavioral affinities among groups (e.g., elderly vs. youngsters). In this context we investigate a bio-inspired approach for promoting the autonomous emergence of grouping behaviors among the robots in the same class.

To each agent a we associate an internal, time-variant state $w_a \in \mathbb{R}$ – an adimensional quantity which we refer to as *wellness*. The wellness of an agent represents its *emotional state* in relation to the navigation and the presence of agents of the same or different groups. The perception of wellness modulates the obstacle-avoidance behavior of an agents, and depends on how well, in terms of class membership, the robot fits within its neighborhood. Let us define

$$\theta(a, b) = \begin{cases} +1 & \text{if } a \text{ and } b \text{ belong to the same class} \\ -1 & \text{otherwise} \end{cases} \quad (4)$$

then, w_a is computed as:

$$w_a = \sum_{b \in \text{visible neighbors}} \theta(a, b) \, e^{-\frac{d(a,b)}{g}}, \quad (5)$$

where $d(a, b)$ denotes the distance between the agents a and b, and g defines the spatial scale of the influence of the neighbors on the agent's wellness.

An agent with a large w_a value *feels* well and safe, since most of its closest visible neighbors belong to its own class. On the contrary, agents whose closest visible neighbors belong to a different class are associated to negative wellness values. A lonely agent feels neutral ($w_a = 0$). w_a may alter different aspects of the agent's sensing and/or behavior as to mimic human brain neuromodulation [13]. In the following, we let w_a modulate the agent's optimal speed (v_{opt} in Eq. (2)) such that a robot, when it does not feels safe, will tend to be more cautious and move slower. In Eq. (2), we replace v_{opt} with $v_{\text{opt}} - \Delta(w_a)$. $\Delta(w_a)$ is defined as follows:

$$\Delta(w_a) = \begin{cases} 0 & \text{if } w_a \geq 0, \\ \Delta_{\max} & \text{if } -kw_a > \Delta_{\max}, \\ -kw_a & \text{otherwise}, \end{cases} \quad (6)$$

where $\Delta_{\max} \geq 0$ (measured in m/s) is a parameter bounding the maximum effect of wellness on the optimal speed, and $k \geq 0$ (also measured in m/s) is a parameter controlling how much effect the agent's wellness has on its own speed.

Section 5.5 shows that this modeling results in robots of the same class clustering together, led by cautious group leaders.

4   Implementation on Real and Simulated Robots

The navigation algorithm described in Sect. 3 has been implemented on real robots (the *foot-bots*) as well as in simulations.

4.1   The Foot-Bot Real Robot Platform

The *foot-bot* robot is a small mobile platform, directly derived from the *marXbot* [2], specifically designed for swarm robotics [22]. The robot is 17 cm wide and 30 cm tall, and is based on an on-board ARM-11 processor programmed in a Linux-based operating environment. Differential-driven motorized tracks allow mobility at speeds up to 0.3 m/s.

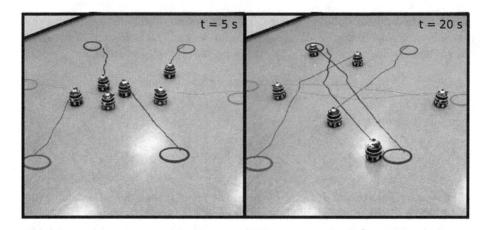

Fig. 2. Six *foot-bot* robots implementing our algorithms, moving towards opposite sides of a circle without collisions. Full video at http://bit.ly/WkJ0ld

In the context of this work, foot-bots sense neighbors by means of a forward-facing camera with a $2\phi = 50°$ field of view and a down-sampled resolution of 128×92 px, which is used for localizing humans, navigation targets, and walls at 25 frames-per-second (Fig. 2).

Since our main focus is on navigation algorithms and not on sensing, we use straightforward techniques for processing camera images: entities of interest, (e.g., landmarks used to identify a destination point, or humans) are marked with differently colored bands at a known height from the floor. Robots convert each frame to the HSV color space, and segment pixels corresponding to each object. After performing connected component analysis, this results in a set of binary blobs. From the image coordinates of each blob's centroid, the robot computes distance and bearing of the corresponding entity by means of an homography transform, which can be estimated in advance given that the camera parameters and height of each entity are known. The velocity of humans is estimated by finite differencing, after smoothing position readings with a moving average filter defined over a period of 0.5 s. Note that the position of targets (i.e., destination points) is sensed online through vision, and not given by an external observer or a higher-level path planning algorithm.

Robot controllers operate on a 0.1 s timestep and are not synchronized with each other. At each timestep, rules in Eqs. (1) and (2) yield the desired values for heading (α_{des}) and speed (v_{des}), respectively. Both in simulation and in the implementation on the foot-bots, we use a mobility model similar to Eq. (3) that takes into account the robots constraints and independently controls the speed of the two differential driven track wheels.

4.2   Robot Simulation

We developed a custom simulator for the efficient and accurate simulation of large foot-bot swarms. In the simulation experiments, the observed position and

speed of neighboring agents (i.e., the readings of simulated sensors) are arti-
ficially corrupted by random localization errors approximating the statistical
properties of errors observed in the real implementation. This means a precise
and uniform bearing resolution but large uncertainty in depth estimation, which
increases for objects farther away. More specifically, given an obstacle whose
ground truth relative position is expressed in robot-centered polar coordinates
as (ρ, θ), the observed position (ρ', θ') is given by $(\theta' = \theta + \phi e; \ \rho' = \rho + \gamma \rho \phi e)$.
In the formulae, $e \sim \mathcal{N}(0, \sigma)$ models the localization error in the normalized
image space, ϕ denotes the camera field of view, and γ is a constant depending
on the characteristics of the depth estimation approach. In the following, we set
$\sigma = 1/128$ (i.e., 1 pixel on a 128×96 sensor) and $\gamma = 10$. In both the real
and simulated robot implementations, velocity vectors are estimated by finite
differencing.

5   Experimental Results

In this section, we report the results of experiments aimed at studying the effect
of the proposed algorithm in terms of: (i) efficiency of individual trajectories
(Sect. 5.2), and (ii) emergence of macroscopic behavioral patterns (Sects. 5.3, 5.4
and 5.5). These aspects are investigated in two different experimental settings,
denoted as *cross* and *periodic corridor*, respectively.

In the *cross* setting, we consider four target destinations at the vertices of a
square with an edge of 4 m. N robots are divided in two equally-sized groups:
robots of each group travel back and forth between two opposite vertices, thus
creating a busy crossroad in the middle (see Fig. 3). In this setting, we use

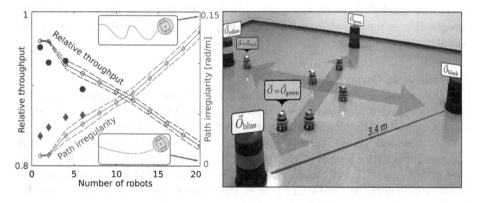

Fig. 3. Experimental results (left) for scalability in the *cross* scenario (pictured on the
right). Large filled markers correspond to results measured on real robots. Blue thick
line with round markers denotes relative throughput. Green thin line with diamond
markers shows path irregularity. Dashed lines delimit ± standard deviation over 100
randomly initialized replicas. In this experiment, both real and simulated robots use
360° range-and-bearing sensing, $H = 3$ m, $m_s = 10$ cm (Color figure online).

infrared range and bearing sensors for robot-robot detection both in real robot implementation and in simulation. In the *periodic corridor* setting, $N = 60$ robots travel along a straight corridor with a given width w_C and fixed length $l_C = 16$ m, whose opposite ends "wrap around" as if the corridor was the lateral surface of a cylinder. This setting is often considered in the crowd analysis literature, and allows us to observe emerging macroscopic behaviors.

For each experiment we study the effects of one or two parameters. For each setting of the parameters we perform 100 simulation runs (replicas), each lasting 360 s of simulated time. For each run, robots are positioned in randomly chosen initial locations.

5.1   Performance Metrics

We define several metrics, for quantifying the quality of individual trajectories (*relative throughput* and *path irregularity*), and for observing the emergence of macroscopic behaviors in the *periodic corridor* scenario (*line order*, *group order*, and *number of clusters*).

Trajectory Quality Metrics

Relative throughput is defined as the number of targets that a robot has reached during the simulation, expressed as a fraction of the number of targets that the robot could reach if traveling along straight paths and ignoring collisions (which is an ideal upper bound for throughput). The measure is adimensional and is averaged over all the robots in the simulation.

Path irregularity is defined as the amount of *unnecessary turning* per unit path length performed by a robot, where *unnecessary turning* corresponds to the total amount of robot rotation minus the minimum amount of rotation which would be needed to reach the same targets with the most direct path. Path irregularity is measured in rad/m, and is averaged over all the robots in the simulation.

Macroscopic Order Metrics

For the *periodic corridor* scenario, following [1], we extend the approach in [23] and define three macroscopic order metrics quantifying interesting characteristics of the agents' spatial configuration at a given moment in time.

Let $\mathcal{L}$ denote the set of all longitudinal corridor bands with width Δ_L and spanning the entire length of the corridor. Let $L(x) = [x, x + \Delta_L] \times [0, l_C] \in \mathcal{L}$ denote one specific band. Similarly, let $\mathcal{T}$ denote the set of all transversal corridor bands with length Δ_T, and let $T(y) = [0, w_C] \times [y, y + \Delta_T] \in \mathcal{T}$ denote one specific band.

Let $n_k(B)$ be the number of agents of class k whose center lies inside a transversal or longitudinal band $B \in \mathcal{L} \cup \mathcal{T}$. In the 2-class case, the Yamori's band index $Y(B)$ [23] measures the prevalence of any class in B, and is defined as:

$$Y(B) = \frac{|n_1(B) - n_2(B)|}{n_1(B) + n_2(B)}$$

Line order: the line order O_L is defined as the average Yamori index of longitudinal bands of width $\Delta_L = 0.3\,\text{m}$ over $\mathcal{L}$, and measures how agents of the same class tend to position themselves along ordered longitudinal lines. O_L is bounded between 0 (random configuration) and 1 (representing perfect organization of the swarm classes in longitudinal lines).

$$O_L = \langle Y(B) \rangle_{B \in \mathcal{L}}$$

Group order: the group order O_G is defined as the average Yamori index of transversal bands of length $\Delta_T = 0.6\,\text{m}$ over $\mathcal{T}$, and quantifies how agents of the same class tend to group themselves in compact clusters. $0 \leq O_G \leq 1$, with $O_G = 1$ meaning perfect organization of the swarm classes in clusters.

$$O_G = \langle Y(B) \rangle_{B \in \mathcal{T}}$$

Number of clusters: the number of clusters N_G is computed as follows. Let $\{T_1, T_2\}$ denote a pair of adjacent transversal bands of length $\Delta_T = 0.6\,\text{m}$. N_G is defined as the number of such pairs where the majority class in T_1 differs from the majority class in T_2.

5.2 Algorithm Scalability and Trajectory Efficiency

Within scenario *cross*, we initially verify the scalability of the algorithm versus an increasing number of agents, and validate simulation results by comparison with the performance measured on foot-bot robots. Results are reported in Fig. 3. We can observe that the results obtained with real robots in the same conditions closely match simulations. As the swarm size grows, relative throughput decreases and path irregularity increases, because robots must follow longer and more curvy trajectories in order to avoid collisions. Performance scales well, since, even in very dense scenarios, paths remain efficient, smooth, and predictable.

In the real robot implementation, despite the severe hardware limitations, the navigation controller requires invariably less than 20 ms of computation time per timestep. In simulation, we also tested robustness to timesteps longer than 0.1 s, and found that in all considered scenarios, performance begins to degrade only when the timestep exceeds 0.4 s.

5.3 Formation of Ordered Lines of Opposite Flow

For the remaining experiments, we consider the *periodic corridor* scenario. We observe that robots traveling in opposite directions, just like pedestrians, tend to form stable ordered lines of flow from initial random settings. This macroscopic collective behavior results in increased efficiency, since agents can follow more direct trajectories. In Fig. 4 we illustrate this process, and show that there is a critical corridor width below which robots are too dense to reach an ordered formation. Such critical density is also influenced by robot parameters. Figure 4ⓔ shows that the critical width depends on the value of v_{opt}: faster robots need larger maneuvering spaces, and therefore require a larger corridor for reaching an ordered configuration.

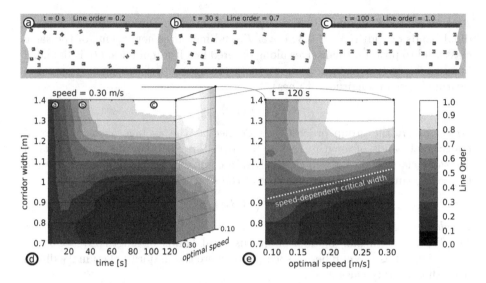

Fig. 4. Formation of flow lines when the corridor is shared by 60 robots traveling in different directions (30 dark blue: to the left; 30 bright yellow: to the right). ⓐ–ⓒ: configuration at $t = \{0, 30, 100\}$ s. ⓓ: line order (color value) as function of time (x axis) and corridor width (y axis), for robot speed $v_{\mathrm{opt}} = 0.3$ m/s. Datapoints corresponding to configurations a–c are marked. ⓔ: line order at $t = 120$ s, as function of v_{opt} (x axis) and corridor width (y axis) (Color figure online).

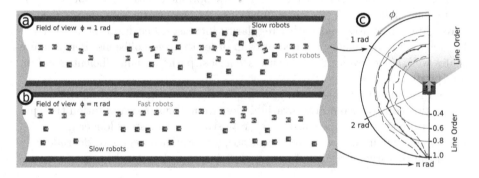

Fig. 5. 30 slow (dark blue) and 30 fast (bright yellow) robots sharing the corridor ($w_c = 2$ m) traveling in the same direction (right), for different values of the half-field of view ϕ of slow robots. ⓐ, ⓑ: configuration at $t = 300$ s, for $\phi = 1$ rad and $\phi = \pi$ rad, respectively. ⓒ: line order at $t = 300$ s as a function of ϕ. Dashed lines represent ± 1 standard deviations over 100 replicas (Color figure online).

5.4   Effects of Sensing on Line Order

In Fig. 5 we study how the sensing methodology used for localizing neighbors affects swarm's macroscopic behaviors. Unlike humans, which exhibit very little variations in their ability to sense the environment, different types of robots may

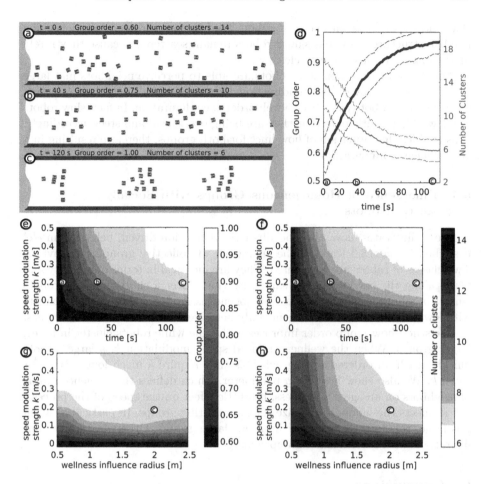

Fig. 6. Two classes of 30 robots (with wellness-controlled speed) sharing the corridor (width $w_C = 2\,\mathrm{m}$), traveling in the same direction (right). ⓐ–ⓒ: configuration at $t = \{0, 40, 120\}\,\mathrm{s}$, with $g = 2\,\mathrm{m}$, $k = 0.2\,\mathrm{m}$; robot colors are mapped to their wellness value w_a (dark red: low; bright green: high); note formation of groups led by cautious robots with low w_a. ⓓ: group order (thick blue line) and number of clusters (thin green line) against time (x axis). ⓔ: group order (color value) against time (x axis) and value of speed modulation strength k (y axis), with $g = 2\,\mathrm{m}$. ⓕ: number of clusters in the same context. ⓖ: group order (color value) against g (x axis) and k (y axis) at $t = 120\,\mathrm{s}$. ⓗ: number of clusters in the same context (Color figure online).

feature very different sensing subsystems. We consider the case in which two types of robots – *fast* and *slow* – move along the corridor in the same direction, with different speeds, equal to $0.1\,\mathrm{m/s}$ and $0.3\,\mathrm{m/s}$. We study how the field of view ϕ of the slow robots affects the ability to reach an ordered configuration.

With low ϕ values, slow robots cannot perceive and react to fast robots approaching behind them. In contrast, fast robots frequently need to steer around slow ones. As a consequence, fast robots tend to form ordered but zigzagging line-like structures, whereas slower robots remain scattered because they rarely need to navigate around obstacles.

On the contrary, when slow robots are able to perceive neighbors in a large field of view (e.g., by using omnidirectional cameras or additional back-pointing sensors), both robot types reach a well-ordered configuration. In fact, slow robots are now able to anticipate that they are being overtaken, and steer accordingly. This enables the formation of flow lines for both groups, therefore resulting in a very efficient configuration.

5.5   Emergence of Homogeneous Groups with Socially Active Robots

Figure 6 illustrates how the socially-active, emotion-driven behavior model described in Sect. 3.3 leads to the emergence of collective grouping behaviors. Two classes of robots are considered. They are identical in terms of mobility and travel along the corridor in the same direction. After a short time, the robots cluster in a limited number of class-homogeneous groups, led by cautions agents with low wellness values: such agents only perceive neighbors of a different class.

We study how group order improves with time while robots cluster in fewer, larger groups. When the wellness-induced speed modulation k is large, robots quickly reach an ordered setting, whereas they require a much longer time for small k. We also show how the g parameter, which defines the influence radius of neighbors on an agent's wellness, has little effect on the speed of the process, but it affects the number (and size) of the resulting groups. In particular, when g is large, robots tend to cluster in a few, large groups. Otherwise, stable configurations with several small groups emerge.

6   Conclusions

We introduced a novel obstacle avoidance algorithm for robot swarms based on a heuristic that closely models the behavior of human pedestrians. We have presented an implementation on real robots as well as on a simulator capable to handle several different scenarios of interest. We studied the characteristics and the efficiency of the obtained navigation trajectories in a number of extensive simulation tests (validated by real robot tests).

We observed that robot swarms implementing our algorithm exhibit several macroscopic behaviors which can also be observed in human crowds, such as the formation of ordered lines of flow in corridors. We investigated how different environment properties or algorithm parameters affect such emergent behaviors, and introduced a simple neuro-modulation/emotional model which promotes the emergence of new behaviors, such as the ability to form groups in heterogeneous swarms.

We are currently advancing the implementation on real robots in order to test the system in scenarios in which the space is shared between humans and robots, and/or among robots with very different mobility characteristics. Moreover, we are running simulations comparing our approach with state-of-the art obstacle avoidance algorithms. In addition, we are designing experiments for quantitatively evaluating our main objective, i.e. the predictability and perceived friendliness of robots' trajectories by humans.

References

1. Moussaïd, M., Helbing, D., Theraulaz, G.: How simple rules determine pedestrian behavior and crowd disasters. Proc. Natl. Acad. Sci. USA **108**(17), 6884–6888 (2011)
2. Bonani, M., Longchamp, V., Magnenat, S., Rétornaz, P., Burnier, D., Roulet, G., Vaussard, F., Bleuler, H., Mondada, F.: The marXbot, a miniature mobile robot opening new perspectives for the collective-robotic research. In: Proceedings of the IEEE/RSJ International Conference on Intelligent Robots and Systems (IROS), pp. 4187–4193 (2010)
3. Shillert, Z., Fiorini, P.: Motion planning in dynamic environments using velocity obstacles. Int. J. Robot. Res. **17**(7), 760–772 (1998)
4. Kluge, B., Prassler, E.: Recursive agent modeling with probabilistic velocity obstacles for mobile robot navigation among humans. In: Proceedings of IEEE/RSJ International Conference on Intelligent Robots and Systems, vol. 1, pp. 376–381 (2003)
5. van den Berg, J., Manocha, D., Lin, M.: Reciprocal velocity obstacles for real-time multi-agent navigation. In: Proceedings of IEEE International Conference on Robotics and Automation, pp. 1928–1935 (2008)
6. Snape, J., van den Berg, J., Guy, S.J., Manocha, D.: The hybrid reciprocal velocity obstacle. IEEE Trans. Robot. **27**(4), 696–706 (2011)
7. Guy, S.J., Lin, M.C., Manocha, D.: Modeling collision avoidance behavior for virtual humans. In: Proceedings of the International Conference on Autonomous Agents and Multiagent Systems, pp. 575–582 (2010)
8. Gibson, J.J.: Visually controlled locomotion and visual orientation in animals. Br. J. Psychol. **49**(3), 182–194 (1958)
9. François, M., Morice, A.H., Bootsma, R.J., Montagne, G.: Visual control of walking velocity. Neurosci. Res. **70**(2), 214–219 (2011)
10. Bootsma, R.J., Craig, C.M.: Information used in detecting upcoming collision. Perception **32**(5), 525–544 (2003)
11. Lee, D.N.: A theory of visual control of braking based on information about time-to-collision. Perception **5**(4), 437–459 (1976)
12. Coull, J.T., Vidal, F., Goulon, C., Nazarian, B., Craig, C.: Using time-to-contact information to assess potential collision modulates both visual and temporal prediction networks. Front. Hum. Neurosci. **2**, 10 (2008)
13. Krichmar, J.L.: The neuromodulatory system: a framework for survival and adaptive behavior in a challenging world. Adapt. Behav. **16**(6), 385–399 (2008)
14. Cox, B.R., Krichmar, J.L.: Neuromodulation as a robot controller: a brain inspired strategy for controlling autonomous robots. IEEE Robot. Autom. Mag. **16**(3), 1–25 (2009)
15. Krichmar, J.L.: A biologically inspired action selection algorithm based on principles of neuromodulation. In: Proceedings of IEEE World Congress on Computational Intelligence, pp. 10–15 (2012)

16. Hall, E.T.: A system for the notation of proxemic behavior. Am. Anthropol. **65**(5), 1003–1026 (1963)
17. Helbing, D., Farkas, I., Vicsek, T.: Simulating dynamical features of escape panic. Nature **407**(6803), 487–490 (2000)
18. Johansson, A.: Constant-net-time headway as key mechanism behind pedestrian flow dynamics. Phys. Rev. E **80**, 026120 (2009)
19. Moussaïd, M., Guillot, E.G., Moreau, M., Fehrenbach, J., Chabiron, O., Lemercier, S., Pettré, J., Appert-Rolland, C., Degond, P., Theraulaz, G.: Traffic instabilities in self-organized pedestrian crowds. PLoS Comput. Biol. **8**(3), e1002442 (2012)
20. Helbing, D., Molnár, P., Farkas, I.J., Bolay, K.: Self-organizing pedestrian movement. Environ. Plan. B Plan. Des. **28**(3), 361–383 (2001)
21. Moussaïd, M., Helbing, D., Garnier, S., Johansson, A., Combe, M., Theraulaz, G.: Experimental study of the behavioural mechanisms underlying self-organization in human crowds. Proc. R. Soc. B **276**(1668), 2755–2762 (2009)
22. Dorigo, M., Floreano, D., Gambardella, L.M., Mondada, F., Nolfi, S., Baaboura, T., Birattari, M., Bonani, M., Brambilla, M., Brutschy, A., Burnier, D., Campo, A., Christensen, A.L., Decugnière, A., Di Caro, G.A., Ducatelle, F., Ferrante, E., Förster, A., Martinez Gonzales, J., Guzzi, J., Longchamp, V., Magnenat, S., Mathews, N., Montes de Oca, M., O'Grady, R., Pinciroli, C., Pini, G., Rétornaz, P., Roberts, J., Sperati, V., Stirling, T., Stranieri, A., Stützle, T., Trianni, V., Tuci, E., Turgut, A.E., Vaussard, F.: Swarmanoid: a novel concept for the study of heterogeneous robotic swarms. IEEE Robot. Autom. Mag. (2012, to appear)
23. Yamori, K.: Going with the flow: micro–macro dynamics in the macrobehavioral patterns of pedestrian crowds. Psychol. Rev. **105**(3), 530–557 (1998)

Evolving Networks Processing Signals with a Mixed Paradigm, Inspired by Gene Regulatory Networks and Spiking Neurons

Borys Wróbel[1,2,3]([⊠]), Ahmed Abdelmotaleb[2], and Michał Joachimczak[3]

[1] Institute of Neuroinformatics, University of Zurich/ETHZ, Zurich, Switzerland
[2] Evolutionary Systems Laboratory, Adam Mickiewicz University in Poznan, Poznan, Poland
[3] Systems Modelling Laboratory, IOPAS in Sopot, Sopot, Poland
wrobel@evosys.org

Abstract. In this paper we extend our artificial life platform, called GReaNs (for Genetic Regulatory evolving artificial Networks) to allow evolution of spiking neural networks performing simple computational tasks. GReaNs has been previously used to model evolution of gene regulatory networks for processing signals, and also for controlling the behaviour of unicellular animats and the development of multicellular structures in two and three dimensions. The connectivity of the regulatory network in GReaNs is encoded in a linear genome. No explicit restrictions are set for the size of the genome or the size of the network. In our previous work, the way the nodes in the regulatory network worked was inspired by biological transcriptional units. In the extension presented here we modify the equations governing the behaviour of the units so that they describe spiking neurons: either leaky integrate and fire neurons with a fixed threshold or adaptive-exponential integrate and fire neurons. As a proof-of-principle, we report the evolution of spiking networks that match desired spiking patterns.

Keywords: Evolutionary algorithms · Gene regulatory networks · Spiking neural networks · Signal processing · Leaky integrate and fire neurons · Adaptive-exponential neurons

1 Introduction

Drawing from biology when building artificial systems aims to distil the crucial features of natural processes which allow in Nature for the existence and constant generation of extremely complex, efficient entities – biological organisms. Research in one of the fields using this approach, Artificial Life, often centres around software (or, more rarely, hardware) platforms which are based on a particular biologically-inspired paradigm. We are building one such platform, based on the paradigm of the encoding of regulatory network in a linear genome. This platform, GReaNs (which stands for Gene Regulatory evolving

G.A. Di Caro and G. Theraulaz (Eds.): BIONETICS 2012, LNICST 134, pp. 135–149, 2014.
DOI: 10.1007/978-3-319-06944-9_10, © Institute for Computer Sciences, Social Informatics and Telecommunications Engineering 2014

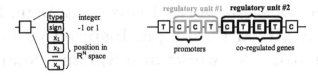

Fig. 1. Schematic structure of the genome in GReaNs. The left part shows the internal structure of a genetic element which consists of an integer specifying the type (*cis-regulator*, C; *trans-regulator*, T; or *external element*, E), a sign field (1 or −1, which determines if a trans-cis interaction is inhibitory or excitatory), and real numbers that specify a point in space (which determines trans-cis affinity). The genome (left) is a series of genetic elements that build regulatory units, which correspond to nodes in the regulatory network. A regulatory unit is at least one cis-regulator followed by at least one trans-regulator. External elements correspond to the inputs and outputs of the network. In the work presented here, the activity of the input is driven by a random Poisson spike source, while the spiking pattern of the output is used to measure the fitness of the network in the genetic algorithm.

artificial Networks) has been previously used to evolve regulatory networks to control development of multicellular structures in two dimensions [1, 2, which also introduced a method to transform the structures into soft-bodied animats swimming in a fluid-like environment] and three dimensions [3–5], and to investigate the ability of single cells to forage for resources in an artificial environment [6, 7] and to process signals [8].

In our previous work we were inspired by the way transcriptional units work in biology to formulate the rules governing the nodes in the regulatory network. The structure of the network in GReaNs is encoded in a linear genome (Fig. 1). The genome consists of "regulatory units": series of "genetic elements" of type C, followed by a series of elements of type T. C stands for "cis-regulators" (it is common to use the term promoters in the field, although this does not follow strictly the biological nomenclature). T elements code for "products", some of which act as "trans-regulators". Trans-regulators have "affinity" to cis-regulators. This affinity determines the computational properties of the network. All products can change concentration in each simulation step. All products coded in the same regulatory unit have the same concentration. In addition to the elements of type C and T, there are elements of type E. This letter stands for "external": the elements of type E allow for external inputs and outputs of the network.

Each genetic element is by itself a series of numbers: an integer specifying the type, a bit specifying the sign (signs determine if a particular cis-trans interaction is inhibitory or excitatory), and real numbers (coordinates) that specify a point in space (affinity is a function of the Euclidean distance between the corresponding points). Because the connection is always from a product-coding element (a "gene") to the cis-regulator, the connections between regulatory units are asymmetric. The net regulatory effect of a one unit (let us call it A) on the other unit (B) results from the combined effect of all trans-regulators coded in

A on all cis-elements of the unit B. An important feature of our encoding is that the topology of the network is not restricted, although in most experiments (also here) we do not allow for direct connections between the input and the output. However, we do not limit in any way the size of the genome, and thus of the network. There is, of course, a limit imposed by the computer memory, but it is never reached in practice.

In our previous papers [1–8] each regulatory unit in GReaNs was considered to be an analogue of a node in a biological gene regulatory network. But they could equally well be seen as computationally equivalent to neurons (or groups of neurons) in a neural network. Indeed, the recurrent networks explored previously in GReaNs can be seen as networks of discretely or continuously variable artificial neurons (cf. [9,10]). In this paper we explore the analogy further by analysing the evolvability of a model in which linear genomes encode the structure of a spiking neural network (SNN), in other words, when the equations governing each computational unit produce a spiking behaviour. We have recently introduced two models in GReaNs: the leaky integrate and fire model with a fixed threshold (LIF) [11,12] and the adaptive exponential model (AdEx) [11–14]. Both models are computationally simple, but the less simple one, AdEx, can show much richer and biophysically plausible behaviour [11,13,14].

In biology the encoding of the neural structures in the genome is much more indirect, so at this point our model has little to do with biological reality. However, the introduction of spiking neuron models in GReaNs is the first step towards a more biologically plausible model. Furthermore, this introduction allows us to investigate the evolvability of this encoding, and its potential for possible practical applications. The work presented here extends our previous preliminary results (presented as an extended abstract, [15]).

2   Evolving Spiking Neural Networks in GReaNs

To provide more detail to the description above, the connectivity of the network is specified in GReaNs by trans-cis affinity, using a function that relates the weights of the connections in an inverse exponential way to the Euclidean distance between points in the abstract space of interactions, with a cut-off value to prevent full connectivity.

Once the topology of the network and the weights are determined, every regulatory unit is transformed into a single neuron, the value of the concentration of the products in the unit to the neuron membrane potential, and the connections between the regulatory units, to synapses. If both elements determining the connection have the same value of the sign field, the connection is excitatory. If the signs differ, it is inhibitory.

In the work presented here, we allow for one input and one output of the network. The first external element in the genome whose type field specifies it is an input is connected to a Poisson spike source. In other words, its spiking pattern is determined externally to the network, and its connectivity to other neurons (interneurons) determines the activation of the network. We do not allow for a direct connection between the nodes in the network that correspond to the input and output. The connectivity of the output unit is determined by the first element in the genome whose type specifies it is an output. Its spiking pattern is used in the calculation of the value of the fitness function for a particular network.

In the experiments described here, units in the network work as either of LIF or AdEx neurons. In the former case, their behaviour is described by the equation:

$$\frac{dV}{dt} = \frac{g_L(V_R - V) + g_E(E_{rev,E} - V) + g_I(E_{rev,I} - V)}{C} \tag{1}$$

where V is the membrane potential, $V_R = -65.0\,\text{mV}$ is the resting potential, $g_L = 0.05\,\mu\text{S}$ is the leak conductance, g_E is the conductance of the excitatory synapses, g_I is the conductance of the inhibitory synapses, $E_{rev,E} = 0\,\text{mV}$ is the reversal potential of the excitatory input, $E_{rev,I} = -70.0\,\text{mV}$ is the reversal potential of inhibitory input, and $C = 1\,\text{nF}$ is the capacitance of the membrane. When the membrane potential of a LIF neuron crosses the threshold ($V_{threshold} = -50\,\text{mV}$), a stereotypical spike is created: the membrane potential is first forced to reach a maximum value, and then it decreased to the reset voltage value ($V \leftarrow V_{reset}$, $V_{reset} = -70.0\,\text{mV}$).

The AdEx neurons are described by two equations:

$$\frac{dV}{dt} = \frac{g_L(E_L - V) + \delta e^{\frac{V - V_T}{\delta}} + g_E(E_E - V) + g_I(E_I - V) - W}{C} \tag{2}$$

$$\frac{dW}{dt} = \frac{a(V - E_L) - W}{\tau_W} \tag{3}$$

where V is the membrane potential, $E_L = -65.0\,\text{mV}$ is the leak reversal potential, $\delta = 2.0\,\text{mV}$ is the slope factor, $V_T = 50.0\,\text{mV}$ is the threshold potential, $g_L = 0.05\,\text{S}$ is the leak conductance, g_E (or g_I) is the conductance of the excitatory (inhibitory) synapses, $E_E = 0\,\text{mV}$ ($E_I = -70.0\,\text{mV}$) is the reversal potential of the excitatory (inhibitory) input, W is the adaptation variable, $C = 1\,\text{nF}$ is the capacitance of the membrane, $a = 4.0\,\text{nS}$ is the adaptation coupling parameter, and $\tau_w = 40.0\,\text{ms}$ is the adaptation time constant.

In contrast to the LIF neurons, AdEx neurons do not have a fixed threshold for spike generation, but when the neuron is activated enough, its membrane potential will diverge to infinity. It is necessary, therefore, to specify the maximum value of the potential. When this value is reached ($V_{spike} = -40.0\,\text{mV}$), both the potential and the adaptation variable are decreased or increased, respectively, in a discontinuous manner ($V \leftarrow V_{reset}$, $w \leftarrow w + b$, where $V_{reset} = -65.0\,\text{mV}$, and $b = 0.0805\,\text{nA}$). It has been shown that the behaviour of the AdEx model does not depend on the value of the maximum potential. In other words, the behaviour would be essentially the same if a more biologically realistic value (e.g., $+40\,\text{mv}$) was used for V_{spike} [14].

AdEx neurons, compared with LIF, while remaining computationally simple, have a more biophysically plausible behaviour, including, for example, spike latency, and rebound spiking/bursting, phasic/tonic spiking/bursting, initial bursting, or even chaotic behaviour [11,13,14], depending on the value of the parameters. The parameters used in this work allow for phasic spiking.

In GReaNs, as in other implementations of SNNs, a spike is marked at time t when the threshold is crossed ($V = V_{threshold}$) for the LIF neuron, or $V = V_{spike}$, for AdEx. This spike will arrive at all the connected neurons at the next simulation time step ($t + 1$), and the membrane potential of these neurons will be affected by this spike starting from the next step still ($t + 2$). The way it is affected depends on the weight of the synaptic connection (determined by the Euclidean distance between the points specified by the genetic elements, in this paper we use two coordinates for both LIF and AdEx) and its sign (positive for excitatory, negative for inhibitory; a product of the sign of the genetic elements).

The conductance of the synapses was modelled using decaying exponential functions:

$$\frac{dg_E}{dt} = \frac{-g_E}{\tau_E} \quad \text{and} \quad \frac{dg_E}{dt} = \frac{-g_I}{\tau_I} \tag{4}$$

where $\tau_E = 5.0$ ms and $\tau_I = 5.0$ ms are the decay time constants of the excitatory and inhibitory synaptic conductance, respectively. Each arriving spike produces an increase of conductance which is directly proportional to the weight of the synapse.

In this paper we use a genetic algorithm to evolve SNNs which are able to generate a specific spike pattern if presented with a specific input. The population size was kept constant (300 individuals). The genetic algorithm used the tournament selection (pick two, select the best). The genomes in the initial population were generated randomly (with random signs and coordinates of the elements), with 5 regulatory units, but a random number of C and T elements in each. At each generation, a new population was formed by taking the best 5 individuals from the previous one without mutation (elitism), 100 with crossover and mutation, and 195 with mutation only. Mutations could change the element type, sign, coordinates, or consist of a deletion of a random number of elements or a duplication of a random number at a random position, so the number of regulatory units (and neurons in the network) could change over generations from the initial 5.

While we use GReaNs for the time-integration of the network, we have allowed for the export of the network description using PyNN [16] which allows testing the network with other simulation backends, e.g. Brian simulator [12] (we have used IF_cond_exp class for LIF and EIF_cond_exp_isfa_ista class for AdEx in pyNN, and Euler integration with 1 ms integration time step in both Brian and GReaNs).

The genetic algorithm used a fitness function which rewards the match between the output of the recurrent network in GReaNs with the desired output:

$$f_{fit} = \frac{\alpha |S_{desired} - S_{GReaNs}| + \beta(S_{desired} - m_{GReaNs})}{S_{desired}} \tag{5}$$

where $S_{desired}$ is the desired number of spikes, S_{GReaNs} is the number of spikes generated by a network under evaluation, m_{GReaNs} is the number of matching

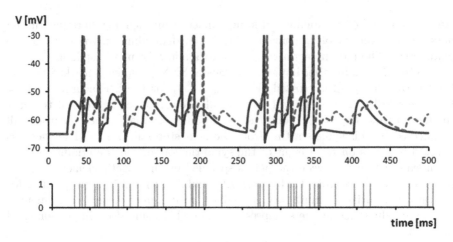

Fig. 2. The membrane potential of the output neuron of the best network (in terms of generalization, in 10 independent evolutionary runs) of LIF neurons evolved with GRe-aNs (blue line) to match the spikes of one AdEx neuron (red line), shifted by 20 ms in response to the Poisson spike train (green) used during evolution. The evolved network had 1 input neuron, 1 output neuron and 5 interneurons, with recurrent connections (Color figure online).

spikes in these two spike trains, while α and β are constant fractions ($\alpha + \beta = 1$, we used $\alpha = 0.3$ and $\beta = 0.7$ in this work). We have searched for matching spikes over the window $[t_{desired} - 9, t_{desired} + 9]$ and weighted them using a Gaussian function, $f_{Gaussian} = e^{(t_{desired} - t_{GReaNs})^2 / 15}$, where $t_{desired}$ is the spike time in the desired output, and t_{GReaNs} is the spike time in GreaNs. Only one specific spike train was used for evaluation during evolution (Figs. 2, 3, 4, 5).

3   Results

In this paper we used GReaNs to evolve recurrent SNNs able to perform two types of tasks. In one task the network generates a specific spike train in response to a specific input, generated by a Poisson spike source with 100 Hz spike frequency. The target spike train was generated by one LIF or AdEx neuron (modelled, for convenience, using Brian, with the same parameters as the LIF/AdEx neurons in the evolving network, with the exception of using $V_{reset} = -65.0$ mV for the LIF neuron in Brian).

Because direct connection of the input to the output was not allowed in the evolving networks, and because synaptic connection introduce delays, it is not possible to match the desired output exactly unless the spikes are "shifted". Therefore, in this type of task we evolved networks which produced an output shifted (in comparison to one LIF or AdEx neuron) by 20 ms. In our previous preliminary report [15], we investigated also shorter shift times: 5 ms and 10 ms; these tasks were much easier.

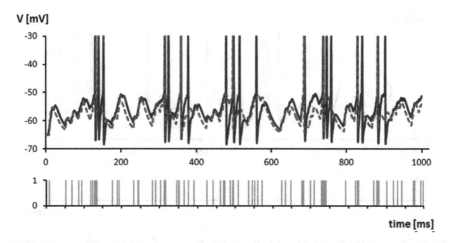

Fig. 3. The membrane potential of the output neuron of the best network (in terms of generalization, in 10 independent evolutionary runs) of LIF neurons evolved with GReaNs (blue line) to double the spikes of one AdEx neuron (red line), shifted by 5 ms in response to the Poisson spike train (green) used during evolution. The evolved network had 1 input neuron, 1 output neuron and 8 interneurons, with recurrent connections (Color figure online).

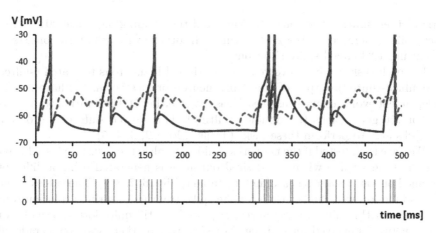

Fig. 4. The membrane potential of the output neuron of the best network (in terms of generalization, in 10 independent evolutionary runs) of AdEx neurons evolved with GReaNs (blue line) to match the spikes of one LIF neuron (red line), shifted by 20 ms in response to the Poisson spike train (green) used during evolution. The evolved network had 1 input neuron, 1 output neuron and 12 interneurons, with recurrent connections (Color figure online).

The second task consists also of producing a specific output, but here the target is created by first presenting the input (also generated by a Poisson source, with 75 Hz frequency) to, again, one neuron (same as above, modelled using

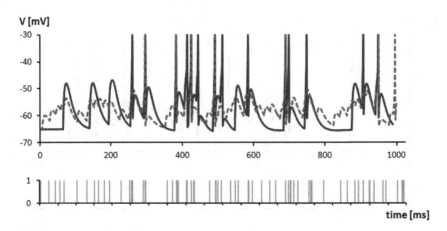

Fig. 5. The membrane potential of the output neuron of the best network (in terms of generalization, in 10 independent evolutionary runs) of AdEx neurons evolved with GReaNs (blue line) to double the spikes of one LIF neuron (red line), shifted by 5 ms in response to the Poisson spike train (green) used during evolution. The evolved network had 1 input neuron, 1 output neuron and 9 interneurons, with recurrent connections (Color figure online).

Brian), then shifting the spikes by 5 ms, and then adding one spike 20 ms after each spike generated by the Brian neuron. In other words, the task consists of producing a "doubled-shifted" output.

For each task, we have evolved a network or LIF neurons to match (shifted or doubled-shifted) output of one AdEx neuron or vice versa. We have run 10 independent evolutionary runs in each case, with either 500 generations (for shifting) or 750 generations (for doubled-shifting). This proved sufficient to obtain networks able to perform these tasks (Table 1, Figs. 2, 3, 4, 5).

We have then tested if the networks could generalize the tasks: we tested what was the fitness value when the desired output was generated using a different specific Poisson spike train as an input (Table 1, Figs. 6, 7, 8, 9). The values of the fitness function were higher than for the spike train used during evolution (our genetic algorithm aims to minimize the f_{fit}), but still quite low, especially for AdEx networks evolved for shifting. Not surprisingly, there was some trade-off between the value of f_{fit} that a particular network has reached during evolution (this can be seen as the performance for the training set) and its generalization. For example, the best shifting network of LIF neurons in 10 runs ($f_{fit} = 0.146$) had only $f_{fit} = 0.652$ when tested on a different input, while the best generalizing network ($f_{fit} = 0.517$ for generalization, Figs. 2, 6) had $f_{fit} = 0.285$ for the input used during evolution. The situation was similar for doubled-shifting LIF networks (the best network: $f_{fit} = 0.112$ for "training", and $f_{fit} = 0.327$ for generalization; the best generalizing network: $f_{fit} = 0.240$, and $f_{fit} = 0.290$, respectively). For the AdEx networks, the best shifting network ($f_{fit} = 0.006$ for "training") had also the best fitness for a different input ($f_{fit} = 0.142$),

Table 1. The average ± standard deviation and range (square brackets) of the fitness function and size of the networks evolved with GReaNs for the best networks in each of 10 independent evolutionary runs for each task, both for the input spike train used during evolution and for another Poisson spike train (generalization).

Type of the network and task for evolution	f_{fit}, input used in evolution	f_{fit}, another input	Network size
LIF network matching a shifted output of one AdEx neuron	0.240 ± 0.043 [0.146, 0.309]	0.730 ± 0.130 [0.517, 0.963]	8.3 ± 3.2 [5, 15]
AdEx network matching a shifted output of one LIF neuron	0.087 ± 0.062 [0.006, 0.158]	0.390 ± 0.175 [0.142, 0.625]	9.6 ± 4.1 [5, 16]
LIF network matching a doubled-shifted output of one AdEx neuron	0.235 ± 0.057 [0.112, 0.300]	0.379 ± 0.055 [0.290, 0.434]	11.2 ± 5.5 [4, 26]
AdEx network matching a doubled-shifted output of one LIF neuron	0.258 ± 0.042 [0.171, 0.301]	0.478 ± 0.042 [0.373, 0.541]	8.1 ± 3.7 [4, 17]

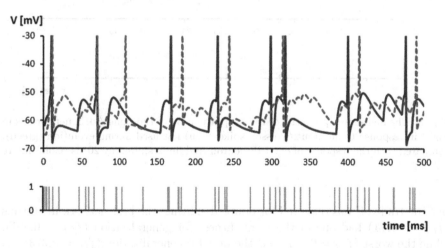

Fig. 6. The membrane potential of the output neuron of the same LIF network as in Fig. 2 in response to an input (Poisson spike train) not used during evolution (green), compared with the response of one AdEx neuron (red line), shifted by 20 ms (Color figure online).

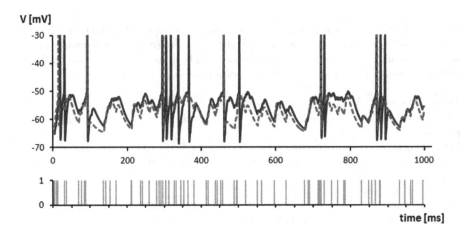

Fig. 7. The membrane potential of the output neuron of the same LIF network as in Fig. 3 in response to an input (Poisson spike train) not used during evolution (green), compared with the response of one AdEx neuron (red line), shifted by 5 ms. The output network was evolved to double-shift the spikes (Color figure online).

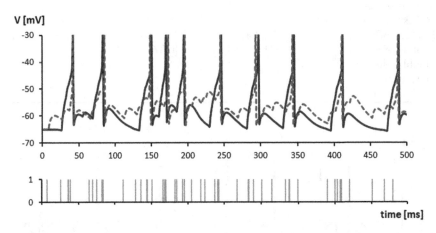

Fig. 8. The membrane potential of the output neuron of the same AdEx network as in Fig. 4 in response to an input (Poisson spike train) not used during evolution (green), compared with the response of one LIF neuron (red line), shifted by 20 ms (Color figure online).

but the trade-off was observed for doubled-shifting: the best network in 10 runs ($f_{fit} = 0.171$) had one of the worst fitness for generalization ($f_{fit} = 0.501$), while the worst ($f_{fit} = 0.301$) was the best for generalization ($f_{fit} = 0.373$).

To sum up, evolving a network of AdEx neurons to match a shifted output of one LIF neuron proved the easiest task, and generalized the best. In contrast, evolving a LIF network to match an output of one AdEx neuron proved more difficult, and the evolved networks did not generalize well. Doubled-shifting was

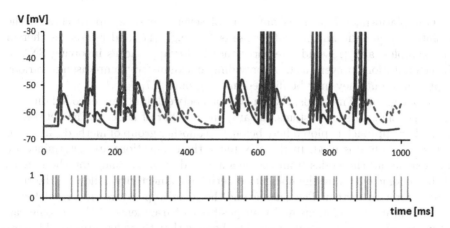

Fig. 9. The membrane potential of the output neuron of the same AdEx network as in Fig. 5 in response to an input (Poisson spike train) not used during evolution (green), compared with the response of one LIF neuron (red line), shifted by 5 ms. The output network was evolved to double-shift the spikes (Color figure online).

also difficult, regardless of the type of the network, but the solutions generalized quite well.

Shifting could in principle be accomplished by a feed-forward network in which the first layer, connected directly to the input, has one neuron (or several) which would generate the desired output without any shift (after all, it was generated by one neuron), while the neurons in subsequent layers would generate a spike for every spike received (to provide the shift). Doubled-shifting could be achieved in a similar fashion, using two feed-forward networks (of different sizes) working in parallel. However, we did not observe any such trivial topology. Instead, all the best networks were heavily interconnected, with recurrent connections, so (as is typical in the field of evolving networks) it is difficult to find out how they actually work. The analysis of the spiking pattern of the interneurons did, however, reveal that there is sometimes a match between their spiking pattern and the target pattern. Perhaps even more interestingly, we have observed the presence of neurons spiking with very high frequency in the doubled-shifting networks. We plan to investigate this latter issue in more detail in further work.

4 Discussion and Future Work

We have described here a biologically-inspired (even if not biologically plausible) system for evolution of SNNs. In our system the networks are encoded in a linear genome without imposing any limitations on the size of the network or on the size of the genome. We use this encoding and the genetic operators inspired by the biological mutations (point mutations, deletions, duplications; the last two may also result from cross-over) to evolve recurrent SNNs able to perform simple processing tasks.

Our platform, GReaNs, is only one of several existing approaches to the evolution of genetic and neural networks (see e.g., [17] and references therein for examples, and [18] and references therein for approaches involving SNNs). Space limitations prevent us from providing an exhaustive comparison of various approaches, but they may be divided into 3 groups [17].

In one group of methods, the connections between units are encoded directly in the genome. In such systems, usually, the number of nodes in the network is fixed. The approach explored here belongs to another group of methods, in which implicit encoding is used. In these systems, the connections are determined by first extracting the nodes from the genome, and then by analysing the regions of the genome encoding the nodes to determine connectivity, following the rules specific for each system. The genome and the genetic operators are formulated so that duplications, deletions and transpositions of fragments of the genome can allow for network complexification. The hope is that these features would allow for higher evolvability. The same hope [19,20] drives the work on another still group of systems in which the mapping between the genome and the structure of the neural network is even more indirect, and inspired by biological development, with simulated cell division and growth of neurites (see [21] for one of the earliest work using this approach).

We have explored 2- and 3-dimensional development in GReaNs before [1–5], but the possibility for the cells to be converted into neurons and to establish connections has not been so far included in our platform. We plan to do so, and we hope that the implementation of various methods for specifying connectivity of the networks in one platform will make their comparison easier. Such comparison lies within our interests, but it is a daunting task: it requires finding appropriate benchmarks with selection pressures, and optimal parameters for each approach.

The task of comparing particular platforms is also daunting because even systems which share paradigms differ in design principles and details. For example, when compared to Analog Genetic Encoding [17], GReaNs uses an entirely different way of specifying the connection weights, and the number of connections per unit is not pre-specified. On the other hand, Analog Genetic Encoding uses a mechanism inspired by biological homologous recombination for the cross-over during the genetic algorithm. Such a mechanism may allow for better evolvability, and we plan to include it in our future extensions of the GReaNs platform.

As we have already mentioned, the investigation of the evolvability of the system requires a set of tasks (benchmarks). The networks evolved in this paper are able to perform quite simple (but not trivial) tasks. We are currently working on other signal processing tasks and also on using the spiking neurons to control the behaviour of animats in a simulated environment, a subject we have explored previously using evolving gene regulatory networks in GReaNs [6,7]. The results presented here can be seen as a proof-of-principle; they demonstrate that after the extension of our system with SNN models, it remains evolvable at least for simple tasks.

The possible advantage of using the spiking nodes in the network over non-spiking units is that the evolved networks could be implemented in neuromorphic hardware [22], or even evolved using such hardware. Both models used here, LIF and AdEx, have been implemented in neuromorphic silicon neuron circuits (both analogue circuits and digital very large scale integration circuits, and implementations are available also for multi-core based architectures, based on multiple ARM cores and GPUs) [18, and references therein]. Such circuits might allow for a speed-up of the evolutionary process, especially for larger networks. This is one the directions of research we are already pursuing, allowing for the possibility of interfacing through different software/hardware platforms using Python scripts (such as PyNN [16]).

In simulations reported here we have investigated the evolution of SNNs with homogeneous nodes: all nodes had the same, specific values of the parameters in the equations describing their behaviour. We are currently investigating how the values of these parameters affect the ability of the system to evolve networks performing the tasks, and if there are tasks which can be best approached with some type of nodes (for example, with tonic as opposed to phasic, or with bursting or chaotic response spiking [13,14]). We also plan to explore if there is any advantage of including learning mechanisms (synaptic plasticity) in the model (this may be of particular interest for animat control), and of using heterogenous spiking networks, in which nodes would have different parameters. Regarding the latter issue, the approach we would like to explore first is not to evolve the parameters in a continuous fashion, but rather to specify several stereotypical neurons (with particular behaviour). In other words, one mutation in the system would change the type of the neuron (setting in one step another set of values specific for another stereotype). We believe that this approach can keep the size of the search space small enough for the genetic algorithm to explore efficiently the search space of solutions.

Acknowledgement. This work was supported by the Polish Ministry of Science and Higher Education (project 2011/03/B/ST6/00399 to BW); BW acknowledges the support of the Swiss-Polish Research Fund, AA the support of the Foundation for Polish Science, co-financed by EU Regional Development Fund (Innovative Economy Operational Programme 2007–2013). We are also grateful to Volker Steuber for discussions.

References

1. Joachimczak, M., Kowaliw, T., Doursat, R., Wróbel, B.: Brainless bodies: Controlling the development and behavior of multicellular animats by gene regulation and diffusive signals. In: Artificial Life XIII: Proceedings of the 13th International Conference on the Simulation and Synthesis of Living Systems, MIT Press (2012) 349–356
2. Joachimczak, M., Wróbel, B.: Co-evolution of morphology and control of soft-bodied multicellular animats. In: Proceedings of the 14th International Conference on Genetic and Evolutionary Computation. GECCO '12, pp. 561–568. ACM (2012)

3. Joachimczak, M., Wróbel, B.: Evo-devo *in silico*: a model of a gene network regulating multicellular development in 3D space with artificial physics. In: Artificial Life XI: Proceedings of the 11th International Conference on the Simulation and Synthesis of Living Systems, pp. 297–304. MIT Press (2008)
4. Joachimczak, M., Wróbel, B.: Evolution of the morphology and patterning of artificial embryos: scaling the tricolour problem to the third dimension. In: Kampis, G., Karsai, I., Szathmáry, E. (eds.) ECAL 2009, Part I. LNCS, vol. 5777, pp. 35–43. Springer, Heidelberg (2011)
5. Joachimczak, M., Wróbel, B.: Open ended evolution of 3d multicellular development controlled by gene regulatory networks. In: Artificial Life XIII: Proceedings of the 13th International Conference on the Simulation and Synthesis of Living Systems, pp. 67–74. MIT Press (2012)
6. Joachimczak, M., Wróbel, B.: Evolving gene regulatory networks for real time control of foraging behaviours. In: Artificial Life XII: Proceedings of the 12th International Conference on the Simulation and Synthesis of Living Systems, pp. 348–355. MIT Press (2010)
7. Wróbel, B., Joachimczak, M., Montebelli, A., Lowe, R.: The search for beauty: evolution of minimal cognition in an animat controlled by a gene regulatory network and powered by a metabolic system. In: Ziemke, T., Balkenius, C., Hallam, J. (eds.) SAB 2012. LNCS, vol. 7426, pp. 198–208. Springer, Heidelberg (2012)
8. Joachimczak, M., Wróbel, B.: Processing signals with evolving artificial gene regulatory networks. In: Artificial Life XII: Proceedings of the 12th International Conference on the Simulation and Synthesis of Living Systems, pp. 203–210. MIT Press (2010)
9. Flood, I., Kartam, N.: Artificial Neural Networks for Civil Engineers: Advanced Features and Applications. American Society of Civil Engineers, New York (1998)
10. Beer, R.D.: On the dynamics of small continuous-time recurrent neural networks. Adapt. Behav. **3**(4), 469–509 (1995)
11. Izhikevich, E.: Dynamical Systems in Neuroscience: The Geometry of Excitability and Bursting. MIT press, Cambridge (2007)
12. Goodman, D., Brette, R.: Brian: a simulator for spiking neural networks in python. Front. Neuroinform. **2**, 5 (2008)
13. Brette, R., Gerstner, W.: Adaptive exponential integrate-and-fire model as an effective description of neuronal activity. J. Neurophysiol. **94**(5), 3637–3642 (2005)
14. Touboul, J.: Bifurcation analysis of a general class of nonlinear integrate-and-fire neurons. SIAM J. Appl. Math. **68**(4), 1045–1079 (2008)
15. Wróbel, B., Abdelmotaleb, A., Joachimczak, M.: Evolving spiking neural networks in the GReaNs (gene regulatory evolving artificial networks) platform. EvoNet2012: Evolving Networks, from Systems/Synthetic Biology to Computational Neuroscience Workshop at Artificial Life XIII, pp. 19–22 (2008)
16. Davison, A.P., Brüderle, D., Eppler, J., Kremkow, J., Muller, E., Pecevski, D., Perrinet, L., Yger, P.: PyNN: a common interface for neuronal network simulators. Front. Neuroinform. **2**, 11 (2008)
17. Mattiussi, C., Floreano, D.: Analog genetic encoding for the evolution of circuits and networks. Trans. Evol. Comput. **11**(5), 596–607 (2007)
18. Veredas, F.J., Vico, F.J., Alonso, J.M.: Evolving networks of integrate-and-fire neurons. Neurocomputing **69**(13–15), 1561–1569 (2006)
19. Stanley, K.O., Miikkulainen, R.: A taxonomy for artificial embryogeny. Artif. Life **9**(2), 93–130 (2003)

20. Tufte, G.: Phenotypic, developmental and computational resources: scaling in arti-ficial development. In: GECCO '08: Proceedings of the 10th Annual Conference on Genetic and Evolutionary Computation, pp. 859–866. ACM (2008)
21. Jakobi, N.: Harnessing morphogenesis. In: Proceedings of Information Processing in Cells and Tissues, pp. 29–41 (1995)
22. Indiveri, G., Linares-Barranco, B., Julia, T., van Schaik, A., Etienne-Cummings, R., Delbruck, T., Liu, S.C.C., Dudek, P., Häfliger, P., Renaud, S., Schemmel, J., Cauwenberghs, G., Arthur, J., Hynna, K., Folowosele, F., Saighi, S., Serrano-Gotarredona, T., Wijekoon, J., Wang, Y., Boahen, K.: Neuromorphic silicon neuron circuits. Front. Neurosci. 5, 73 (2011)

Algorithmically Transitive Network: Learning Padé Networks for Regression

Hideaki Suzuki[✉]

National Institute of Information and Communications Technology,
588-2, Iwaoka, Iwaoka-cho, Nishi-ku, Kobe 651-2492, Japan
hsuzuki@nict.go.jp

Abstract. The learning capability of a network-based computation model named "Algorithmically Transitive Network (ATN)" is extensively studied using symbolic regression problems. To represent a variety of functions uniformly, the ATN's topological structure is designed in the form of a truncated power series or a Padé approximant. Since the Padé approximation has better convergence properties than the Taylor expansion, the ATN with the Padé can construct an algebraic function with a relatively small number of parameters. The ATN learns with the standard back-propagation algorithm which optimizes intra-network parameters by the steepest descent method. Numerical experiments with benchmark problems show that the ATN in the form of a Padé approximant has better learning capability than linear regression analysis in a power series, the standard multi-layered neural network with the back-propagation learning, the support vector machine using the radial basis function as kernel, or the simple genetic programming.

1 Introduction

"Algorithmically Transitive Network (ATN)" [32–35] is a novel model for computation and learning which is a kind of combination of 'data-flow computer' [6,28] and 'artificial neural network' (ANN). Like the data-flow computer, a program (algorithm) is represented by a data-flow network whose nodes represent arithmetic/logical operations and whose edges transmit data tokens for variables. After propagating tokens forward in the network for calculation, the ATN propagates tokens backward in the netowrk and revises its internal parameters using the same mechanisms as the back-propagation learning in the ANN [24,25,41]. Moreover, the network topology (algorithm) can be modified/revised by the execution of movable agents in the network.

In [35], Suzuki *et al.* altered the ATN by both the parameter learning and topological alteration and succeeded in organizing networks for some simple arithmetic functions. These results, however, are unsatisfactory from the following points. First, the learning and topological alteration are mixed with each other, and no information for the assessment/improvement of each operation was provided. Second, the implemented agent operations were far from being

G.A. Di Caro and G. Theraulaz (Eds.): BIONETICS 2012, LNICST 134, pp. 150–166, 2014.
DOI: 10.1007/978-3-319-06944-9_11, © Institute for Computer Sciences, Social Informatics and Telecommunications Engineering 2014

optimized, and as a consequence, the demonstrated ability to produce functions was only shown for a few elementary arithmetic functions.

To remedy these problems and examine the ATN's ability for regression, Suzuki focused on the learning aspect of the ATN and conducted some preliminary experiments [36]. The ATNs describing two universal formulas, truncated Taylor expansion and Padé approximant [2,42], were prepared (manually designed) and fixed, and their coefficient parameters were optimized by the supervised learning. The present paper considerably extends this former study, and fully examines the ATN's regression ability using standard benchmark problems with up to three input variables. Some other conventional models are also applied to the same problems and compared, and it is demonstrated that the ATN in the form of the Padé approximant is a more powerful regression tool than linear regression analysis, the ANN, the support vector machine, and simple Genetic Programming. By visualizing the results for one-dimensional problems in input-output (sensor-answer) planes, characteristics of the tested methods are also clarified.

In the following, Sect. 2 presents the model description. After the experimental methods and numerical results are presented in Sect. 3, the results are discussed in Sect. 4. Possible future research directions are argued in Sect. 5, and concluding remarks are given in Sect. 6.

2   The Model

2.1   ATN Architecture

The ATN was originally invented by the combination of a network-based computational model for artificial chemistry [31] and 'active network' in the communication network [38]. By propagating agents/packets and delivering various functions to the processing units/routers, both methods enable flexible emulation/control of complex chemical/communication systems. Borrowing this mechanism, the ATN propagates computational or learning tokens in a dataflow network (namely, a program) and revises the program by using the learning tokens' information while executing computation.

In actual operations of the ATN, the computational tokens are initially created at the variable/constant nodes and are propagated forward to 'answer' nodes. When the answer nodes receive those tokens, the nodes evaluate energy error between the answer values and teaching signals given from the outside, and generate the learning tokens which are propagated backward towards the constant nodes. Like the original back-propagation learning in the ANN, the learning tokens include information for the derivatives of the energy error, and when they arrive at the constant nodes, the constants' values are updated so that the energy error might be minimized by the steepest descent method. See [35] for more detailed implementation of the ATN.

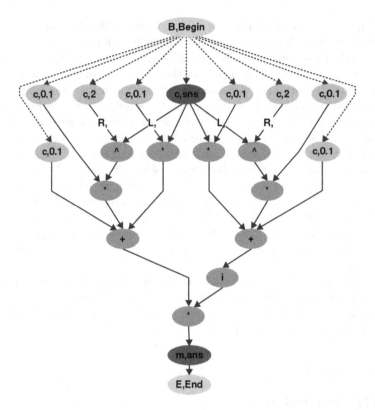

Fig. 1. ATN for the [2/2]th-order Padé approximant of a single variable. Here, shown inside of the nodes are the single-character operation codes (B, c, ^, etc.) and the operation names (sns='sensor', ans='answer', Begin, and End) or values (2, 0.1). Values written in the form of real numbers (e.g., 0.1) are learnable (changeable), and values in the form of integers (e.g., 2) are unlearnable (unchangeable). The graph is drawn by a graph visualization software named *aiSee* [1].

2.2 Topological Design

By the learning, the ATN can optimize any form of mathematical formula comprising arithmetic and logical operations. To apply this scheme to regression, here we assume two types of universal functions and translate them to the ATNs. The ATN topology is fixed to those forms and is never changed during a simulation run. Let M be the number of variables (sensor nodes), s and $s \equiv \{s_m\}$ be the sensor value and sensor vector, respectively, $k \equiv^t (k_1, \cdots, k_M)$ be the index vector, a be the answer value, and $c \equiv \{c_k\}$ or $c \equiv \{c_k\}$, $p \equiv \{p_k\}$ or $p \equiv \{p_k\}$, and $q \equiv \{q_{k'}\}$ or $q \equiv \{q_{k'}\}$ be the coefficient vectors to be optimized. The universal forms we take in this paper are the Kth-order power series (truncated Taylor expansion)

$$a = \sum_{k=0}^{K} c_k s^k = c_0 + c_1 s + \cdots + c_K s^K \qquad \text{(single variable's case),} \quad (1a)$$

$$a = \sum_{k=0}^{K} c_k s_1^{k_1} \cdots s_M^{k_M} \qquad \text{(multiple variables' case),} \quad (1b)$$

or the $[K_0/K_1]$th-order Padé approximants

$$a = \frac{\sum_{k=0}^{K_0} p_k s^k}{\sum_{k'=0}^{K_1} q_{k'} s^{k'}} = \frac{p_0 + p_1 s + \cdots + p_{K_0} s^{K_0}}{q_0 + q_1 s + \cdots + q_{K_1} s^{K_1}} \text{(single variable's case),} \quad (2a)$$

$$a = \frac{\sum_{k=0}^{K_0} p_k s_1^{k_1} \cdots s_M^{k_M}}{\sum_{k'=0}^{K_1} q_{k'} s_1^{k_1'} \cdots s_M^{k_M'}} \qquad \text{(multiple variables' case).} \quad (2b)$$

Here, the summation $\sum_{k=0}^{K}$ in Eq. (1b) is over all the integer vectors satisfying $1 \le \forall m \le M$, $0 \le k_m \le K$ and $k_1 + \cdots + k_M \le K$. The number of such integer vectors, i.e., the number of terms in the expansion of Eq. (1b) is given by $\sum_{k=0}^{K} {}_M H_k = \binom{M+K}{M}$. The same argument applies to the numerator/denominator of Eq. (2b) as well.

ATN for the single-variable, $[2/2]$th-order Padé approximant is shown in Fig. 1. In this figure, the coefficients' values expressed as real numbers ('c,0.1') are optimized through the ATN learning, and the index values expressed as integers ('c,2') are kept constant. In this way, we are able to construct an ATN for the Padé approximant of arbitrary order, which is automated by a program and is used in the subsequent experiments.

3 Experiments

The methods we take in this paper for experimental comparison with regression problems are as follows:

Method (1): Linear regression analysis of a power series,
Method (2): Linear regression analysis of a Padé approximant,
Method (3): ATN learning with a power series,
Method (4): ATN learning with a Padé approximant,
Method (5): Layered ANN with the back-propagation (BP) learning,
Method (6): Support Vector Regression with soft margin (ν-SVR) [4, 26], and
Method (7): Genetic Programming (GP) [15].

Method (1) uses a polynomial formula Eqs. (1a) or (1b). Because these formulas are linear in terms of the coefficients c, we can analytically calculate the coefficients that minimize the sum-of-squares error

$$E = \frac{1}{2} \sum_{j} \{t_j - a(s_j, c)\}^2 + \frac{\lambda}{2} \|c\|^2 \qquad (3)$$

by the standard least square method. Here, j is the sample number, s_j and t_j are the jth sample's sensor vector and teaching value, respectively, $a_j \equiv a(s_j, c)$ is the jth answer value, and λ is the regularization coefficient. The summation for j is over all the sample pairs (s_j-t_j pairs) of the training data. The regularization term in Eq. (3) is necessary to suppress 'over-fitting' and enhance robustness (an extensive argument on this issue is found, for example, in [4]). Method (1) is an analytic but powerful method to represent functions. Actually, if we take the order K of Eqs. (1a) or (1b) to be sufficiently large, Method (1) can be regarded as a superset of STROGANOFF [10, 21], which automatically constructs a polynomial function using evolutionary operations. The STROGANOFF is a kind of 'offspring' research of the GP (Method (7)), and it is reported in [10] that the STROGANOFF's regression ability is comparable to or higher than the original GP.

Similarly, Method (2) uses a Padé formula Eqs. (2a) or (2b), but this time, we minimize the error function defined as

$$E = \frac{1}{2} \sum_j \left\{ t_j \sum_{k'} q_{k'} s^{k'} - \sum_k p_k s_j^k \right\}^2 + \frac{\lambda}{2} \left(\|p\|^2 + \|q\|^2 \right), \qquad (4)$$

in the single variable's case. Since the expression in the curly brackets is linear in terms of the coefficients p and q, we can apply the standard least square method to Eq. (4) and calculate the coefficients.

Methods (3) and (4) use the polynomial formula Eqs. (1a)/(1b) and the Padé formula Eqs. (2a)/(2b), respectively, but the both methods try to determine the coefficients not analytically but numerically using the learning of the ATN *without* regularization term. Some arguments on the generalization in this and other methods are described later.

Based upon some preliminary experiments, we take the following 'optimized' simulation constants/conditions: $\lambda = 0.01$ in Method (1), $\lambda = 0.1$ in Method (2), $\eta_{1p} = 0.1$ and the initial coefficient values 0.1 uniformly in Methods (3) and (4), the learning coefficient 3.0, the hidden layer number 3, the neuron number per a hidden layer 20 (total $3 \times 20 = 60$ hidden neurons), and the initial edge weights randomly chosen within $[-0.5, 0.5]$ in Method (5), the kernel function in the form of a radial basis function $\exp(-\beta\|x - y\|^2)$ (where β is the reciprocal of the input variable number), the ν parameter 0.8, and the C parameter 10000 in Method (6), and the population size 200, the number of node functions 4 ($+$, $-$, $*$, and $/$), the number of terminal constants 110 (randomly generated within $[-5, 5]$), the maximum depth of initial inidividuals 5, the maximum node number in an inidividual 10000, the tournament size of selection 2, the probability of creating new individuals via crossover 0.9, and the mutation probability (per node) 0.05 in Method (7). The simulation program is implemented in Java and is run on a standard desktop computer with Intel Duo processor (1.86 GHz).

3.1   Single Variable Problems

In regression with a single input variable, the problems (s_j-t_j pairs) and results (s_j-a_j pairs) can be plotted in a two-dimensional space. Figures 2 and 3 are

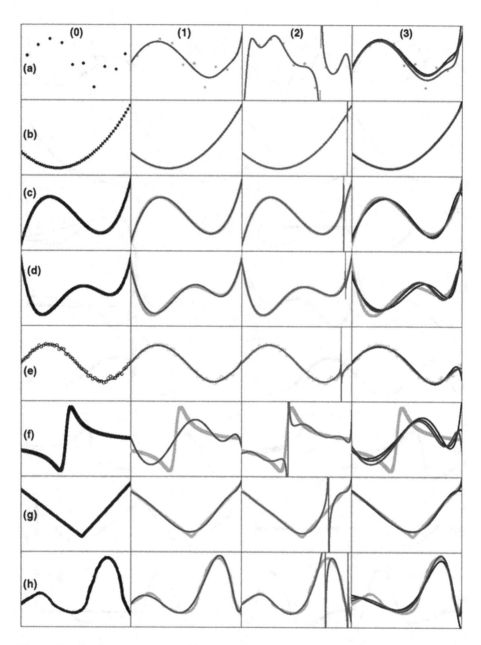

Fig. 2. Results for regression problems with a single sensor variable. (0) Original training image, and the results for Methods (1)~(3) are plotted. Here and in the next figure, integers are column indices, and alphabets are row indices. To specify a particular cell, we sometimes use such notation as (a3) in text.

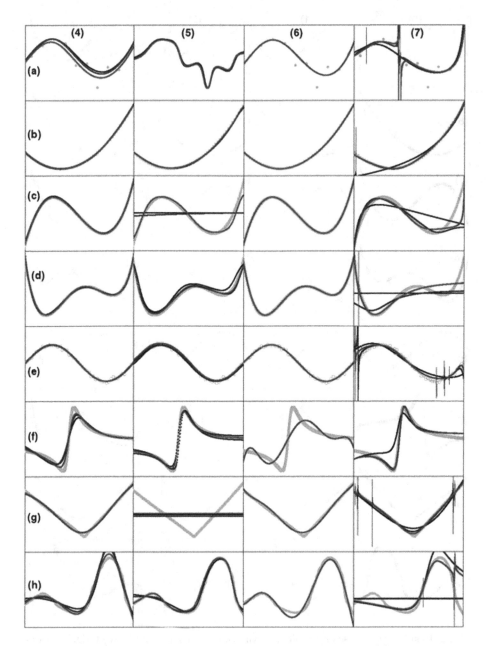

Fig. 3. Results for regression problems with a single sensor variable. (0) Original training image, and the results for Methods (4)∼(7) are plotted.

matrices of the two-dimensional planes, where sample values $((s_j, t_j)s)$ are plotted in black (in the (0)th column) or gray (in the other columns). Each plane's horizontal and vertical domain is $[0, 1]$. We prepare the eight different problems, which are made up of (a) 457 samples/pixels representing scattered dotts, (b) 2023 samples/pixels representing a quadratic function, (c) 3958 samples/pixels representing a cubic function, (d) 3977 samples/pixels representing a quartic function, (e) 1960 samples/pixels representing a sin function, (f) 4744 samples/pixels representing a 'fraction' function defined as $t = \dfrac{4.4 - 24.8s + 35.2s^2}{13 - 64s + 80s^2}$, (g) 3255 samples/pixels representing a polyline function, and (h) 3908 samples/pixels representing a man-made curve.

In the experiments of this subsection, we focus on the basic representation ability of each method, and do not distinguish between the training and the test data. Methods (1), (2), and (6) take all the samples to calculate coefficients, and Methods (3), (4), and (5) randomly choose a training sample for one-pass (one return) learning out of all the black pixels in the figures. Only Method (7) needs a set of samples to evaluate fitness of an individual, in which case we take 30 representative pixels distributed at (almost) even intervals within the domain $[0, 1]$ and calculate $[\text{fitness}] \equiv - \sum_j |t_j - a_j|$ for each fitness case.

From some preliminary experiments, the orders of the polynomials/Padé approximants were taken to be optimal values, $K = 19$ for Methods (1) and (3), and $K_0 = K_1 = 9$ for Methods (2) and (4). These numbers are fairly large, but the problems in Figs. 2 and 3 have so many samples (pixels) as the training data that we do not have to worry about the 'over-fitting' problem here [4].

The columns (1)∼(7) in Figs. 2 and 3 show the results for Methods (1)∼(7), respectively. We conducted 1,000,000-time iterations for Methods (3), (4), and (5) and fifty trials of 100-generation runs with different random number sequences for Method (7), during which functions created at one-third, two-third, and three-third of the run or the best run (in the GP) are plotted in black, blue, and red, respectively. The columns (1), (2), and (6) plot the analytic results in red. To obtain these results, Methods (3), (4), and (5) read total 1,000,000 samples for the training, and Method (7) evaluates individual fitness $50 \times 100 \times 200 = 1,000,000$ times. Note that since the GP uses 30 representative samples per a single fitness evaluation, the GP actually uses 30 times 1,000,000 of sample evaluations in total. This means that the GP uses much more samples to obtain the results, but as discussed later, the GP's results are not better than those of the others.

See the columns (1)∼(4) in Figs. 2 and 3. According to these figures, polynomial-based Methods (1) and (3) produces similar results; both are not able to precisely approximate (a), (d), (f), and (h). Specifically, the fraction function (f) is difficult for the polynomials, which means that a truncated power series has only limited ability to represent functions with strong nonlinearity. This situation is much improved when we replace the power series with a Padé approximant. Figure 3(4) shows good agreement between the Method (4)'s results and the original images with almost all the problems. The only (a) is difficult for (4), which suggests that the ATN in the Padé tries to avoid over-fitting to some extent even without

Table 1. Padé coefficients for the quartic function

Order	(2)			(4)	
	Numerator	Denominator		Numerator	Denominator
0	0.835	1.0		0.861 (0.0136)	1.0 (0.0157)
1	−4.47	2.904		−6.82 (−0.107)	3.44 (0.0543)
2	−29.08	28.68		18.2 (0.287)	−2.66 (−0.0420)
3	391.11	−391.24		3.46 (0.0546)	3.36 (0.0531)
4	−1624.30	1624.27		−7.53 (−0.119)	6.71 (0.106)
5	3570.65	−3570.64		−10.1 (−0.160)	6.39 (0.101)
6	−4664.94	4664.98		−6.90 (−0.109)	3.90 (0.0616)
7	3656.76	−3656.71		−0.798 (−0.0126)	0.281 (0.00444)
8	−1594.09	1594.15		6.46 (0.102)	−3.63 (−0.0574)
9	297.50	−297.44		13.8 (0.218)	−7.41 (−0.117)

Coefficients of the Padé approximants obtained in Figs. 2(d2) and 3(d4). The numbers in the parentheses of column (d4) are raw data by the ATN's learning. For comparison with column (2), those numbers are normalized in order that the order-0 coefficient of the denominator might become 1.0. The larger magnitude of coefficients by Method (4) is more than hundred times as small as those by Method (2)

regularization. This is also supported by the results in Fig. 2(1, 3), where the ATN without regularization gives the same result as the analysis with regularization.

According to Figs. 2(2) and 3(4), however, the analysis and the ATN do not give the same results for the Padé approximant's identification. The primary reason is in the existance of poles (zeros of the denominator) of the Padé by the linear regression analysis. See Table 1 which shows typical coefficient sets of the Padé obtained by the the analysis (Method (2)) and the learning (Method (4)). As evident from this table, the ATN's learning makes the Padé with much smaller coefficients than the analysis, which surely helps create a fitting curve with no singular points. The analysis, on the other hand, tries to find a rigorous function by calculating an inverse matrix, which finally creates a kind of 'brittle' function with some divergent points.

Columns (5)∼(7) in Fig. 3 are the results for representative other models in machine learning and evolutionary computation. As was theoretically proved [8] and demonstrated by a number of experimental studies [9, 27], the ANN with a sufficient number of hidden neurons is able to represent *any* function. The results in Fig. 3(5) support this fact. Figure 3(g5) might seem to be an exception; but in this case, we tested a run of 2, 000, 000-time (two times larger) iterations and observed that the ANN eventually fits the polyline almost perfectly. When we are given sufficient time to repeat learning iterations, the ANN with the BP can be a powerful method to construct a function.

By contrast, as exemplified by Fig. 3(6), the support vector machine has limited ability to represent functions. This highly artificial method is reported to be parameter-sensitive [5, 30], and when the sample number is fairly large, it is said to be difficult to choose an appropriate kernel set to construct a function. In the current optimized parameter setting, the ν-SVR is unable to represent the fraction function (f) like the polynomial-based methods (1) and (3). Also, the SVM's intrinsic ability in generalization makes the result in Fig. 3(a6) similar to those in Figs. 2(a1), (a3), and 3(a4).

As compared to these learning methods, the GP is much worse with the tested regression problems. As mentioned before, to create Fig. 3(7), the GP uses thirty times as many sample evaluations as the other methods; nevertheless, none of the functions obtained by the GP fits the training data except for the quadratic function (b) and the polyline function (g). We can say that the GP which is unable to adjust constants in the model has poor performmace with regression.

3.2 Multiple Variable Problems

In the case of regression on two or more input (sensor) variables, the variables' domain is so large that it is practically impossible to prepare the training data covering the entire domain. With such problems, we wish to construct a 'generalized' function that produces good answers for both the training and test data only by teaching the training data. To examine this ability with Methods (1) and (4) to (7), here we take several difficult target functions from [10, 14, 23, 39, 40] (which are listed in Table 2), and prepare the training data (30 s_j-t_j pairs) and test data (300 s_j-t_j pairs) separately by randomly chosing sensor values within the domain.

The iteration numbers and the orders of the polynomials/Padé approximants are determined from the criterion of genaralization. In regression, the generalization (i.e., avoidance of over-fitting) is one of the most well-studied topics in machine learning [4], and now we have a number of options to incorporate it in a model: (A) reducing the model's degree of freedom (taking a small order in Methods (1)~(4), a small number of hidden neurons in Method (5), and supressing the size of a program tree in Method (7)); (B) preparing a large number of training samples; (C) adding a regularization term to the error function (e.g., the λ-term in Eq. (3) or Eq. (4)); (D) stopping the learing when the error with the test data begins to increase (early stopping); (E) estimating the model parameters using the maximum likelihood method by assuming the Gaussian noise and a prior probability.

In Sect. 3.1, we had so many training samples (450 to 4700 pixels in Fig. 2(0)) that the generalization was semi-automatically met by option (B). In this subsection, however, we just have 30 training samples and implement/rely on other methods for generalization: (C) in Model (1), and (D) in Models (4), (5), and (7). We intentionally avoided (A) because the optimal value of the degree of freedom is problem-dependent and difficult to identify in advance.

See Fig. 4 which shows a typical result for the effects of the order and iteration number in Method (4). We can say from this figure that the mean square error with the training data is always smaller than the error with the test data, and that the former monotonously decreases with an increase in the iteration number,

Table 2. Results for regression problems with multiple sensor variables

	(4)		(5)	
(i)	◎ 6.29E-5 / 0.00246	(960,000)	× 5.65E-4 / 0.128	(800,000)
(j)	◎ 3.10E-7 / 1.18E-6	(960,000)	△ 2.65E-6 / 1.92E-5	(970,000)
(k)	○ 0.00333 / 0.00969	(640000)	△ 0.0127 / 0.150	(980,000)
(l)	○ 0.0261 / 0.145	(20,000)	◎ 9.36E-4 / 0.0497	(460,000)
(m)	○ 0.0372 / 0.142	(20,000)	◎ 1.30E-5 / 0.0725	(920,000)
(n)	◎ 1.43E-5 / 5.18E-5	(100,000)	△ 9.31E-6 / 1.69E-4	(970,000)
(o)	○ 3.66E-7 / 1.48E-6	(650,000)	× 5.27E-6 / 2.71E-5	(970,000)
(p)	△ 2.52E-6 / 6.94E-6	(970000)	× 3.53E-6 / 2.71E-5	(970,000)
(q)	◎ 0.174 / 0.497	(50,000)	○ 0.0934 / 1.20	(220,000)
(r)	◎ 9.05E-5 / 2.70E-4	(140,000)	○ 1.05E-5 / 5.74E-4	(140,000)
(s)	◎ 3.39E-4 / 0.00803	(30,000)	○ 5.72E-5 / 0.0199	(580,000)
(t)	◎ 7.08E-6 / 1.78E-4	(10,000)	○ 4.51E-6 / 2.53E-4	(610,000)
	(1)	(6)	(7)	
(i)	× 0.0260 / 0.192	× 0.0248 / 0.156	× 0.497 / 0.403	
(j)	× 4.94E-5 / 7.49E-5	○ 1.05E-7 / 1.25E-6	× 0.00162 / 0.00175	
(k)	× 0.616 / 1.34	◎ 9.19E-4 / 0.00692	× 0.291 / 1.17	
(l)	△ 0.169 / 0.769	△ 0.181 / 0.421	× 2.77 / 1.69	
(m)	○ 0.0285 / 0.186	○ 0.0333 / 0.156	○ 0.0349 / 0.151	
(n)	× 5.45E-4 / 0.00184	△ 6.13E-6 / 4.13E-4	× 5.70E-4 / 0.00259	
(o)	× 5.67E 5 / 2.05E-4	◎ 2.82E-7 / 6.08E-7	× 0.00110 / 0.00162	
(p)	× 1.26E-4 / 5.74E-4	◎ 1.67E-7 / 5.62E-7	× 2.68E-4 / 5.47E-4	
(q)	○ 0.191 / 0.679	○ 0.202 / 1.26	△ 8.79 / 8.84	
(r)	× 0.00203 / 0.0145	△ 2.41E-4 / 0.00566	× 0.00641 / 0.0177	
(s)	△ 0.00198 / 0.0653	○ 4.17E-6 / 0.00932	× 0.152 / 0.323	
(t)	○ 9.25E-6 / 3.90E-4	○ 3.94E-7 / 1.88E-4	△ 3.33E-4 / 7.43E-4	

The number on the left-hand side of a '/' is the mean square error with the training samples, the number on the right-hand side of a '/' is the mean square error with the test samples, and the number in a pair of parentheses is the iteration number at which the smallest test error was observed in Methods (4) and (5). Using the sensor variables s_1, s_2, and s_3, the target functions are written as (i) $8/(2 + s_1^2 + s_2^2)$ $[-3,3]$, (j) $s_1^4 - s_1^3 + s_2^2/2 - s_2$ $[0,1]$, (k) $s_1^3/5 - s_2^3/2 - s_2 - s_1$ $[-3,3]$, (l) $\dfrac{(s_1 - 3)^4 + (s_2 - 3)^3 - (s_2 - 3)}{(s_2 - 2)^4 + 10}$ $[0,6]$, (m) $\dfrac{1}{1 + s_1^{-4}} + \dfrac{1}{1 + s_2^{-4}}$ $[-5,5]$, (n) $s_1^{s_2}$ $[0,1]$, (o) $\sin(s_1) + \sin(s_2^2)$ $[0,1]$, (p) $2\sin(s_1)\cos(s_2)$ $[0,1]$, (q) $s_1 s_2 + \sin((s_1 - 1)(s_2 - 1))$ $[-3,3]$, (r) $\dfrac{\exp(-(s_1 - 1)^2)}{1.2 + (s_2 - 2.5)^2}$ $[0,4]$, (s) $120\dfrac{(s_1 - 1)(s_3 - 1)}{(s_2 + 2)^2(s_1 - 10)}$ $[0,2]$, and (t) $\sqrt{(s_1 + s_2 + s_3)(-s_1 + s_2 + s_3)(s_1 - s_2 + s_3)(s_1 + s_2 - s_3)}/16$ (Heron's formula) $[0,1]$, where each formula's subsequent value pair in the square brackets are the domains of the variables. The functions (i) to (r) have two input variables, and the functions (s) and (t) have three input variables. In each row, comparing values with the test data, the best (smallest) entry is marked with ◎, entries not larger than three times the best are marked with ○, entries not larger than thirty times the best are marked with △, and the others are marked with ×.

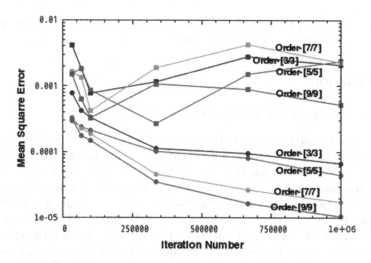

Fig. 4. Order- and iteration number-dependency of Method (4)'s results with Problem (r) (see Table 2). Mean square errors by the order-[3/3] (blue), [5/5] (red), [7/7] (green), and [9/9] (pink) Padé approximants are plotted as a function of the iteration number. The circular/square dots are values for the training/test data, respectively (Color figure online).

whereas the latter, beyond some threshold, begins to increase with the iteration number. The threshold is smaller with a larger order of the Padé approximant ($333, 333$ with $K_0 = K_1 = 5$, but $100,000$ with $K_0 = K_1 = 7$ and 9), but the minimal mean square errors at the thresholds make little difference if the order is large enough to express the target function.

From these observations, to make the models powerful enough to represent functions and accelerate the learning, we take fairly large orders, $K = 19$ in Methods (1) and $K_0 = K_1 = 9$ in Method (4) (the same numbers as in the previous subsection). During a run in Methods (4) and (5), we intermittently (every 10,000 iterations) observe the mean square error for the test data, and stop the run if the mean square error becomes three times as large as the minimal error or the iteration number exceeds a very large limitation 1,000,000. The final error values with the training/test data are taken from the values at the threshold iteration number. Methods (1) and (6) calculate the model parameters with the 30 training samples, and Method (7) evaluates the fitness with the 30 training samples. As in the previous subsection, Method (7) conducts fifty runs (i.e., 1,000,000 fitness evaluations) for each problem and the best of them is taken as the result.

The results are shown in Table 2. According to this table, we can say that Methods (4), (5), and (6) are far better than Methods (1) or (7). Specifically, Method (4) gives the best results with seven out of twelve problems, which is better than the second best (Method (6)) or the third best (Method (5)). We can also see that the threshold iteration numbers in the early stopping are much larger in Method (5) than in Method (4) with many problems, namely Method (5) is slower in convergence than Method (4).

4   Discussion

4.1   Completeness and Compactness

When we try to solve regression in a parametric way, the model function we prepare and optimize is the most important factor that determines the learning performance. The model function should satisfy the two antithetic conditions, *completeness* and *compactness*; in this sense, the Padé approximant is one of the most ideal functions we can utilize. The Padé is able to approximate any function (i.e., complete) because the Padé is a superset of the truncated Taylor expansion, and the Padé needs far fewer terms than the Taylor expansion to represent a function (i.e., compact) since the Padé approximant's error is suppressed in much wider variable domain than the Taylor expansion. (The Taylor expansion is known to swiftly magnify the convergence error with an increase in the distance from the center.)

This paper first translated the Padé approximant to a data-flow network and succeeded in optimizing it through the supervised learning.

4.2   Learning Speed, Stability, and Generalization

As shown in the results in Table 2, the ATN learns much faster than the ANN in general. This is rooted in the difference of network structures between the ATN and ANN. An ANN nodes' output function which is typically sigmoidal does not usually respond to a change in the weight/input strongly unless the weighted sum of the inputs is close to zero. This helps stabilize the ANN's output, but at the same time, makes the ANN go into a long period of stasis from time to time, decelerating the ANN's learning.

In the ATN without such a strong nonlinear factor, on the other hand, the situation is very different. In the ATN, every slight change in a parameter/input directly affects the output value, which causes incessant changes of the ATN's answer during the learning. This surely helps accelerate the learning, but at the same time, makes the ATN unable to converge without adjusting the learning coefficient (η) appropriately depending upon the error function between the answer and teaching values. The normalization scheme of η (Eq. (6) in [35]) is devised to mitigate this instability. Without this scheme, the ATN cannot have properties of both high speed learning and convergence stability.

From the generalization's point of view, these two methods are also contrasting. The ANN with a large number of strongly nonlinear elements is able to represent any functions precisely (see Fig. 3(a5, h5)). Without such elements, the Padé has limited ability to represent functions (see Fig. 3(a4, h4)), but this limitation surely enhances the ATN-Padé's ability for generalization as evident from the results in Table 2.

5   Future Directions

5.1   Combining Evolution and Suprevised Learning

As shown in Fig. 3 and Table 2, out of the seven methods tested in this paper, the simple GP exhibits the worst performance in regression. This is primarily because the simple GP cannot adjust constants in a program tree.

Of course, up to the present, some researchers have proposed optimizing constants within the framework of the GP. These works include Group Method of Data Handling (GMDH) [12], Adaptive Transformation Network [3], Adaptive Learning Networks [29], STROGANOFF [10,21] which limit the program search space to polynomials and optimize constants with the least square method (i.e., Method (1)), and Iba *et al.*'s approaches [7,11,17] which apply reinforcement learning (Q-learning) to adjusting node parameters in a GP program tree.

Along this line of research, the ATN implemented more powerful supervised learning in a GP-like program graph and achieved high ability in functional regression. The paper proposed constructing a data-flow network in the form of the Padé approximant and making it learn, but the ATN's learning scheme itself is topology-free and applicable to any network with differentiable node functions. The author surmises that incorporating the ATN's learning scheme into a GP program tree or other program graphs (PADO [37], Cartesian GP [19,20], GNP [17], or whatever) would be an interesting research topic to be tackled in the future. This would also provide a way to solve an intrinsic problem of the Padé approximant, whose term number $\binom{M+K_0}{M} + \binom{M+K_1}{M}$ increases in a higher polynomial order with the input number M.

5.2   Applicability to Motion Control

On the other hand, the ATN in the present form can be directly applied to control which requires fine estimation/modeling of the controlled object with limited actuators. The regression method established in this paper (Method (4)) has the following advantages in this problem domain.

First, the Padé function obtained through the ATN's learning has a kind of 'mild' coefficients with no pole (see Figs. 2(2), 3(4), and Table 1). Regression is called 'system identification' in control, and the above result suggests that if we apply Method (4) to identifying the transfer function (whose form is the same as the single-variable Padé), we would be able to construct a transfer function with much more stablilty than by the classical system identification method utilizing an (iterative) least square method [16].

Moreover, the Padé's algebraic form gives another advantage to the method. As is well known, in system identification for motion control, we want to construct an 'inverse model' from a kind of 'forward' information observed from the controlled object. Here, the forward information consists of actuator-to-sensor value pairs, and the inverse model calculates the actuator signal from the sensor signals. Several strategies have been proposed: The 'feedback error learning' [13] prepares a conventional feedback controller and an inverse model (typically, implemented with the ANN) and gradually replaces the former's answer with the latter's answer through the learning. Since the inverse model by the ANN is difficult (time-consuming) to modify, to make the controller adaptable to environmental changes, some hybrid methods, for example, a hybrid of forward model and Kalman filter [18] and a hybrid of inverse model and iterative control (which uses the Newton method in a wide sense) [22], have been also proposed. When applied to motion control, on the other hand, the ATN first constructs a forward model in the form of the Padé approximant (whose input number, i.e., actuator's number is typically very small) with the supervised learning. After or during the learning, we can solve the Pad'e formula (algebraic function) inversely with the Newton method in a wide sense, and obtain the actuator signal. In other words, the forward model can be used as the inverse model as well. Because the ATN's learning is faster than the ANN (Table 2), this controller will be able to modify the model itself faster than the inverse controller by the ANN.

6   Conclusion

Focusing on function regression problems, the learning capability of Algorithmically Transitive Network (ATN) was extensively studied. The ATN's topology is manually designed in the form of the truncated Taylor expansion or the Padé approximant, and its coefficients were optimized through the supervised learning that brings about forward and backward propagation of tokens in the network. From the experiments using a number of benchmark functions, it was demonstrated that the ATN with the Padé is able to identify and represent a variety of target functions on up to three input variables better than linear regression analysis with regularization, layered Artificial Neural Networks (ANN) with the BP learning, Support Vector Regression with soft margin (ν-SVR), or simple Genetic Programming (GP).

References

1. aiSee: Commercial software for visualizing graphs with various algorithms such as rubberband. http://www.aisee.com/
2. Baker, G.A.: Essentials of Padé Approximants. Academic Press, New York (1975)
3. Barron, R.L.: Adaptive transformation networks for modelling, prediction, and control. In: IEEE Systems, Man, and Cybernetics Group Annual Symposium Record, pp. 254–263 (1971)
4. Bishop, C.M.: Pattern Recognition And Machine Learning: Information Science and Statistics. Springer, Berlin (2006)
5. Burges, C.: A tutorial on support vector machines for pattern recognition. Data Min. Knowl. Disc. 2, 21–167 (1998)
6. Dennis, J.B.: Data flow supercomputer. IEEE Comput. 13(11), 48–56 (1980)
7. Downing, K.L.: Reinforced genetic programming. Genet. Program. Evolvable Mach. 2(3), 259–288 (2001)
8. Funahashi, K.: On the approximate realization of continuous mappings by neural networks. Neural Netw. 2(3), 183–192 (1989)

9. Haykin, S.: Neural networks and learning machines. Prentice-Hall, Upper Saddle River (2009)
10. Iba, H., de Garis, H., Sato, T.: A numerical approach to genetic programming for system identification. Evol. Comput. **3**(4), 417–452 (1995)
11. Iba, H.: Multi-agent reinforcement learning with genetic programming. In: Koza, J.R., et al. (eds): Genetic Programming 1998: Proceedings of the Third Annual Conference (GP-98), pp. 167–172 (1998)
12. Ivakhnenko, A.G.: Polynomial theory of complex systems. IEEE Trans. Sys. Man Cyber. **SMC-1**, 364–378 (1971)
13. Kawato, M., Furukawa, K., Suzuki, R.: A hierarchical neural-network model for control and learning of voluntary movement. Biol. Cybern. **57**, 169–185 (1987)
14. Keijzer, M.: Improving symbolic regression with interval arithmetic and linear scaling. In: Ryan, C., Soule, T., Keijzer, M., Tsang, E.P.K., Poli, R., Costa, E. (eds.) EuroGP 2003. LNCS, vol. 2610, pp. 70–82. Springer, Heidelberg (2003)
15. Koza, J.R.: Genetic Programming: on the Programming of Computers by Means of Natural Selection. MIT Press, Boston (1992)
16. Ljung, L.: System Identification: Theory for the User. Prentice Hall, Upper Saddle River (1998)
17. Mabu, S., Hirasawa, K., Hu, J.: A graph-based evolutionary algorithm: genetic network programming (GNP) and its extension using reinforcement learning. Evol. Comput. **15**(3), 369–398 (2007)
18. Mehta, B., Schaal, S.: Forward models in visuomotor control. J. Neurophysiol. **88**, 942–953 (2002)
19. Miller, J.F.: An empirical study of the efficiency of learning boolean functions using a cartesian genetic programming approach. In: Banzhaf, W., et al. (eds.) Proceedings of the Genetic and Evolutionary Computation Conference, vol. 2, pp. 1135–1142. Morgan Kaufmann (1999)
20. Miller, J.F., Smith, S.L.: Redundancy and computational efficiency in cartesian genetic programming. IEEE Trans. Evol. Comput. **10**(2), 167–174 (2006)
21. Nikolaev, N.Y., Iba, H.: Regularization approach to inductive genetic programming. IEEE Trans. Evol. Comput. **5**(4), 359–375 (2001)
22. Otani, M., Taji, K., Uno, Y.: Motion adaptation with iterative control using inverse-dynamics model. (Japan.) Inst. Electron. Inf. Commun. Eng. J. D **J95-D**(2), 305–313 (2012)
23. Pagie, L., Hogeweg, P.: Evolutionary consequences of coevolving targets. Evol. Comput. **5**(4), 401–418 (1997)
24. Rumelhart, D.E., Hinton, G.E., Williams, R.J.: Learning representations by back-propagating errors. Nature **323**, 533–536 (1986)
25. Rumelhart, D.E., Hinton, G.E., Williams, R.J.: Learning internal representations by error propagation. In: McClelland, J.L., Rumelhart, D.E. (eds.) The PDP Research Group: Parallel Distributed Processing, vol. 1, pp. 45–76. MIT Press, Cambridge (1986)
26. Schölkopf, B., Smola, A.J., Williamson, R.C., Bartlett, P.L.: New support vector algorithms. Neural Comput. **12**(5), 1207–1245 (2000)
27. Sejnowski, T.J., Rosenberg, C.R.: Parallel networks that learn to pronounce English text. Complex Syst. **1**, 145–168 (1987)
28. Sharp, J.A. (ed.): Data Flow Computing: Theory and Practice. Ablex Publishing Corp, Norwood (1992)
29. Skomorokhov, A.O.: Adaptive learning networks in APL2. In: Proceedings of the international conference on APL (APL '93), pp. 219–229. ACM, New York (1993)

30. Soares, C., Brazdil, P.B.: Selecting parameters of SVM using meta-learning and kernel matrix-based meta-features. In: Proceedings of the 2006 ACM Symposium on Applied Computing (SAC '06), pp. 564–568. ACM, New York (2006)

31. Suzuki, H.: A network cell with molecular agents that divides from centrosome signals. BioSystems **94**, 118–125 (2008)

32. Suzuki, H., Ohsaki, H., Sawai, H.: A network-based computational model with learning. In: Calude, C.S., Hagiya, M., Morita, K., Rozenberg, G., Timmis, J. (eds.) Unconventional Computation. LNCS, vol. 6079, pp. 193–193. Springer, Heidelberg (2010)

33. Suzuki, H., Ohsaki, H., Sawai H.: Algorithmically transitive network: a new computing model that combines artificial chemistry and information-communication engineering. In: Proceedings of the 24th Annual Conference of Japanese Society for Artificial Intelligence (JSAI), 2H1-OS4-5 (Japanese, 2010)

34. Suzuki, H., Ohsaki, H., Sawai, H.: An agent-based neural computational model with learning. Frontiers in Neuroscience. Conference Abstract: Neuroinformatics (2010). doi:10.3389/conf.fnins.2010.13.00021

35. Suzuki, H., Ohsaki, H., Sawai, H.: Algorithmically transitive network: a self-organizing data-flow network with learning. In: Suzuki, J., Nakano, T. (eds.) BIO-NETICS 2010. LNICST, vol. 87, pp. 59–73. Springer, Heidelberg (2012)

36. Suzuki, H.: Padé network: a parametric machine Learning system for regression. In: Proceedings of the 22nd Annual Conference of Japanese Neural Network Society, P2–20 (2012)

37. Teller, A., Veloso, M.: PADO: Learning tree-structured algorithm for orchestration into an object recognition system. Carnegie Mellon University Technical report, CMU-CS-95-101 (1995)

38. Tennenhouse, D.L., Wetherall, D.J.: Towards an active network architecture. ACM Comput. Commun. Rev. **26**(2), 5–18 (1996)

39. Uy, N.Q., Hoai, N.X., O'Neill, M., McKay, R.I., Galván-López, E.: Semantically-based crossover in genetic programming: application to real-valued symbolic regression. Genet. Program. Evolvable Mach. **12**(2), 91–119 (2011)

40. Vladislavleva, E.J., Smits, G.F., den Hertog, D.: Order of nonlinearity as a complexity measure for models generated by symbolic regression via pareto genetic programming. IEEE Trans. Evol. Comput. **13**(2), 333–349 (2009)

41. Werbos, P.J.: The roots of backpropagation: From ordered derivatives to neural networks and political forecasting. Adaptive and Learning Systems for Signal Processing, Communications and Control Series. Wiley-Interscience, New York (1994)

42. Wuytack, L. (ed.): Padé Approximation and Its Applications: Proceedings of a Conference Held in Antwerp, Belgium. Lecture Notes in Mathematics. Springer, Heidelberg (1979)

Molecular Scale and Bioinformatics

Investigation of Developmental Mechanisms in Common Developmental Genomes

Konstantinos Antonakopoulos$^{(\boxtimes)}$ and Gunnar Tufte

Department of Computer and Information Science,
Norwegian University of Science and Technology, Sem Sælandsvei 7-9, 7491
Trondheim, Norway
kostas@idi.ntnu.no

Abstract. The potentiality of using a common developmental mapping to develop not a specific, but different classes of architectures (i.e., species), holding different structural and/or computational phenotypic properties is an active area of research in the field of bio-inspired systems. To be able to develop such species, there is a need to understand the governing properties and the constraints involved for their development. In this work, we investigate how common developmental genomes influence evolution and how they push the developmental process in directions where it would have been impossible to achieve with ordinary genomes. Relations between mutation and evolution along with a comprehensive study of developmental mechanisms involved in development are worked out. The results are promising as they unveil that common developmental genomes perform better in more complex and random environments.

Keywords: Common developmental genomes · Evolvability · Cellular automata · Boolean network · L-systems

1 Introduction

The importance of development in life is crucial since it enables multicellular organisms to grow in well-defined stages. Besides 'direct-development', which is the simplest form of development, one can also find 'indirect development' through which the organism changes radically. These changes impose a new kind of adult organism over a period of evolutionary generations [1]. In the artificial analog, one can find a "simplified" artifact comprised of a genotype targeting a special phenotypic structure or other computational goal (a.k.a., Evo-Devo systems).

Taking a step further, there is a possibility of creating a genome (i.e., a genetic representation), which can be common for more than one phenotypes. In previous work, we studied whether the same mapping (i.e., common developmental genomes), can favor the evolvability of different computational architectures under the same (single-cell) environment (i) with limited resources [2], and (ii) in problems with increasing complexity [3].

G.A. Di Caro and G. Theraulaz (Eds.): BIONETICS 2012, LNICST 134, pp. 169–183, 2014.
DOI: 10.1007/978-3-319-06944-9_12, © Institute for Computer Sciences, Social Informatics and Telecommunications Engineering 2014

The ability of common developmental genomes to drive evolution lies in the developmental process and has a positive influence in directing evolution, as it has been seen so far. In other words, common developmental genomes may be the essence for what is called *developmental drive* [4]. Though, it is not yet clear how common developmental genomes influence evolution and how they push the developmental process in directions where it would have been impossible to achieve with ordinary genomes (i.e., genomes evolved separately for each architecture at hand). Also, it is still unclear whether and in what way the environment affects the ontogenetic pathways of development. Therefore, further research is needed towards three aspects. First, how genetic operators (i.e., mutation) affect evolution (i.e., selection) in common developmental genomes, as they are part of the "orienting mechanism" of both short-term and long-term evolution. Second, whether development and the developmental dynamics of common developmental genomes prescribe a certain pathway for evolution. Third, it is interesting to see if the external environment is at all important to the final phenotype.

The motivation for this work is to identify potential relations between mutation and selection in the underlying genetic process, discover inherent ontogenetic directionalities during the stages of evolution, and see whether environmental conditions affect the outcome in some way.

The rest of the article is laid out as follows. Section 2 describes the challenges involved in designing such a developmental model. The common genetic representation is given in Sect. 3. Detailed explanation of the genetic representation and developmental model can be found in [5]. The definition and structure of the environment is given in Sect. 4. Experiments come in Sect. 5 with the conclusion and future work in Sect. 6.

2   Development for Sparsely-Connected Computational Structures

The developmental goal is to be able to generate not a specific, but different classes of structures (i.e., species), using the same developmental model. This should be achieved through the same developmental approach. Such developmental approach requires sufficient knowledge of the targeted computational architectures and of their governing properties. That is, for the 2-dimensional cellular automata (CA) architecture, the properties of dimensionality and neighborhood must be defined, where the connectivity is predetermined (i.e., the Euclidean space). For boolean networks (BN), the connectivity (i.e., the node connections of the network), must be determined. The problem just described can be better expressed as *three-challenge* problem: (a) the *genome* challenge, (b) the *developmental processes* involved in the model, and (c) the *developmental model* challenge.

2.1   The Genome Challenge

Based on the properties of a 2D-CA, the genome contains information about the cells at each developmental step, in order to place them on a 2D-CA lattice structure. The wiring of the cell is given by the CA's neighborhood. At the same time and based on the properties of a boolean network, the genome contains enough information to feed the developmental model to develop a boolean network, at each developmental step.

2.2   The Developmental Processes Challenge

The resulting structure is able to grow, alter the functionality of a cell/node, and shrink. These processes are introduced in the developmental mapping through *growth*, *differentiation*, and *apoptosis* (i.e., the death of the cell/node). Having these properties in mind, our genome incorporates the notion of *chromosomes* - inspired by biology. Each chromosome contains respective information about the structural and/or functional requirements. More specifically, a chromosome will contain the information required for the cell/node creation (i.e., for the CAs and BNs), where another chromosome will contain the information required for wiring the nodes (i.e., for BNs). The notion of chromosomes allows us to exploit the genome in a modular way in the sense that if an additional computational architecture needs to be described through the same genome, more chromosomes can be added to it. To better illustrate how these processes will influence the developing structure, Fig. 1 shows the three developmental processes as applied to a developing cellular automata.

2.3   The Developmental Model Challenge

The developmental model is able to develop these structures, taking into account the special properties employed by each architecture. Figures 2 and 3, illustrate this requirement. The developmental model receives the same genome as input, regardless of the target architecture. Then, it is possible – depending on some properties of the genome – to discriminate whether it will develop a CA or a BN.

   Figure 2, visualizes this by showing step-by-step the development of a cellular automata from the developmental model. At DS 0, the first cell of the cellular automata is created. At DS 1, the cellular automata grows in size and a new cell is added. At DS 2, the architecture grows again by adding one more cell to

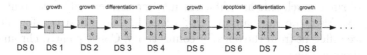

Fig. 1. The developmental model should be able to incorporate the processes of growth, differentiation and apoptosis. Here, each of the processes are illustrated as the model develops step-by-step a 2D cellular automata.

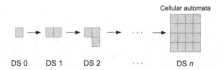

Fig. 2. The developmental model should be able to develop a cellular automata. Here, the development of the automata is shown from developmental step (DS) 0 until the final developmental step (DS) n.

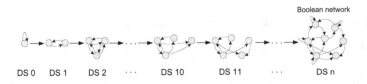

Fig. 3. The developmental model should be able to develop a boolean network. Here, the development stage of a BN is shown from developmental step (DS) 0, until the last developmental step (DS) n.

the cellular automata. At DS n, development has stop and the cellular automata has its final structure (adult organism).

Similarly, Fig. 3 presents step-by-step the development of a boolean network with the same developmental model. At DS 0, the first node with its self-connections is created. At DS 1, the boolean network grows in size and a new node is added to the network. This will cause new connections to be created for all the nodes existing in the network. At DS 2, the network adds another node and new connections are created for the existing nodes. This algorithm continues until the boolean network has created all the nodes and the connections for the existing nodes at DS n).

3   The Common Genetic Representation

In biology, a specie is often used as the basic unit for biological classification and for taxonomic ranking [6]. As such, an organism with unifying properties and same characteristics can be of the same specie. Figure 4, shows how the genome looks like. The genome is split into two parts (*chromosomes*). The first chromosome is responsible for creating the cells/nodes. The second chromosome is responsible for creating the connectivity (i.e., for the BNs). Each chromosome is built out of rules. Each rule has sufficient information for cell/node creation and connectivity. Also, the rules are of certain length. Those destined for cell/node creation are different from the ones for connectivity. Consequently, chromosomes contain different rules.

Fig. 4. This is how the genome looks like with the genome split into two (*chromosomes*). The first chromosome is responsible for generating the cells/nodes whereas the second chromosome is responsible for generating the connectivity of the network

3.1 An L-system for the Genetic Representation

A rewriting approach was chosen due to the ease of defining a specific rule set, that can target to rewrite specific features of a structure, e.g., connections or node functions that enable a way of splitting genetic information into separate information carrying units (i.e., chromosomes).

A prominent model is L-systems. They are rewriting grammars, able to describe developmental systems, simulate biological processes [7], and describe computational machines [8]. Since there are different types of rules in the two chromosomes, there is a need for two separate L-systems. The first L-system processes the rules of the first chromosome, while the second L-system deals with the connectivity rules of the second chromosome.

3.2 The L-system for the First Chromosome

The L-system used here is context-sensitive. As such, development is using the strict predecessor/ancestor to determine the applicable production rule. The rules are able to incorporate all the cell processes. Table 1(a), shows the type of symbols used by the L-system of the first chromosome. Some cells perform special cell processes and influence the intermediate and final phenotypes. Symbol a is the *axiom*. Apart from the symbols a, b, and c, which perform *growth* of the phenotype, symbol d performs *apoptosis*, leading to the deletion of the current rule (i.e., cell/node), of the intermediate phenotype. Additionally, symbols X and Y, are responsible for *differentiation*, leading to the replacement of the predecessor cell/node (i.e., if X→Y the outcome will be Y, whereas, if Y→X the outcome will be X). The length of each rule is 4 symbols (i.e., 4x8bits=32bits). For node/cell generation the L-system runs for 100 timesteps and then it stops. As such, the intermediate phenotypes generated by development are of variable size.

Figure 5(a), gives an example of a L-system for the first chromosome. A simple example with step-by-step development of a 2D-CA architecture is illustrated in Fig. 6. Development starts with the axiom (a) representing a cell at developmental step (DS) 0. Since the axiom is found in the L-system rules, development continues and the next rule triggered is the a→bX. This rule will create two more cells b and X, resulting in growth of the CA, at DS 1. The next rule triggered is bX→Y. Since X→Y denotes differentiation, the symbol X is replaced by Y, at DS 2. For differentiation to occur, the rules should either be X→Y, or Y→X. Next,

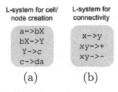

Fig. 5. (a) Example of L-system rules for the first chromosome, (b) Example of L-system rules for the second chromosome

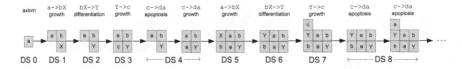

Fig. 6. A step-by-step development of a 2D-CA architecture based on the example L-system for the first chromosome

rule Y→c triggers causing again growth of the CA, at DS 3. At DS 4, the rule c→da is triggered, causing the death of the cell c and the growth of the CA with the cell a. From DS 5 up to DS 8, the rules are being triggered once more in the same sequence.

3.3   The L-system for the Second Chromosome

The rules are able to generate the connections necessary for the wiring of the nodes. They contain symbols which when executed by the L-system, result in creating a connection forward or backwards from the current node. Each node in the network has unique numbering; the current node has always the number zero and any nodes starting from the current node forward have positive numbering, where nodes that exist from the current node backwards, have negative numbering. So, there is a need to differentiate between the current and the next node, using different symbols and also a need to describe whether a connection will be created forward or backward from the current node.

The rules involved for connectivity are not as complex as the ones found in the first chromosome. The length of the rules here is also 4 symbols/rule. Also, there is a need to assure that the chromosome will have sufficient information for the developmental processes (i.e., growth, differentiation and apoptosis). The L-system uses a D0L (i.e., with zero-sided interactions). An example L-system for the second chromosome is shown in Fig. 5(b), and the symbols used are explained at Table 1(b). The *axiom* rule for the second chromosome is x→y. It means that development initially searches if the axiom exists. If so, development continues and looks for rules of type xy→+value, or xy→-value. In short, these two rules imply that if two different (i.e., distinct) nodes are found (x≠y), then it creates a connection forward (if the rule includes a '+'), or backwards (if the rule includes a '-'). The field value is encoded in the genotype and denotes the

Table 1. (a) Symbol table for node generation, (b) Symbol table for connectivity generation

(a)			(b)	
Symbol	Description		Symbol	Description
a (AXIOM)	Add (growth)		x	Node (different from y)
b	Add (growth)		y	Node (different from x)
c	Add (growth)		+	Connect forward
d	Delete (apoptosis)		−	Connect backwards
X	Substitute (differentiation)		→	Production
Y	Substitute (differentiation)			
→	Production			

node number for the generated connection. For example, rule xy→+3 denotes that a connection will be created from the current node (node 0), to the one being three nodes forward. Similarly, rule xy→-3, denotes that a connection will be created starting from the current node (node 0), to the one that is three nodes backwards. If value=0, a self-connection is created to the current node. A step-by-step development of a boolean network based on the chromosomes of Table 1(a) and (b), can be found in [5] and is not shown here due to page limitation. The modularity of the genome, gives the possibility to develop itself to enable or disable parts of it (chromosomes), when this is required and driven by the goal set. For example, if the target architecture is a 2D-CA, the second chromosome (i.e., connectivity) is disabled, since connectivity is predetermined. Similarly for BN development, both chromosomes are enabled (i.e., nodes and connectivity).

3.4   The Genetic Algorithm for the Common Genetic Representation

A genetic algorithm is used to generate and evolve the rules found in the genome (i.e., in the chromosomes). Since there are two separate L-systems involved in development, the evolutionary process comprise of two phases: node and connectivity generation phases. Mutation and single-point crossover were used as genetic operators. Mutation may happen anywhere inside the 4-symbol rule, ensuring that the production symbol (→) is not distorted by mutation. In short, we want to make sure that after mutation, the production symbol is still in the rule (i.e., the rule is valid). Single-point crossover between two parents is executed at the location of the production symbol, ensuring that a valid rule is created as offspring. The evolutionary cycle ends after a predetermined number of generations.

4   Definition of the Environment

Here, we define the 'environment' in a consistent way. In the literature, the environment has been used in various levels, depending on the system. In [9] for example,

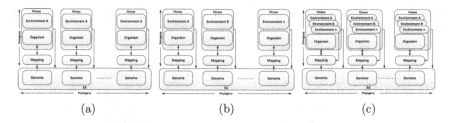

Fig. 7. The three different environmental setups. (a) Single-cell environment/genome evaluation, (b) Random environment/genome evaluation, (c) Multiple-random environments/genome evaluation.

the environment is used within the cell itself (i.e., the *cell's metabolism* [9,10]). In most models, environment refers to inter-cell communication, where cells can communicate their protein levels [11] or chemical levels and cell types [10]. The neighborhood of the underlying architecture (i.e,. cellular automata) can also become the environment, as in [10] with a 2D von Neumann neighborhood. Herein, we consider only an external environment, where the emerging organism needs to survive in it. Also, the behavior of the organism - interpretation of the state plot, may be distinctly different for the same organism when developed in two different environments [9]. As such, we introduce different external environments and give the possibility to the developing organism to adapt its behavior.

Figure 7 shows the various external environmental setups applied to the developing organisms. In the first subfigure, the individuals are evaluated based on environment A. In the second subfigure, the individuals are still being evaluated in a single environment, but the environment can be different (i.e., environment A, environment B, etc.). In the third subfigure, each individual is being assessed on a set of different environments. The environment in Fig. 7(a), is a single-cell environment where everywhere except in the first node/cell has a value of zero. The first cell/node holds initially the value of one. The environments in Fig. 7(b) and (c) are random. Feeding evolution with different information (i.e., environments), is expected to affect the genome, intermediate phenotypes and the final phenotype, not only in terms of cell types (i.e., for CA, BN), but also in terms of connectivity (i.e., for the BN).

5   Experiments

The motivation in Sect. 1, dictates a need to identify whether common developmental genomes have an inherent advantage over the separately developed genomes. We take different approaches to be able to draw solid conclusions. First, the influence mutations may have to the final phenotype and in driving evolution in different environmental conditions (Subsect. 5.1). Second, we try to discover inherent ontogenetic trends of common developmental genomes, by focusing on the developmental mechanisms and how these are deployed under different environments. This is studied in Subsects. 5.2 and 5.3.

5.1   Influence of Mutation Over Evolution

In this experiment, we study whether mutations help to determine the direction of phenotypic evolution. That is, try to identify specific patterns by simply counting the positive, neutral or negative influence a mutation has over the phenotype for each generation. The new phenotype after a mutation is compared to the phenotype from the previous generation. If the fitness of the new phenotype is bigger, then the mutation is considered positive. Neutral mutation is when both phenotypes have the same fitness where negative mutation will have a destructive influence to the phenotype.

5.2   Influence of Developmental Processes Over Evolution

Here, the mechanisms involved in development, i.e., the developmental processes during evolution, are investigated. More specifically, the appearance rate of growth, apoptosis and differentiation per individual are measured for each generation. In this way, we hope to get a better understanding of how development works for the separate and common genomes respectively.

5.3   Influence of Conditional Developmental Processes Over Evolution

Taking one step further, we capture conditional appearance for each process, given a certain process has appeared earlier during development. For example, we measure the number of growths after an apoptosis has occurred ($growth|$ $apoptosis$) or after a differentiation has occurred ($growth|differentiation$). Given we have three different developmental processes and each process can be in one of the three different conditional cases, we conclude to a total of 9 conditional cases for evaluation (Table 2).

5.4   Experimental Setup

The experiments were performed both for separate genomes and the common genome cases. Each experiment is based on the different settings shown in Fig. 7. For each setup, 20 runs were performed resulting in 20 different organisms. The developmental process was apportioned of 1000 state steps. The fitness function gives credit for cycle attractors between 2 and 800; the best score 100 is assigned to individuals with a cycle attractor of 400. The fitness scores have a

Table 2. Conditional appearance of the developmental processes

Cond. case 1	Cond. case 2	Cond. case 3
$growth|growth$	$growth|apoptosis$	$growth|differentiation$
$apoptosis|growth$	$apoptosis|apoptosis$	$apoptosis|differentiation$
$differentiation|growth$	$differentiation|apoptosis$	$differentiation|differentiation$

Table 3. Cell types with their functionality

Cell type	Function name
a	NAND
b	OR
c	AND
d	IDENTITY CELL
X	XOR
Y	NOT

bell-curve "distribution" with the worst score 4 being assigned at limit values. A total number of 70 rules for node generation (i.e., total size `70x32=2248bits`). For the second chromosome, we use the same number of rules (i.e., a size of `70x32=2248bits`). In this setup, each rule can be used more than once during development. For the common developmental genomes case, fitness is the average of CA and BN fitnesses. In the multiple-random environment case, fitness is the average over 10 evaluations (i.e., different external environments). The evaluation of CA and BN phenotypes was based on the cell types of Table 3.

The 2D-CA is non-uniform of size `6x6`. The BN network has a maximum size of $N = 36$ nodes. The reason for this choice is that we want the architecture to have the same state space. The number of outgoing connections per node is $K = 5$. For inputs more than 5, a self-connection to the originating node is created instead. *Generational mixing* protocol was used as the GA's global selection mechanism and *Rank selection* for parental selection. For CA development, the mutation rate was set to 0.005 and crossover rate at 0.001. The population size is 20. The GA was set to 10000 generation/Run.

5.5   Results

The result figures were generated after sampling the data every 100 values. To present the influence mutation has over evolution, an average number for each mutation type (i.e., positive, neutral and deleterious), is drawn across all runs along with the standard deviation per generation. Common developmental genomes show higher ratio of positive mutation during evolution for the single-cell environment (Fig. 8(a), (b) and (c)). Neutrality levels seems to be less throughout evolution. Deleterious mutations follow positive mutations across all species. In the random environment case, common genome holds a constant level of neutrality which is slightly higher than the separate genomes for CA and BN (Fig. 8(d), (e) and (f)).

Higher neutrality ratio gives genomes the ability to choose amongst larger span of potential trajectories in the fitness landscape, and greater ability to cope with uncertain and random environments. The same applies for the multiple-random environment, where common genome has a higher constant ratio on positive and neutral levels as compared to the other architectures (Fig. 8(g), (h) and (i)). Also, common genomes appear to have lower standard deviation, for the random and multiple-random environment cases.

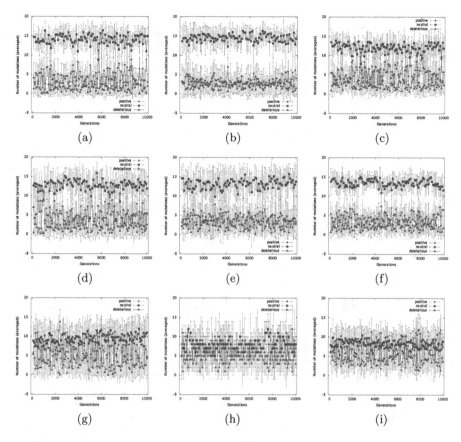

Fig. 8. Mutational analysis. Single-cell environment: (a) CA, (b) BN, (c) Common developmental genome. Random environment; (d) CA, (e) BN, (f) Common developmental genome. Multiple-random environment: (g) CA, (h) BN, (i) Common developmental genome.

To present the influence of developmental processes over evolution, an average for each developmental process (i.e., growth, apoptosis and differentiation), is drawn along with the standard deviation per generation. Common developmental genomes hold higher growth and differentiation ratios for the single-cell, random and multiple-random environments (Fig. 9(c), (f) and (i)). Growth and differentiation are crucial components of a genome towards evolvability [2]. Here, it is obvious that as the environment becomes more difficult (i.e., from the single-cell to the multiple-random), common genomes acquire higher ratios of growth and differentiation, if compared to the other architectures. Apoptosis levels are close to zero in all cases. Higher standard deviation is shown for the multiple-random environment case.

Each conditional developmental process is averaged across all 10 runs and the standard deviation is shown per generation. The conditional developmental

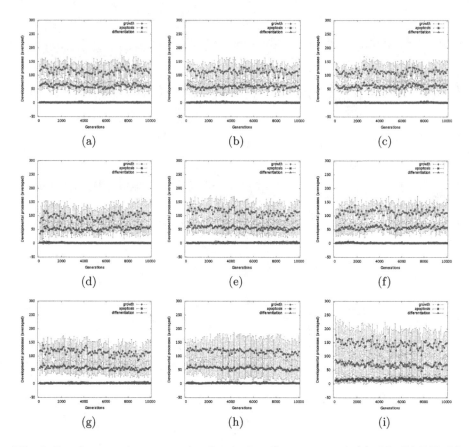

Fig. 9. Developmental processes for the single-cell environment: (a) CA, (b) BN, (c) Common developmental genome. Random environment; (d) CA, (e) BN, (f) Common developmental genome. Multiple-random environment: (g) CA, (h) BN, (i) Common developmental genome.

processes ratios for the single-cell environment seem homogeneous in all architectures (not shown). What is worth noting is that the ratio of the *apoptosis|growth* conditional process for the common genome looks like a convolution of the other two architectures (Fig. 10(a), (b), and (c)), with values being spread out over a larger range (i.e., high standard deviation). It is not yet clear why and what effect this has for common genomes development. The conditional growth and differentiation processes show similar behavior across all architectures (not shown).

The same result is obtained also for the random environment. The overall pattern is similar across all architectures (not shown). The ratio of the *apoptosis|growth* conditional process for the common genome looks (again) like a convolution of the other two architectures (Fig. 11(a), (b) and (c)).

The last set of experiments on the multiple-random environment, contributes clearly to the overall findings. For the common genome case, there is higher

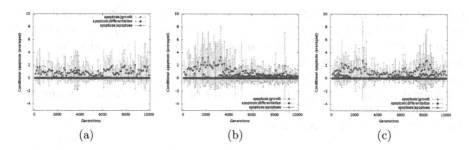

Fig. 10. Conditional apoptosis process for the single-cell environment. CA (a), BN (b), Common developmental genomes (c).

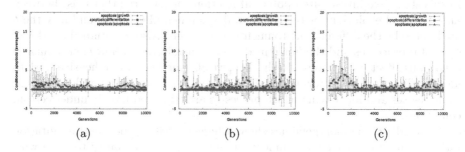

Fig. 11. Conditional apoptosis process on a random environment. CA (a), BN (b), Common developmental genomes (c).

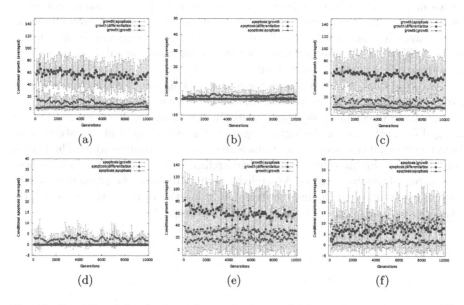

Fig. 12. Conditional developmental processes on multiple-random environment. CA: Conditional growth (a), apoptosis (b). BN: Conditional growth (c), apoptosis (d). Common developmental genomes: Conditional growth (e), apoptosis (f).

exploitation of *growth|differentiation* and *growth|apoptosis* conditional processes (Fig. 12(e)). Also, a higher exploitation ratio of *apoptosis|growth* and *apoptosis|differentiation* conditional processes (Fig. 12(f)), and higher standard deviation is observed. No *apoptosis|differentiation* conditional processes were observed for any of the architectures (not shown). Higher ratios of these conditional processes dictates the complexity of the multiple-random environment and the effect it applies on the common genomes as they need to respond and better adapt to the environment.

6   Conclusion

In this study, we investigated potential relations or patterns that exist between mutation and evolution for the common developmental genomes and how these are limited by the different environmental conditions. Also, we studied the developmental processes and the inherent dynamics involved and how environmental conditions affect the outcome. In this line, we also made a comprehensive work to identify directionalities during ontogeny by looking at conditional developmental processes and identifying which processes are triggered mostly under certain conditions.

Concluding, common genomes showed higher positive and neutral mutation ratios in more complex environments giving an inherent ability to cope with such environments. Also, they acquired higher ratios of growth and differentiation processes as compared to the other architectures. Lastly, high exploitation ratios of *growth|differentiation* and *growth|apoptosis* conditional processes, offered to common genomes the ability to perform in random environments. They have also shown a larger plurality of conditional processes during development. More work is needed towards conditional developmental processes and how these affect evolution and the final phenotype.

As future work, we shall change the way of looking into the architectures, i.e., instead of looking at them as different species, we can consider them as organs of a common developing biological entity. In this case, architectures need to be merged (as is the case in biological organs). The overall goal of this study is to target more adaptive scalable systems able of complex computation. The exploration of these architectures (i.e., hybrid architectures) with common genomes, the current developmental model and the ability to shape the phenotype of the system as modules, to change the dynamic properties of the entire system (i.e., *phenotypic shaping*), seem promising as future research.

References

1. Wallace, A.: Evolution: A Developmental Approach. Wiley-Blackwell, Oxford (2010)
2. Antonakopoulos, K., Tufte, G.: On the evolvability of different computational architectures using a common developmental genome. In: Rosa, A., Dourado, A., Madani, K., Filipe, J., Kacprzyk J. (eds.) IJCCI 2012, pp. 122–129. SciTePress Publishing (2012)
3. Antonakopoulos, K., Tufte, G.: Is common developmental genome a panacea towards more complex problems? In: 13th IEEE International Symposium on Computational Intelligence and Informatics (CINTI2012), pp. 55–61 (2012)
4. Arthur, W.: Developmental drive: an important determinant of the direction of phenotypic evolution. Evol. Dev. **3**(4), 271–278 (2001)
5. Antonakopoulos, K., Tufte, G.: A common genetic representation capable of developing distinct computational architectures. In: IEEE Congress on Evolutionary Computation (CEC2011), pp. 1264–1271 (2011)
6. Wilkins, J.: What is a species? essences and generation. Theor. Biosci. **129**(2), 141–148 (2010)
7. Lindenmayer, A., Prusinkiewicz, P.: Developmental models of multicellular organisms: a computer graphics perspective. In: Langton, C.G. (ed.) Proceedings of ALife. pp. 221–249. Addison-Wesley Publishing (1989)
8. Stauffer, A., Sipper, M.: Modeling cellular development using L-systems. In: Sipper, M., Mange, D., Pérez-Uribe, A. (eds.) ICES 1998. LNCS, vol. 1478, pp. 196–205. Springer, Heidelberg (1998)
9. Tufte, G., Haddow, P.C.: Achieving environmental tolerance through the initiation and exploitation of external information. In: IEEE Congress on Evolutionary Computation (CEC2007), pp. 2485–2492 (2007)
10. Federici, D.: Evolving a neurocontroller through a process of embryogeny. In: Eigth International Conference on Adaptive Behaviour, pp. 373–384 (2004)
11. Gordon, T.G.W., Bentley, P.J.: Development brings scalability to hardware evolution. In: 2005 NASA/DoD Conference on Evolvable Hardware, pp. 272–279 (2005)

Improving Diffusion-Based Molecular Communication with Unanchored Enzymes

Adam Noel[(✉)], Karen Cheung, and Robert Schober

Department of Electrical and Computer Engineering,
University of British Columbia, Vancouver, BC, Canada
adamn@ece.ubc.ca

Abstract. In this paper, we propose adding enzymes to the propagation environment of a diffusive molecular communication system as a strategy for mitigating intersymbol interference. The enzymes form reaction intermediates with information molecules and then degrade them so that they have a smaller chance of interfering with future transmissions. We present the reaction-diffusion dynamics of this proposed system and derive a lower bound expression for the expected number of molecules observed at the receiver. We justify a particle-based simulation framework, and present simulation results that show both the accuracy of our expression and the potential for enzymes to improve communication performance.

Keywords: Molecular communication · Reaction-diffusion system · Intersymbol interference · Nanonetwork

1 Introduction

Molecular communication is the use of molecules emitted by a transmitter into its surrounding environment to carry information to an intended receiver. This strategy has recently emerged as a popular choice for the design of new communication networks where devices with nanoscale components need to communicate with each other, i.e., nanonetworks. Molecular communication is suitable because its inherent biocompatibility can facilitate implementation inside of a living organism; many mechanisms in cells, organisms, and subcellular structures already rely on the transmission of molecules for communication, as described in [1, Chap. 16]. It is envisioned, as in [2,3], that by using bio-hybrid components (such as synthesized proteins or genetically-modified cells), we can take advantage of these mechanisms for a range of applications that can include health monitoring, targeted drug delivery, and nanotechnology in general.

The design of molecular communication systems should reflect both the limited capabilities of small individual transceivers and the physical environment

The first author was supported by the Natural Sciences and Engineering Research Council of Canada, and a Walter C. Sumner Memorial Fellowship. Computing resources were provided by WestGrid and Compute/Calcul Canada.

G.A. Di Caro and G. Theraulaz (Eds.): BIONETICS 2012, LNICST 134, pp. 184–198, 2014.
DOI: 10.1007/978-3-319-06944-9_13, © Institute for Computer Sciences, Social Informatics and Telecommunications Engineering 2014

in which they operate. The state-of-the-art has only begun to take advantage of the unique characteristics of molecular communication systems and their operational environments. The simplest and arguably most popular molecular communication scheme proposed has been communication via diffusion. Diffusion is a naturally-occurring process where free molecules tend to disperse through a medium over time. Diffusion requires no added energy and can be very fast over short distances; bacterial cells, many of which are on the order of one micron in diameter, can rely on diffusion for all of their internal transport requirements; see [4, Chap. 4]. By adopting diffusion, network designers do not need to worry about the development of the infrastructure required for active methods such as the molecular motors described in [5].

The major drawbacks of using diffusion are the need for a large number of information molecules to send a single message, long propagation times over larger distances, and the intersymbol interference (ISI) due to molecules taking a long time to diffuse away. Fortunately, biological systems commonly store large numbers of molecules for release at specific instances, such as the storage of Calcium ions in cellular vesicles until they are needed for signalling or secretion, as described in [1, Chap. 16]. Thus, delay and ISI become the performance bottlenecks. Strategies in the literature for mitigating ISI have been limited to making the transmitter wait sufficiently long for the presence of previously-emitted molecules to become negligible, as in [6–8]. The primary drawback of this strategy is a reduced transmission rate.

We propose adding reactive molecules to the propagation environment to significantly decrease the ISI in a molecular communication link when a single type of information molecule is used. The reactive molecules transform the information molecules so that they are no longer recognized by the receiver. If using chemical reactants, then they must be provided in stoichiometric excess relative to the information molecules, otherwise their capacity to transform those molecules may be limited over time. However, a catalyst lowers the activation energy for a specific biochemical reaction but does not appear in the stoichiometric expression of the complete reaction so (unlike a reactant) is not consumed.

An enzyme is a biomolecule that acts as a catalyst, often by providing an active site (a groove or pocket) that encourages a particular molecular conformation; see [1, Chap. 3]. Compared to catalysts in general, enzymes can have the advantage of very high selectivity for their substrates. Thus, we are specifically interested in enzymes as reactive molecules because a single enzyme can be recycled to react many times. Enzymes play a key role in many essential biochemical reactions. For example, acetylcholinesterase is an enzyme present in the neuromuscular junction that hydrolyzes diffusing acetylcholine to prevent continued activation in the post-synaptic membrane because the receptor in the membrane does not recognize acetate or choline, as described in [4, Chap. 12]. Acetylcholine is called the *substrate* for acetylcholinesterase. The physical environment of the neuromuscular junction is referred to as a *reaction-diffusion system* because reaction and diffusion can take place simultaneously. From a purely communications perspective, the enzyme in this example is reducing the ISI of the substrate.

There are many potential benefits for using enzymes to aid in developing new molecular communication systems. The reduction in ISI would enable transmitters to release molecules more often, simultaneously increasing the data rate and decreasing the probability of erroneous transmission. There would also be less interference from neighbouring communication links, so independent sender-receiver pairs could be placed closer together than in an environment dominated by diffusion alone. These gains can be achieved with no additional complexity at the sender or receiver, which is a very useful benefit for the case of individual nanomachines with limited computational capabilities. The enzymatic reaction mechanism could also be coupled to a mechanism that regenerates information molecules once they are degraded so that they are returned to the sender for future use (as is the case for acetylcholinesterase). Of course, it is necessary to select an enzyme-substrate pair that would not otherwise damage the environment where the nanomachines are in operation.

Most existing work in molecular communications, including [9,10], have considered enzymes only at the receiver as part of the reception mechanism. In these cases, the ability for the enzymes to mitigate ISI is limited. Two works that have considered information molecules reacting in the propagation environment are [11,12]. In [11], the spontaneous destruction and duplication of information molecules are treated as noise sources, whereas in [12], information molecules undergo exponential decay in an attempt to mitigate ISI. Papers that have considered reaction-diffusion systems with enzymes in the propagation environment from a biological perspective, such as [13,14] for acetylcholinesterase, have focussed on providing an accurate simulation model for specific biological processes with a particular physical layout and not the manipulation of parameters for the design of new communication systems.

In this paper, we present a model for the analysis of diffusion-based communication systems with enzymes that are present throughout the entire propagation environment. We start with the fundamental dynamics of both diffusion and enzyme kinetics to derive a bound on the expected number of molecules within the volume of an isolated observer placed some distance from the transmitter. In this context, we assume that the reader has a communications background and is not familiar with reaction-diffusion dynamics. We justify a particle-based simulation framework to assess the accuracy of our analytical results, and show that adding enzymes drastically reduces the "tail" created by relying on diffusion alone.

The rest of this paper is organized as follows. In Sect. 2, we introduce our model for transmission between a single transmitter and receiver. This model is based on both reaction and diffusion. In Sect. 3, we derive the number of information molecules expected at the receiver. We present the simulation framework in Sect. 4 before giving numerical and simulation results in Sect. 5. In Sect. 6, we present conclusions and discuss the on-going and future direction of our analysis.

Unless otherwise noted, we use meters (m) for distance, seconds (s) for time, and molecules per m^3 for concentrations (concentrations are typically given in

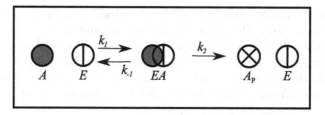

Fig. 1. The Michaelis-Menten reaction mechanism. Substrate molecule A can react with enzyme molecule E if they collide with sufficient energy and in the correct orientation. The reaction produces an intermediate EA that can either return to its original constituents or degrade particle A into A_P. The enzyme is not degraded by this process so it can react with multiple A molecules. An A molecule, once degraded, cannot be returned to its original state via this mechanism.

moles per litre, but molecules per m^3 makes our analysis easier to follow by limiting the number of conversions).

2   System Model

We consider an unbounded 3-dimensional aqueous environment. There is a sender fixed at the origin, treated analytically as a point source but as a sphere in simulation. The receiver is a fixed spherical volume of radius r_{obs} and size V_{obs}, centered at the point defined by $\mathbf{r_0} = \{x_0, y_0, z_0\}$. The receiver acts as a passive observer by not disturbing the diffusion of any molecules in the environment. This is not a strong assumption, since many small molecules are able to diffuse freely through cells and other objects if the molecules are non-polar or if there are protein channels specific to the molecules in the cell's plasma membrane; see [1, Chap. 12]. The immobility of both the sender and receiver is generally impractical at the nanoscale unless they are anchored to larger objects, but here we assume immobility for ease of analysis.

Before describing our communication process, we must overview the environment's chemical dynamics. There are three mobile species (types of molecules) in the system that we are interested in: A molecules, E molecules, and EA molecules. The number of molecules of species S is given by N_S where $S \in \{A, E, EA\}$. A molecules are the information molecules that are released by the sender. These molecules have a natural degradation rate that is negligible over the time scale of interest, but they are able to act as substrates with enzyme E molecules. We assume that A and E molecules react according to the following Michaelis-Menten reaction mechanism (which is generally accepted as the fundamental mechanism for enzymatic reactions; see [4,15], and Fig. 1):

$$E + A \xrightarrow{k_1} EA, \tag{1}$$

$$EA \xrightarrow{k_{-1}} E + A, \tag{2}$$

$$EA \xrightarrow{k_2} E + A_P, \tag{3}$$

where EA is the intermediate formed by the binding of an A molecule to an enzyme molecule, A_P is the degraded A molecule, and k_1, k_{-1}, and k_2 are the reaction rates for the reactions as shown with units $\text{molecule}^{-1}\text{m}^3\text{s}^{-1}$, s^{-1}, and s^{-1}, respectively. We see that A molecules are irreversibly degraded by reaction (3) while the enzymes are released intact so that they can participate in future reactions. We are not interested in the A_P molecules once they are formed because they cannot participate in future reactions.

We assume that every molecule of each species S diffuses independently of all other molecules, unless they are bound together. We assume that all free molecules are spherical in shape so that we can state that each molecule diffuses with diffusion constant D_S, found using the Einstein relation as [4, Eq. 4.16]

$$D_S = \frac{k_B T}{6\pi\eta R_S},\qquad(4)$$

where k_B is the Boltzmann constant ($k_B = 1.38 \times 10^{-23}\,\text{J/K}$), T is the temperature in kelvin, η is the viscosity of the medium in which the particle is diffusing ($\eta \approx 10^{-3}\,\text{kg}\,\text{m}^{-1}\text{s}^{-1}$ for water at room temperature), and R_S is the molecule radius. Thus, the units for D_S are m^2/s. The diffusion of a single molecule along one dimension has variance $2D_S t$, where t is the diffusing time [4, Eq. 4.6].

We note that reaction rate constants are experimentally measured for specific reactions under specific environmental conditions (i.e., temperature, pH, etc.) using large populations of each reactant. By "large", we mean sufficiently large for the rate of change of species concentrations to be deterministic. Diffusion also becomes deterministic with sufficiently large populations. We are not interested in such large populations due to the size of our system. However, the rate constants also describe the stochastic affinity of reactions in single-molecule detail, as proven in [16]. It is impossible to precisely predict where a specific molecule will diffuse and if or when it will react with other molecules, but the diffusion and reaction rate constants will be used to generate random variables when executing stochastic simulations of system behavior.

We can now describe the communication process. The sender emits impulses of N_A A molecules, which is a common emission scheme in the molecular communication literature; see, for example, [6]. We deploy binary modulation with constant bit interval T_B, where N_A molecules are released at the start of the interval for binary 1 and no molecules are released for binary 0 (there have been works studying the use of different T_Bs depending on the values of the current and previous bit, as in [8], since, for example, consecutive 0s can be transmitted with less risk of intersymbol interference). N_E E molecules are randomly (uniformly) distributed throughout a finite cubic volume V_{enz} that includes both the sender (TX) and receiver (RX), as shown in Fig. 2. V_{enz} is impermeable to E molecules (so that we can simulate using a finite number of E molecules) but not A molecules (in simulation, we make EA molecules decompose to their constituents if they hit the boundary). Therefore, the total concentration of the free and bound enzyme is constant. V_{enz} is sufficiently large to assume that it is infinite in size, such that there would be negligible change in observations at the receiver if V_{enz} were also impermeable to A molecules.

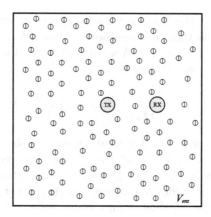

Fig. 2. The bounded space V_{enz} in 2-dimensions showing the initial uniform distribution of enzyme E. V_{enz} inhibits the passage of E so that the total concentration of free and bound E remains constant. A molecules can diffuse beyond V_{enz}.

The receiver counts the number of free (unbound) A molecules that are within the receiver volume, without disturbing those molecules. For a practical bio-hybrid system, the A molecules would need to bind to receptors on either the receiver surface or within the receiver's volume, but we assume perfect passive counting in order to focus on the propagation environment. We also assume that the degraded A_P molecules were modified in such a way that they cannot be detected by the receiver, so A_P molecules can be ignored.

3   Observations at the Receiver

Generally, the spatial-temporal behavior of the three mobile species can be described using a system of reaction-diffusion partial differential equations. Even though these equations are deterministic, we noted in Sect. 2 that they will enable stochastic simulation. In this section, we use the deterministic partial differential equations to derive the expected number of information molecules at the receiver.

3.1   Diffusion Only

For comparison, we first consider the dynamics when there is no enzyme present, i.e., $N_E = 0$. So, we only consider the diffusion of A molecules in the unbounded environment. By Fick's Second Law we have [4, Chap. 4]

$$\frac{\partial C_S(\mathbf{r}, t)}{\partial t} = D_S \nabla^2 C_S(\mathbf{r}, t), \tag{5}$$

where $C_S(\mathbf{r}, t)$ is the point concentration of species S at time t and location $\mathbf{r}$. Closed-form analytical solutions for partial differential equations are not always possible and depend on the boundary conditions that are imposed. Here, we have no E molecules, so there are also no EA molecules, and we immediately have $C_E(\mathbf{r}, t) = C_{EA}(\mathbf{r}, t) = 0 \, \forall \mathbf{r}, t$. Assuming that the A molecules are released from the origin at $t = 0$, we then have [4, Eq. 4.28]

$$C_A(\mathbf{r}, t) = \frac{N_A}{(4\pi D_A t)^{3/2}} \exp\left(\frac{-|\mathbf{r}|^2}{4 D_A t}\right). \tag{6}$$

Equation (6) is the form that is typically used in molecular communications to describe the local concentration at the receiver; the receiver is assumed to be a point observer, as in [7,17], or the concentration throughout the receiver volume is assumed to be uniform and equal to that expected in the center, as in [6]. Equation (6) is the baseline against which we will evaluate our proposed system design.

It has been noted that Fick's second law violates the theory of special relativity, since there is no bound on how far a single particle can travel within a given time. A finite propagation speed can be added as a correction, as in [18], but we assume that Fick's second law is sufficiently accurate without this correction.

3.2 Reaction-Diffusion

We now include active enzymes in our analysis. If we write $C_S(\mathbf{r}, t) = C_S$, $S \in \{A, E, EA\}$, for compactness, then the general reaction-diffusion equation is [19, Eq. 8.12.1]

$$\frac{\partial C_S}{\partial t} = D_S \nabla^2 C_S + f(C_S, \mathbf{r}, t), \tag{7}$$

where $f(\cdot)$ is the reaction term. Using the principles of chemical kinetics (see [15, Chap. 9]), we write the complete partial differential equations for the species in our system as

$$\frac{\partial C_A}{\partial t} = D_A \nabla^2 C_A - k_1 C_A C_E + k_{-1} C_{EA}, \tag{8}$$

$$\frac{\partial C_E}{\partial t} = D_E \nabla^2 C_E - k_1 C_A C_E + k_{-1} C_{EA} + k_2 C_{EA}, \tag{9}$$

$$\frac{\partial C_{EA}}{\partial t} = D_A \nabla^2 C_{EA} + k_1 C_A C_E - k_{-1} C_{EA} - k_2 C_{EA}. \tag{10}$$

This system of equations is highly coupled due to the reaction terms and has no closed-form analytical solution under our boundary conditions. We seek such a solution, so we must make some simplifying assumptions. We first note that the total concentration of enzyme, both free and bound to A, over the entire system is always constant $C_{E_{Tot}} = N_E/V_{enz}$. A common next step for the Michaelis-Menten mechanism in (1)–(3) is to assume that the amount of EA is constant, i.e., $\frac{\partial C_{EA}}{\partial t} = 0$, in order to derive an expression for C_{EA}; see [15, Chap. 10] and its use when considering enzymes at the receiver in [10]. We will use a slightly different assumption to directly derive a lower bound expression. We assume that both C_E and C_{EA} are not time-varying, i.e., C_E and C_{EA} are both constants. It is then straightforward to show that, in our system, (8) has solution

$$C_A \approx \frac{N_A}{(4\pi D_A t)^{3/2}} \exp\left(-k_1 C_E t - \frac{|\mathbf{r}|^2}{4 D_A t}\right) + k_{-1} C_{EA} t, \qquad (11)$$

and we ignore (9) and (10). Next, we assume that the amount of EA at any time is small, such that $k_{-1} C_{EA} \to 0$. If C_{EA} is small, then we can approximate C_E with its upper bound $C_{E_{Tot}}$. All concentrations and rate constants must be non-negative, so we can write the bound

$$C_A \geq \frac{N_A}{(4\pi D_A t)^{3/2}} \exp\left(-k_1 C_{E_{Tot}} t - \frac{|\mathbf{r}|^2}{4 D_A t}\right), \qquad (12)$$

which is intuitively a lower bound because the actual degradation due to enzymes can be no more than if all enzymes were always unbound. In other words, (12) describes the point concentration of A molecules as $k_2 \to \infty$ and $k_{-1} \to 0$. A convenient property of this lower bound is that, while it loses accuracy as EA is initially created ($C_E < C_{E_{Tot}}$), it eventually improves with time as all A molecules are degraded and none remain to bind with the enzyme ($C_A, C_{EA} \to 0$, $C_E \to C_{E_{Tot}}$, as $t \to \infty$). We also note that, had we started with the $\frac{\partial C_{EA}}{\partial t} = 0$ assumption, then we would have arrived at a similar expression to (12), where k_1 is replaced with $k_1 k_2 / (k_{-1} + k_2)$.

Equation (12) can be directly compared with (6). The presence of enzyme results in an additional decaying exponential term. This decaying exponential is what will eliminate the "tail" that is observed under diffusion alone. It can be shown that adding enzymes will always lead to a faster degradation of C_A from its maximum value for a given $\mathbf{r}$ than when not adding enzymes. Furthermore, the maximum value is achieved sooner when enzymes are present, but this value is smaller. These statements are apparent from the results in Sect. 5, and future work will prove these statements analytically.

We have already established that the receiver is able to count the number of free A molecules that are within the receiver volume. Equations (12) and (6) give us the expected point concentrations with and without active enzymes, respectively. We can readily convert these concentrations to the expected number of observed A molecules, $\overline{N_{A obs}}(t) = C_A(\mathbf{r_0}, t) V_{obs}$, if we assume that the concentration throughout the receiver is uniform, as in [6].

4   Simulation Framework

In the previous section, we derived (12) as a lower bound on the local concentration when enzymes are present throughout the propagation environment. We now require an appropriate simulation framework to evaluate the accuracy of (12). This framework will be used to perform stochastic simulations of the system of equations described by (8–10).

4.1 Choice of Framework

Commonly used stochastic reaction-diffusion simulation platforms can be placed into one of two categories. The first are subvolume-based methods, where the reaction environment is divided into one (if diffusion is ignored) or many well-stirred subvolumes. By well-stirred, it is meant that molecules in a specific subvolume are uniformly distributed throughout that subvolume, and that the velocities of those molecules follow the Boltzmann distribution; see [16]. In other words, every subvolume should have more nonreactive molecular collisions than reactive collisions. Stochastic subvolume-based methods are based on the stochastic simulation algorithm, which generates random numbers to determine the time and type of the next reaction in the system; see [20]. We note that these methods, though subvolume-based, still consider discrete species populations. However, the precise locations of individual molecules are not maintained, and diffusion is modeled as transitions of molecules between adjacent subvolumes; see [21].

The second category of simulation platforms use particle-based methods, where the precise locations of all individual molecules are known. Every free molecule diffuses independently along each dimension. These methods require a constant global time step Δt and there is a separation in the simulation of reaction and diffusion; see [22]. First, all free molecules are independently diffused along each dimension by generating normal random variables with variance $2D_S\Delta t$. Next, potential reactions are evaluated to see whether they would have occurred during Δt. For bimolecular reactions, a binding radius r_B is defined as how close the centers of two reactant molecules need to be at the end of Δt in order to assume that the two molecules collided and bound during Δt. For unimolecular reactions, a random number is generated using the rate constant to declare whether the reaction occurred during Δt.

Particle-based methods tend to be less computationally efficient, but they do not have to meet the well-stirred requirement. Our system has an impulse of molecules being released into an environment with highly reactive enzymes (we will discuss specific rate constants in Sect. 5, but for now we note that we are generally interested in large k_1). A general criterion for subvolume size is that the typical diffusion time for each species should be much less than the typical reaction time; see [23]. We cannot guarantee the satisfaction of this criterion for subvolume sizes that make physical sense (i.e., significantly larger than the size of individual molecules), so we adopt a particle-based method.

4.2 Simulating Reactions

Our bimolecular reaction (1) (the binding of E and A to form EA) is reversible, so we must be careful in our choice of binding radius r_B, time step Δt, and what we assume when EA reverts back to E and A molecules. A relevant metric is the root mean square step length, r_{rms}, between E and A molecules, given as [22, Eq. 23]

$$r_{rms} = \sqrt{2\left(D_A + D_E\right)\Delta t}. \tag{13}$$

If reaction (2) occurs, then the root mean square separation of the product molecules A and E along each dimension is r_{rms}. Unless $r_{rms} \gg r_B$, then these two reactants will likely undergo reaction (1) in the next time step. Generally, we need to define an unbinding radius specifying the initial separation of the A and E molecules when reaction (2) occurs. However, in the long time step limit, we can define r_B as [22, Eq. 27]

$$r_B = \left(\frac{3k_1 \Delta t}{4\pi}\right)^{\frac{1}{3}},\tag{14}$$

and this is valid only when $r_{rms} \gg r_B$. Thus, if r_{rms} is much greater than r_B found by (14), which we can impose by our selection of k_1 and Δt, then we do not need to implement an unbinding radius. Also, the evaluation of r_B is much more involved when we are not in the long time step limit and requires generating a lookup table; see [22] for further details. We will select parameters so that $r_{rms} \gg r_B$ is satisfied, even if r_{rms} becomes comparable with the size of the receiver, so we simply use (14).

We have a few additional comments on simulating reaction (1). It does not require the generation of any random values, besides those that are used to diffuse the individual molecules. However, we must check the position of every unbound A molecule with that of every unbound E molecule to see whether they are closer than r_B. For computational efficiency, we create subvolumes so that we only need to check the positions of enzymes in the current and adjacent subvolumes of the current free A molecule. If we find a pair close enough, then we move both of them to the midpoint of the line between their centers and re-label them as a single EA molecule.

Our two unimolecular reactions have the same reactant, EA, so we must consider both of them when calculating the probability of either reaction occuring. For (2), we have [22, Eq. 14]

$$\Pr\{\text{Reaction (2)}\} = \frac{k_{-1}}{k_{-1} + k_2} \left[1 - \exp\left(-\Delta t \left(k_{-1} + k_2\right)\right)\right],\tag{15}$$

where $\Pr\{\cdot\}$ denotes probability and (3) has an analogous expression by switching k_{-1} and k_2. A single random number uniformly distributed between 0 and 1 can then be used to determine whether a given EA molecule reacts. If it does, then we place the products at the same coordinates.

4.3   Simulating the Sender and Receiver

When the sender releases an impulse of N_A A molecules, we enforce an initial separation of $2R_A$ between adjacent A molecules, placing them in a spherical shape centered at the origin. At the receiver, we make observations at integer multiples of time step Δt. When an observation is made, all free A molecules whose centres are within V_{obs} are counted.

4.4   Selecting Component Parameters

We now discuss practical parameter values for the underlying reaction-diffusion system. Specific enzymatic reactions, such as the breakdown of acetylcholine by acetylcholinesterase, are represented by specific molecules and are characterized by specific reaction rate constants. Most enzymes are proteins and are usually on the order of less than 10 nm in diameter; see [1, Chap. 4]. From (4), smaller molecules diffuse faster, so we are most likely to select small molecules as information molecules. Many common small organic molecules, such as glucose, amino acids, and nucleotides, are about 1 nm in diameter. In the limit, single covalent bonds between two atoms are about 0.15 nm long; see [1, Chap. 2].

Higher rate constants correlate to faster reactions. Bimolecular rate constants can be no greater than the collision frequency between the two reactants, i.e., every collision results in a reaction. The largest possible value of k_1 is on the order of 1.66×10^{-19} molecule^{-1}m^{3}s^{-1}; see [15, Chap. 10] where the limiting rate is listed as on the order of 10^8 L/mol/s. k_2 usually varies between 1 and 10^5 s^{-1}, with values as high as 10^7 s^{-1}. In theory, we are not entirely limited to preexisting enzyme-substrate pairs; protein and ribozyme engineering techniques can be used to modify and optimize the enzyme reaction rate, specificity, or thermal stability, or modify enzyme function in the presence of solvents.

5   Results and Discussion

We are now prepared to present results comparing the observed number of A molecules at a receiver with and without the presence of enzymes in the propagation environment. We assume that the environment has a viscosity of 10^{-3} kg m^{-1}s^{-1} and temperature of 25 °C. The sender emits $N_A = 10^4$ A molecules, each having radius $R_A = 0.5$ nm, in a single impulse. V_{enz} is defined as a cube with side length 1 μm and centered at the origin, so its size is on the order of a bacterial cell. $N_E = 2 \times 10^5$ E molecules having radius $R_E = 2.5$ nm are uniformly distributed throughout V_{enz}. For simplicity, we assume that $R_{EA} = R_A + R_E = 3$ nm. In consideration of the limiting values of reaction rate constants, we choose $k_1 = 10^{-19}$ molecule^{-1}m^{3}s^{-1}, $k_{-1} = 10^4$ s^{-1}, and $k_2 = 10^6$ s^{-1}. We also set $\Delta t = 0.5$ μs, resulting in $r_{rms} = 22.9$ nm and $r_B = 2.28$ nm, so that $r_{rms} \gg r_B$ is satisfied.

We compare the number of molecules observed at a receiver due to a single emission from the sender. In Fig. 3, we consider two receivers with radii $r_{obs} = \{25, 45\}$ nm and their centers placed at a distance of $|\mathbf{r_0}| = \{150, 300\}$ nm from the sender, respectively. The expected number of molecules is calculated using either (12) or (6) for enzymes present and absent, respectively. The observed number of A molecules via simulation is averaged over at least 15000 independent emissions by the sender at $t = 0$.

Let us first consider the receiver placed 150 nm from the sender. The maximum number of molecules is received about 8 μs after emission. The maximum value is less than 15 % higher in the absence of active enzymes; over 14 molecules are expected and observed on average via simulation without enzymes, compared

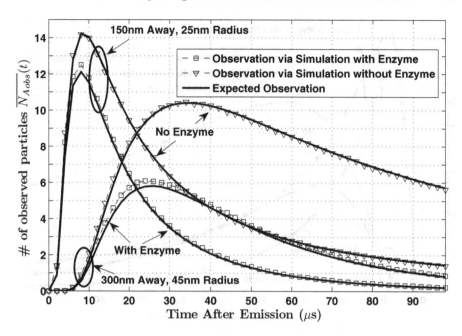

Fig. 3. Number of particles counted by receivers with radii $r_{obs} = \{25, 45\}$ nm that are placed $|\mathbf{r_0}| = \{150, 300\}$ nm from the sender, respectively. The source releases 10^4 molecules at $t = 0$. Simulation and analytical results are shown both with and without active enzymes.

to 12 molecules expected and 12.5 molecules observed with active enzymes. The decay from the maximum value is slower in the absence of active enzymes; 60 μs after emission, 3 molecules are expected and observed without enzymes while 1 molecule is expected and observed with enzymes, a threefold difference. We see that, as previously noted, the expected number of observed A molecules when active enzymes are present is a lower bound on the average number of A molecules observed in simulation, and this is a relatively tight bound.

The simulation and analytical results for the receiver placed 300 nm from the sender follow the same general trends as those for the closer receiver, but with a few noteworthy differences. Obviously, the time elapsed before receiving the maximum number of molecules is greater and the maximum value is less than for the closer receiver, even though the receiver is larger (the receiver being larger accounts for how it is possible for this receiver to observe more molecules than the closer receiver after 23 μs). However, the change in the number of molecules received is much greater in the presence of active enzymes; the peak number of molecules is observed relatively sooner (about 25 μs instead of about 35 μs after emission), but the maximum number of molecules is less than 60 % of that expected without enzymes (about 6 molecules instead of 10.5 molecules). Intuitively, being further from the sender gives more time for the E molecules to bind to and then degrade the A molecules.

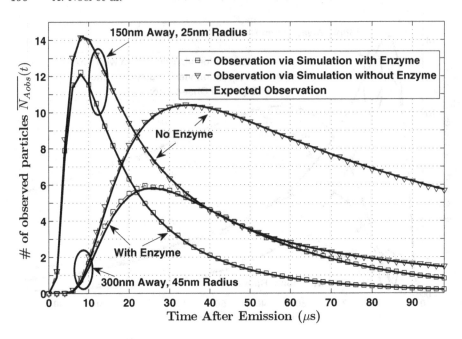

Fig. 4. Number of particles counted by receivers with radii $r_{obs} = \{25, 45\}$ nm that are placed $|\mathbf{r_0}| = \{150, 300\}$ nm from the sender, respectively. The source releases 10^4 molecules from a point at $t = 0$. Simulation and analytical results are shown both with and without active enzymes, where $k_2 = \infty$ and $k_{-1} = 0$. The resultant difference between this figure and Fig. 3 is that here the curves generated via simulation are tighter to the curves generated by the analytical expressions.

Both receivers in Fig. 3 show that adding enzymes decreases the "tail" of diffusion while still providing a peak to be detected at the receiver. It is clear that the sender could emit impulses more often with less risk of ISI. For example, if the criterion for designing the bit interval T_B was the time at which the expected number of particles is some fraction of the maximum expected number, then this time should be shorter in the presence of active enzymes. Alternatively, sender-receiver pairs could be placed closer together with less risk of co-channel interference. We leave formal proofs of these statements for future work, but they are intuitive given the results in Fig. 3.

Finally, we consider in Fig. 4 the limiting case that we used to derive the bound (12), i.e., set $k_2 = \infty$, $k_{-1} = 0$, and co-locate all A molecules at the origin when emitting. In this case, an E molecule binding to an A molecule immediately degrades the A molecule while releasing the E molecule, so all enzymes are always available to react. We otherwise maintain the same parameters that we used for Fig. 3. We see that the average number of particles observed via simulation with active enzymes agrees very well with that expected from (12), and that the average number of particles observed via simulation without active enzymes matches the value expected from (6), even though we are still assuming uniform

C_A throughout the receiver volume. This confirms that the looseness of the lower bound (12) in Fig. 3 comes from both the finite emission volume and the creation of EA molecules. The slight looseness of the lower bound in Fig. 4 for the receiver 300 nm away and when enzymes are present is likely due to having a finite V_{enz}; some A molecules are able to diffuse beyond V_{enz}, where they cannot be degraded, and then enter the receiver volume after returning to V_{enz}. This effect is negligible at the receiver 150 nm away.

6   Conclusions and Future Work

In this paper, we introduced the concept of using enzymes in the propagation environment to improve the performance of a diffusive molecular communication system. Enzymes that break down information molecules are able to reduce the time that a sender must wait before being able to send additional information molecules. There is potential to increase the data rate and to decrease the probability of error. This gain in performance comes with no additional complexity required at either the sender or receiver.

The emphasis in this paper was the description of the underlying reaction-diffusion model and the selection of an appropriate simulation framework, thereby providing a foundation for performance analysis. On-going work includes the derivation of the bit error rate for this binary-coded communication network when multiple emissions are made by the sender given a bit interval T_B and the reception scheme at the receiver. Furthermore, we are currently evaluating the analytical accuracy of the assumption that the concentration observed at the receiver is uniform. We must also consider the ability to choose reaction rate constants based on specific enzymes, as well as the enzyme concentration. In addition, dimensional analysis is useful to arbitrarily scale our system, compare different parameter sets, and derive the looseness of our receiver bound in terms of a dimensionless parameter. We also note that we could forego the use of enzymes altogether and use A molecules with a faster natural degradation rate, as in [12], but without using the number of counted molecules as the amount of information received. This case would allow simpler and accurate analysis though we would have to be concerned with maintaining a stockpile of these molecules at the sender without them degrading before emission. Other relevant problems of interest include interference from nearby sender/receiver pairs and the potential mobility of the sender and receiver.

References

1. Alberts, B., Bray, D., Hopkin, K., Johnson, A., Lewis, J., Raff, M., Roberts, K., Walter, P.: Essential Cell Biology, 3rd edn. Garland Science, New York (2010)
2. Akyildiz, I.F., Brunetti, F., Blazquez, C.: Nanonetworks: a new communication paradigm. Comput. Netw. **52**(12), 2260–2279 (2008)
3. Nakano, T., Moore, M.J., Wei, F., Vasilakos, A.V., Shuai, J.: Molecular communication and networking: opportunities and challenges. IEEE Trans. Nanobiosci. **11**(2), 135–148 (2012)

4. Nelson, P.: Biological Physics: Energy, Information, Life, 1st edn. W. H. Freeman and Company, New York (2008)
5. Hiyama, S., Moritani, Y.: Molecular communication: harnessing biochemical materials to engineer biomimetic communication systems. Nano Commun. Netw. **1**(1), 20–30 (2010)
6. Atakan, B., Akan, O.B.: Deterministic capacity of information flow in molecular nanonetworks. Nano Commun. Netw. **1**(1), 31–42 (2010)
7. Mahfuz, M.U., Makrakis, D., Mouftah, H.T.: Characterization of intersymbol interference in concentration-encoded unicast molecular communication. In: Proceedings of 2011 IEEE CCECE, pp. 164–168, May 2011
8. Einolghozati, A., Sardari, M., Beirami, A., Fekri, F.: Capacity of discrete molecular diffusion channels. In: Proceedings of 2011 IEEE ISIT, pp. 723–727, August 2011
9. Chou, C.T.: Molecular circuits for decoding frequency coded signals in nano-communication networks. Nano Comm. Netw. **3**(1), 46–56 (2012)
10. Nakano, T., Okaie, Y., Vasilakos, A.V.: Throughput and efficiency of molecular communication between nanomachines. In: Proceedings of 2012 IEEE WCNC, pp. 704–708, April 2012
11. Miorandi, D.: A stochastic model for molecular communications. Nano Commun. Netw. **2**(4), 205–212 (2011)
12. Moore, M.J., Suda, T., Oiwa, K.: Molecular communication: modeling noise effects on information rate. IEEE Trans. Nanobiosci. **8**(2), 169–180 (2009)
13. Naka, T., Shiba, K., Sakamoto, N.: A two-dimensional compartment model for the reaction-diffusion system of acetylcholine in the synaptic cleft at the neuromuscular junction. Biosystems **41**(1), 17–27 (1997)
14. Cheng, Y., Suen, J.K., Radi, Z., Bond, S.D., Holst, M.J., McCammon, J.A.: Continuum simulations of acetylcholine diffusion with reaction-determined boundaries in neuromuscular junction models. Biophys. Chem. **127**(3), 129–139 (2007)
15. Chang, R.: Physical Chemistry for the Biosciences. University Science Books, Sausalito (2005)
16. Gillespie, D.T.: A rigorous derivation of the chemical master equation. Phys. A **188**(13), 404–425 (1992)
17. Pierobon, M., Akyildiz, I.F.: Information capacity of diffusion-based molecular communication in nanonetworks. In: Proceedings of 2011 IEEE INFOCOM 2011, pp. 506–510, April 2011
18. Pierobon, M., Akyildiz, I.F.: A physical end-to-end model for molecular communication in nanonetworks. IEEE J. Sel. Areas Commun. **28**(4), 602–611 (2010)
19. Debnath, L.: Nonlinear Partial Differential Equations for Scientists and Engineers, 2nd edn. Birkhaeuser, Boston (2005)
20. Gillespie, D.T.: Stochastic simulation of chemical kinetics. Annu. Rev. Phys. Chem. **58**(1), 35–55 (2007)
21. Iyengar, K.A., Harris, L.A., Clancy, P.: Accurate implementation of leaping in space: the spatial partitioned-leaping algorithm. J. Chem. Phys. **132**(9), 094101 (2010)
22. Andrews, S.S., Bray, D.: Stochastic simulation of chemical reactions with spatial resolution and single molecule detail. Phys. Biol. **1**(3), 137 (2004)
23. Bernstein, D.: Simulating mesoscopic reaction-diffusion systems using the Gillespie algorithm. Phys. Rev. E **71**(4), 041103 (2005)

The Use of Computational Intelligence in the Design of Polymers and in Property Prediction

Xi Chen[1(⊠)], Les M. Sztandera[2], Hugh M. Cartwright[3],
and Stephen Granger-Bevan[3]

[1] Materials Science, Drexel University, Philadelphia, PA 19104, USA
xc37@drexel.edu
[2] Computer Science, Philadelphia University, Philadelphia, PA 19144, USA
sztanderal@philau.edu
[3] Chemistry Department, Oxford University, Oxford OX1 3QZ, UK
hmc@physchem.ox.ac.uk

Abstract. Taking advantage of techniques from the field of Computational Intelligence, the goal of our research is to construct systems that can computationally design polymer optical fiber formulations with specified desirable consumer characteristics and to develop computational tools which can be used to rationalize and predict properties of polymeric materials, such as the glass transition temperature.

Keywords: Computational intelligence · Polymer design · Glass transition temperature · Artificial neural network · Genetic algorithm

1 Introduction

Quantitative Structure Property Relationships (QSPRs) are (generally linear) relationships that are used to correlate the structure of compounds with their physical or molecular properties. These types of relationships show considerable potential in a wide range of scientific areas, including thermodynamics, computational drug design, and material science. Within the last of these areas, material science, the formulation of a new polymer with desired properties usually involves a laborious and expensive trial-and-error procedure, but this may largely be avoided using computer search techniques. Artificial Intelligence (AI) based methods [1], which we consider in this paper, are increasingly used, and their use can avoid a subtle disadvantage of traditional approaches; this is that in the latter the choice of candidate molecules is within the control of the researcher, which can prejudice the nature of the modifications which candidate molecules suffer. Should the optimum polymer formulation exist in

This research was supported by National Textile Center (NTC) grant C05-PH01 through U.S. Department of Commerce. Various aspects appeared in publications in Conference Proceedings and in the International Journal of Intelligent Systems.

G.A. Di Caro and G. Theraulaz (Eds.): BIONETICS 2012, LNICST 134, pp. 199–207, 2014.
DOI: 10.1007/978-3-319-06944-9_14, © Institute for Computer Sciences, Social Informatics and Telecommunications Engineering 2014

an unexpected region of search space, that ideal solution will never be located if the researcher believes investigation of that region will be fruitless. AI methods operate without this sort of bias.

In this research, we are using two promising techniques from the field of Artificial Intelligence: Artificial Neural Networks (ANNs) and Genetic Algorithms (GAs) [2]. The role of an artificial neural network in the first application we describe is to solve the "forward problem", which is the prediction of the properties of a polymer, given its molecular structure. A modified Genetic Algorithm has been prepared to solve the "backward problem", that is, designing the structure of a novel polymer whose properties match as closely as possible those in a list of desirable properties. Various factors in addition to the molecular structure may affect the properties of a finished polymeric product, so we approach the problem first as one of monomer design, and then expand it to include polymerization process conditions, fiber spinning, yarns and fabrics. We report here some results from the ANN approach and also from a further study in which a combined ANN-GA algorithm is used to investigate polymer properties.

2   Polymer Design

Polymers offer considerable potential as the base material for cost effective optical components, such as optical fibers and lenses. If polymer-based products are to be commercially attractive as optical transmission components, it is essential that the polymers be easy to manipulate and sufficiently robust that they can tolerate rough handling during installation; they must also of course possess the required optical properties and reliability. In this paper we focus on one of the parameters which can influence the properties of polymer optical fibers, the glass transition temperature, Tg,. Ballato et al. [3] suggested that performance can be optimized by engineering a polymer that exhibits a lower refractive index and Tg.

Numerous attempts have been made to correlate Tg with polymer structure, including the construction of a variety of group contribution and empirical or semi-empirical models. Van Krevelen [4] for example used a group contribution approach to calculate values not only for the glass transition temperature, but also for other physical and mechanical properties. Following on from Van Krevelen's work, further empirical correlations have been developed. Zuniga and Trevino [5] devised a simpler group contribution scheme to predict T_g. They calculated 66 contribution values of relatively small groups in an approach that was more flexible than that used by Van Krevelen.

In a new group contribution method, Marrero and Gani [6] divided chemical groups into three levels. The first included the simple functional groups required to describe a wide variety of compounds, while the second and third levels included polyfunctional and structural groups carrying information which was not contained in the first order groups. The main shortcoming of such group contribution methods is that they are applicable only when the group contribution values for the polymers are known in advance. A secondary disadvantage is that their predictions of Tg tend to be related only weakly to the position of functional groups, as opposed to the identity of the groups. Consequently, the properties of isomeric polymers which contain the same functional groups but with different geometry are usually predicted to be very similar.

In this paper, we describe how an artificial neural network approach has been employed to attack the glass transition temperature prediction problem. In order to obtain an effective network, descriptors that physical and chemical considerations suggest can be expected to be of relevance were selected by combining the AI methods mentioned above. Contribution information was collected without requiring explicit values for each contribution, thus avoiding the limitation which restricts the applicability of group contribution methods. Descriptors that relate to chemical bonding and to intermolecular bonding were included.

3 Glass Transition Temperature Database

A database of 71 polymers, represented by their repeating units, was compiled. Experimental data for polymer glass transition temperatures were taken from Van Krevelen [4], Zuniga and Trevino [5], Katritzky et al. [7], and suppliers' datasheets. Polymers have a high molecular weight because they contain many monomeric units, so the influence of the pair of end – cap groups is expected to be slight; these groups were therefore neglected in the calculations. Twenty one polymers were selected at random from the database for prediction testing and the remaining 50 polymers were used for network training.

The structure of each polymer was described by its constituent groups in both the main and the side chain. A linear correlation of Tg and constituent groups was developed by multiple linear regression analysis, expressed by:

$$T_g = \frac{\sum Y_i N_i}{\sum M_i N_i} + \text{constant}$$

In this equation, Y_i is the contribution of group i to the glass transition temperature, N_i is the number of group i in the repeating unit and M_i is the molecular weight of group i. The value of T_g calculated from this equation was used as one of the descriptors in the development of a model with 10 descriptors.

As reported in [8], Cao and Lin found that it was the size of the terminal group (R_{ter}) in the side chain rather than the total size of that chain that correlates with the glass transition temperature of polymers, when other factors are excluded. Therefore, two descriptors of the volume of the terminal group in the side chain, $MV(R_{ter})$, and the free length of the side chain (L_F) were introduced to study how the geometry of the side chain affected Tg. Individual and intermolecular bond energies were also considered.

The Free Energy change on melting is related to the mobility of the polymer chain and the intermolecular forces that exist between the chains. The main chain bond was recorded as the type of bond, such as C – C, C – O or C – N, in the backbone of the polymer. Hydrogen bonding is the strongest type of intermolecular bond and, when present, may substantially increase Tg. In the database the number of hydrogen bonds, the hydrogen bond energy and the density of hydrogen bonds were recorded. Nylon is a notable example of the influence of hydrogen bond density on Tg. In naming Nylons, the number after the name is the number of carbon atoms between two nitrogen atoms, which can participate in the formation of hydrogen bonds. As this number increases, the hydrogen bond density decreases, and Tg also decreases, from around 398 K for nylon 6 to roughly 310 K for nylon 12. In our database, aromatic rings in repeating units were recorded as hydrogen bond acceptors, and treated as a participant in a hydrogen bond. The density of hydrogen bonds in the database could then be expressed as:

$$Density = \frac{\min(\text{no. of hydrogen bonding donor, no. of hydrogen bonding acceptor})}{\text{no. of atoms in main chain}}$$

4 Model Development

A model was formed using a back propagation artificial neural network. 28 descriptors were classified into three groups: polymer structure descriptors, main chain bond descriptors and hydrogen bond descriptors. The goal was to find models that required a small number of descriptors, yet generated high R^2 values, in this way identifying the descriptors that are most strongly linked to Tg. This goal was found to be best met by replacing the constituent group descriptors by the pre-calculated Tg, reducing the number of descriptors to 10. This reduction in the total number of descriptors not only leads to a marginal improvement in processing speed, but can also lead to a more robust network.

As an alternative to the empirical approach described here, Principal Component Analysis (PCA) can also be used as an identification tool (Fig. 1) to pick out key descriptors. In this approach the set of descriptors is transformed mathematically from a space in which some descriptors may be correlated, or even redundant, into a new space in which a smaller number of orthogonal, and therefore uncorrelated, vectors exist. If several descriptors are found grouped together after the transformation, they can be replaced by a single new descriptor which accounts for all the variability in the descriptors that have been replaced. The use of PCA to reduce the size of a descriptor

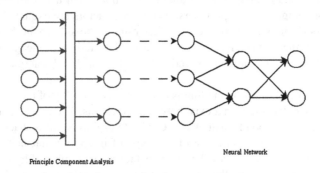

Principle Component Analysis   Neural Network

Fig. 1. Preliminary reduction of a descriptor set using Principal Component Analysis.

set is widespread, especially when the set that forms the starting point for an analysis is large (perhaps as many as several hundred). In the current study the number of starting descriptors is much smaller, so the PCA approach is not essential.

5 Results – Group Contributions Model

A set of randomly selected test polymers was studied. From the database of 71 polymers, 21 polymers (30 %) were selected at random to test the predictability of the neural network trained by the remaining 50 polymers. Two separate models were developed in the case study, one with 28 descriptors and a second with 10 descriptors. R^2 values for predictions were as follows: for 28 descriptors $R^2 = 0.89$, and for 10 descriptors $R^2 = 0.85$.

6 Glass Transition Temperatures for Terpolymers

In related studies, we have also investigated the use of ANNs to predict Tg for a series of terpolymers synthesized from varying ratios of the monomers n-octadecyl acrylate, ethyl acrylate and acrylonitrile. Tg for materials formed by co-polymerisation of these monomers varies significantly with composition, and shows a non-linear correlation with mole fraction, covering a range of 232.9–343.9 K for the polymers in our terpolymer database. Input to the ANN comprised composition of the polymer by mole fraction, while output was the calculated glass transition temperature.

7 Results – Glass Transition Temperatures for Terpolymers

A neural network working with this database initially showed little sign of learning. As Fig. 2 indicates, a negligible reduction in training error was evident over 500 epochs; in addition, during this period the network output showed almost no correlation

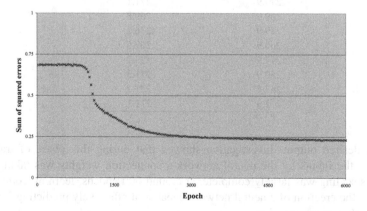

Fig. 2. Variation of network error with epoch.

Table 1. Experimental and predicted glass transition temperatures.

Experimental Tg	Predicted Tg
343.9	319.1
307.9	311.6
301.9	297.2
299.9	316.6
295.9	310.3
309.9	292.2
269.9	270.9
312.9	310.9
285.9	296.2
265.9	272.4
300.9	303.3
276.9	281.8
257.9	260.6
253.9	254.3
289.9	290.4
269.9	268.9
237.9	253.5
256.5	250.1
251.9	244.8
270.2	271.2
285.9	283.6
285.5	282.8
264.9	263.5
244.9	253.4
237.9	248.9
234.9	237.6
279.9	277.6
259.9	260.6
274.9	251.0
232.9	242.6
244.9	272.2
279.9	271.1
269.9	269.8
239.9	236.2
266.9	267.5
277.9	273.2
280.9	271.3
267.4	271.1
277.9	271.1

with the desired output. Investigation showed that during this phase of training a scaling of the inputs by the neural network's connection weights was taking place. Once this scaling was largely complete, at around 600 epochs, learning commenced, leading to the creation of a neural network capable of effectively predicting Tg values from terpolymer composition.

Table 2. Polymer composition predicted from glass transition temperature.

n–octadecyl acrylate fraction	Ethyl acrylate fraction	Acrylonitrile fraction	Error
0.726	0.050	0.223	1.007
0.304	0.664	0.031	1.292
0.326	0.061	0.611	0.175
0.603	0.013	0.383	0.606
0.514	0.160	0.325	0.563
0.404	0.220	0.375	0.387
0.288	0.205	0.505	0.188
0.159	0.364	0.476	0.122
0.652	0.014	0.333	0.603
0.452	0.274	0.272	0.384
0.098	0.624	0.276	0.240
0.179	0.458	0.361	0.154
0.079	0.479	0.441	0.086
0.059	0.284	0.655	0.009
0.363	0.376	0.259	0.213
0.451	0.213	0.335	0.245
0.367	0.351	0.281	0.118
0.378	0.193	0.428	0.084
0.147	0.373	0.479	0.010
0.483	0.287	0.229	0.122
0.033	0.489	0.476	0.031
0.112	0.577	0.310	0.033
0.177	0.113	0.708	0.213
0.288	0.210	0.501	0.088
0.295	0.383	0.321	0.045
0.455	0.464	0.079	0.232
0.183	0.327	0.489	0.118
0.064	0.446	0.489	0.013
0.045	0.512	0.442	0.005
0.051	0.135	0.812	0.259
0.494	0.135	0.369	0.158
0.222	0.295	0.482	0.012
0.372	0.132	0.494	0.489
0.709	0.208	0.081	0.574
0.215	0.753	0.031	0.129
0.246	0.554	0.199	0.023
0.501	0.093	0.405	0.290
0.071	0.250	0.677	0.207
0.716	0.115	0.167	0.351
0.137	0.421	0.440	0.045

A typical set of results is shown in Table 1, in which it is evident that the agreement between experimental and predicted glass transition temperatures is generally good.

The reverse process, in other words, using Tg as the sole input to an ANN and requiring prediction of the composition of the polymer, is considerably harder. This is because the inverse relationship between composition and Tg is not unique: a polymer of specified composition can have only one transition temperature, but by contrast one particular transition temperature may be associated with several polymers of different composition. To investigate the extent to which it might be possible to use Tg values to determine polymer composition, we have used a combined ANN-GA model, with the transition temperature as the sole input, and the terpolymer composition as output. The role of the GA here is to optimize the neural network's architecture (the number of hidden layers and the number of nodes within those layers) and the adjustable parameters that govern network learning, including the learning rate and the type of activation function [2]. The error in the prediction of fractional composition is given by:

$$E = \sum_k \left(prediction_k - target_k \right)^2$$

Some typical results appear in Table 2.

As the table shows the glass transition temperature is adequately predicted for a significant proportion of the samples. As we argued above, it would be unrealistic to expect that the composition of every sample could be predicted correctly from its Tg, no matter what type of model is used. Nevertheless, we believe that calculations of this type may be helpful in identifying those materials that have a glass transition temperature that is strongly influenced by composition (and which, therefore a neural network might reasonably be expected to learn) and other materials which have a Tg which is much less dependent on composition (and for which therefore there is little to guide the network towards the correct composition). In the former case composition prediction should be accurate and the error low, while in the latter case the error will be high.

8 Conclusions

Hydrogen bonding is an important factor in determining Tg of polymers. Thus, introducing hydrophilic groups, such as –OH, into a polymer structure should improve both moisture regain and the glass transition temperature. The optical performance of polymer fibers can be adjusted by engineering a polymer exhibiting a lower refractive index and Tg. As recommended by Ballato [3], it would be beneficial to develop perfluorinated polymers to obtain lower refractive index, since any C-H bonds will lead to increased adsorption. Refractive index can be predicted using the same approach as described in this paper, once refractive index data is included in the ANN training set. The approach used in this study could be used to obtain polymers with both low refractive index and Tg.

It is possible also to use ANNs to establish a direct link between composition and Tg without using group contribution methods. Prediction of Tg from terpolymer composition can be accomplished satisfactorily, while the reverse process, determining composition from a single measurement of Tg, is more difficult. Nevertheless, an analysis of the errors in such a calculation may help identify the regions in which Tg is most sensitive to changes in composition and the use of a GA to optimize the ANN for this calculation can reasonably be expected to improve the likelihood of valid predictions.

References

1. Cartwright, H.M.: Applications of Artificial Intelligence in Chemistry. Oxford University Press, Oxford (1994)
2. Curteanu, S., Cartwright, H.M.: Neural networks applied in chemistry. I. Determination of the optimal topology of multilayer perceptron neural networks. J. Chemom. **25**, 527–549 (2011)
3. Ballato, J., et al.: Theoretical Performance of Polymer Optical Fibers, NTC grant M01-CL01 (2001)
4. Van Krevelen, D.: Properties of Polymers. Elsevier, New York (1990)
5. Camacho-Zuniga, C., Ruiz-Trevino, F.: a new group contribution scheme to estimate the glass transition temperature for polymers and diluents. Ind. Eng. Chem. Res. **42**, 1530–1534 (2003)
6. Marrero, J., Gani, R.: Group-contribution based estimation of pure component properties. Fluid Phase Equilib. **183**, 183–208 (2001)
7. Katritzky, A.R., et al.: Quantitative structure-property relationship (QSPR) correlation of glass transition temperatures of high molecular weight polymers. J. Chem. Inf. Model. **38**, 300–304 (1998)
8. Chen, X., Sztandera, L.M., Cartwright, H.M.: A neural network approach to the prediction of glass transition temperature of polymers. Int. J. Intell. Syst. **23**(1), 22–32 (2008)

A Watermarking Scheme for Coding DNA Sequences Using Codon Circular Code

Suk-Hwan Lee[1(✉)], Eung-Joo Lee[2], Won-Joo Hwang[3],
and Ki-Ryong Kwon[4]

[1] Department of Information Security, Tongmyong University,
535, Yongdang-dong, Busan, South Korea
skylee@tu.ac.kr
[2] Department of Information & Communication Engineering,
Tongmyong University, 535, Yongdang-dong, Busan, South Korea
ejlee@tu.ac.kr
[3] Department of Information & Communications Engineering, Inje University,
197, Inje-Ro, Gimhae, South Korea
ichwang@inje.ac.kr
[4] Department of IT Convergence and Application Engineering,
Pukyong National University, 599-1, Daeyeon-dong, Busan, South Korea
krkwon@pknu.ac.kr

Abstract. This paper discusses DNA watermarking for authentication, privacy protection, and the prevention of illegal copying and mutation of DNA sequences. We propose a DNA watermarking scheme that provides mutation robustness and amino acid preservation. The proposed scheme selects a number of codons as the embedding target at the regular singularity in coding regions. The scheme then embeds the watermark in the watermarked codons such that the original codons are still transcribed as the same amino acids. From in silico experiments using HEXA and ANG sequences, we verified that the proposed scheme is more robust to silent and missense mutations than the conventional scheme, while it also preserves the amino acids of the watermarked codons.

Keywords: DNA watermarking · Coding DNA (cDNA) · Codon coding table · Amino acid preservation · Mutation attack

1 Introduction

The genetic code encoded by DNA contains profound personal information. It may be considered as a personal diary, and its disclosure may be considered as a grave invasion of privacy and violation of human rights. Legal measures have been established to ensure the safety and security of procedures for collecting human genetic information [1–3]. There are ethical laws or guidelines for HGI (human genome information) usage, but security techniques for preventing illegal copying and piracy of HGI are urgently required. DNA is considered as a new biometric storage medium for storing huge amounts of data, because 1 g of DNA can store 108 TB of data. Thus, DNA storage demands that DNA security techniques are addressed.

G.A. Di Caro and G. Theraulaz (Eds.): BIONETICS 2012, LNICST 134, pp. 208–219, 2014.
DOI: 10.1007/978-3-319-06944-9_15, © Institute for Computer Sciences, Social Informatics and Telecommunications Engineering 2014

Recent research in the area of cryptography [4, 5] includes information hiding by steganography and watermarking [7–16] using DNA/RNA sequences, where a character stream of A, G, T(or U), C, encrypts or hides the information of a GMO (genetically-modified organism). These studies were validated by in vivo or in vitro experiments, i.e., with whole living organisms or with isolated components of organisms [4, 5, 7–9, 11, 12], and by in silico experiments, i.e., via computer simulation [10, 13]. The genome contains all of an organism's hereditary information, and it includes coding sequences, which are translated to proteins, and noncoding sequences, which are not translated to proteins. The former is known as coding DNA or cDNA, whereas the latter is known as noncoding DNA or ncDNA. DNA steganography or watermarking methods can be designed differently depending on whether the information is embedded in cDNA [10–13] or ncDNA [7–9].

In this paper, we present a cDNA watermarking method that provides robustness to mutation and amino acid preservation. We analyze the method's performance using in silico-based experiments. Our method allocates all codons to integer values using a codon coding table before re-allocating them to floating numerical values with a circular angle using circular coding. Our method then selects a set of three consecutive codons based on the singularity detection of circular angles and it embeds the watermark into the angle differences of three codons, while preserving codon equivalence. The codon coding table and circular coding were newly designed for cDNA watermarking, unlike conventional genetic codes. The circular coding of numerical sequences makes the numerical transformation easier and it allows symbol errors in arbitrary positions to be estimated, before allocating codons that are translated to the same amino acid using neighboring numerical values. The performance results of our in silico experiments verified that this method ensured amino acid preservation and it was more robust to substitution, insertion, and deletion mutations compared with the conventional method.

2   Related Works

Watermarking can be classified as private and public, depending on the detection method. The first detects a watermark using the original sequence and the watermark, whereas the second detects the watermark without the original sequence and the watermark. Private watermarking has many problems in terms of security, so DNA watermarking must be public. In this paper, we present a public DNA watermarking method with amino acid preservation and robustness to mutations, and we analyze the capacity and security of our method. Research into DNA cryptography or steganography began in the late 1990s and DNA watermarking has been researched since the late 2000s.

Clelland et al. [4] presented the first DNA steganography method using DNA microdots and they implemented it in vitro. This method hides a secret message in DNA microdots based on a simple permutation cipher. However, this method requires the start and end primers to recover the message. Leier et al. [5] mixed binary-encoded plaintext with DNA dummy strands in DNA steganography. This method also required knowledge of the primer sequences. Anam et al. [6] reviewed DNA cryptography methods using PCR (Polymerase Chain Reaction) and DNA steganography methods using DNA or DNA chip technology. They found that the difficulties of DNA cryptography and steganography were due to the lack of a theoretical basis and practical methodologies that could be readily implemented in the field of information security. Further, most existing steganography methods are based on ncDNA sequences, and they cannot be applied to cDNA sequences using the genetic code.

There are several methods for embedding watermarks into ncDNA sequences. Yachie et al. [7] redundantly embedded duplicated data encoded by nucleotide sequences into multiple loci of the Bacillus subtilis genome. Heider et al. [8] applied the DNA-Crypt algorithm [9] to ncDNA sequences using in vivo experiments with the small cytoplasmic RNA I in yeast and the lac promoter region of Escherichia coli. However, their results showed that the watermark can deactivate promoter regions and it further affected the secondary structure of regulatory RNA molecules, although the watermarked RNA and one of the watermarked promoters had no significant differences in the wild type RNA and promoter regions. Therefore, they concluded that watermark embedding is not suitable for regulatory regions such as promoters or regulators.

The following are methods for embedding watermarks into cDNA sequences. Shimanovsky et al. [10] combined codon redundancy with arithmetic encoding and public key cryptography when embedding watermarks into cDNA sequences. They allocated four bases to 2-bit binary and converted the binary sequence to a decimal number from 0 to 1, which was used as the target number in repeated subdivision steps of arithmetic encoding. However, this method failed to accurately extract the target number in each step if any codons were mutated. Arita et al. [11] embedded a short signature into the cDNA sequence of B. subtilis bacteria based on the degenerate genetic code and a permutation cipher function. However, this method required the original sequence to recover the embedded message. Heider et al. [12] inserted encrypted information into cDNA sequences of the yeast S. cerevisiae using the DNA-Crypt algorithm [9]. This method only produced the protein profile because they did not consider the redundancy of the genetic code. Shuhong et al. [13] permuted a third base symbol for codons according to a bit of the watermark bit, which was similar to LSB (Least significant bit) permutation based on Arita's genetic code. This method can only be applied in cases where the cardinality of bases in the amino acid is four. Balado et al. [14–16] used Shannon's theory to model the maximum capacity of data with mutation robustness.

Conventional cDNA watermarking methods have some problems with amino acid preservation and their robustness to substitution, deletion, and insertion mutations.

3   Proposed DNA Watermarking

Figure 1 shows the proposed embedding process, which consists of the numerical mapping of codons using a codon coding table, the selection of target codons by DWT-based singularity detection, bit embedding into target codons, and generation of the watermarked DNA sequence. The index of target codons is used as a key for extracting the watermark. The watermark is extracted using a similar process to

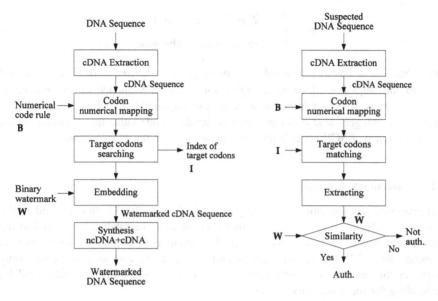

Fig. 1. The proposed process of watermark embedding and extracting.

embedding, as shown in Fig. 1. The main notations used in our paper are as follows. A nucleotide base is $b = \{G, A, C, T\}$ and a codon of triplet bases is $c = b_1b_2b_3$. An amino acid of a codon c is $S = f(c)$ and $f|C \rightarrow S$ is the translation from codons to amino acids. $|S|$ is the number of codons that are translated to an amino acid S, which is known as the cardinality of an amino acid S.

3.1 Numerical Coding

The proposed method uses a codon coding table that is suitable for a watermarking system and it maps 64 codons to integer values of 6-bits based on this table, before converting them to circular angles. First, we set $b = \{G, A, C, T\}$ to $\{0, 1, 2, 3\}$ and we map a codon $c = b_1b_2b_3$ to the integer value n of 6-bits using a polynomial expression.

$$n = 4^2 \times b_1 + 4^1 \times b_2 + 4^0 \times b_3 \tag{1}$$

We then transform n to the circular angle $g(c)$; $g(c) = \frac{2n\pi}{64}$ or $\frac{2n\pi}{64} - 2\pi$. $g(c)$ is allocated to an angle such that the difference from the initial codon angle $g(c_{-1})$ of two angles is small.

$$g(c) = \begin{cases} \frac{2n\pi}{64}, & if \left|\frac{2n\pi}{64} - g(c_{-1})\right| < \left|\frac{2n\pi}{64} - 2\pi - g(c_{-1})\right| \\ \frac{2n\pi}{64} - 2\pi, & if \left|\frac{2n\pi}{64} - g(c_{-1})\right| > \left|\frac{2n\pi}{64} - 2\pi - g(c_{-1})\right| \end{cases} \tag{2}$$

Codon c can be easily recovered from a circular angle $g(c)$ using the inverse mapping. From Eq. (3), the integer value n can be obtained as follows.

$$n = \begin{cases} \frac{64}{2\pi} g(c), & if g(c) > 0 \\ \frac{64}{2\pi} (2\pi + g(c)), & \text{otherwise} \end{cases} \tag{3}$$

Three bases of a codon $c = b_1 b_2 b_3$ are $b_1 = \frac{n}{4^2}$, $b_2 = \frac{n \% 4^2}{4}$, $b_2 = (n \% 4^2) \% 4$. We refer to the sequence of circular angles for a codon sequence as the numerical sequence. Thus, a codon is coded to n by the sequential mapping of three bases. Because the integer value n of the codon is simply a code value that is similar to the decimal point in ASCII code, it has nothing to do with the physical quantity.

3.2   Watermark Embedding

The proposed method embeds a binary watermark $w_i = \{0, 1\}$ into selected triplet codons $C_i = \{c_{k-1}, c_k, c_{k+1}\}$. The triplet codons can be randomly selected as the embedding target. However, we detect the singularity codons in the numerical sequence using DWT hard-thresholding and we use these codons as the embedding target. In this section, we explain two parts of codon searching and individual bit embedding for the watermark embedding.

Target Codon Searching. The numerical sequence of codon circular angles appears to be random noise. The proposed method removes irregular perturbations in the sequence and then searches for codons on regular singularities to locate the embedding target. Thus, we compute 5-level DWT transform coefficients, $W(g(c)) = G(c)$, for the numerical sequence $g(c)$ and we perform hard-thresholding according to Birgé-Massart strategy, which produces numerical sequence without irregular perturbations, $W^{-1}(G(c)) = g'(c)$, using the inverse DWT transform. Daubechies wavelet filter was used for DWT in this study.

We define the local maxima, LM, in the thresholded sequence $g'(c)$ as follows.

$$LM(c_k) = \begin{cases} g'(c_k), & \text{if } (l_k > th \text{ and } r_k > th) g'(c_{k-1}) \neq LM \\ 0, & \text{otherwise} \end{cases} \tag{4}$$

where $th = \alpha \times m$

Thus, the local maxima is the numerical value $g'(c)$ where all cardinalities of amino acids for a codon and its left and right codons are not one, such as $|S(c_{k-1})| > 1$, $|S(c_k)| > 1$, $|S(c_{k+1})| > 1$, while the numerical values of their codons are not local maxima and the numerical differences with left and right codons are above th. Thus, the codon 'TGG' of amino acid $S = 'W'$ of $|S| = 1$ is not selected. A variable m for the threshold th is the average of the numerical differences with neighbor codons.

$$m = \frac{1}{N-2} \sum_{k=2}^{N-1} \frac{l_k + r_k}{2}, \tag{5}$$

where $l_k = |g'(c_k) - g'(c_{k-1})|, r_k = |g'(c_k) - g'(c_{k+1})|$

α is a pre-adjusted factor. The number of watermark bits is determined by adjusting factor α and it is the same as the number of local maxima, $|LM|$.

Triplet codons for the local maxima codon and two neighboring codons are used as the embedding target; $C_i = \{c_{k-1}, c_k, c_{k+1}\}$ where $LM(c_{k-1}) = LM(c_{k+1}) = 0$ and $LM(c_k) = g'(c_k)$.

Bit Embedding. All bits of watermark, $\{w_i | i \in [1, |LM|]\}$, are embedded into the selected triplet codons, $\{C_i = \{c_{k-1}, c_k, c_{k+1}\} | i \in [1, |LM|]\}$, one by one, while ensuring amino acid preservation. In a triplet codon, a center codon c_k is moved to codons with maximum or minimum numerical values in the amino acid $S(c_k)$ of c_k according to the bit w_i. Left and right codons, c_{k-1} and c_{k+1}, are moved to codons with maximum or minimum distances, in clockwise and counterclockwise directions, with numerical values according to the bit w_i.

Circular angles of amino acids of triplet codons $C_i = \{c_{k-1}, c_k, c_{k+1}\}$ are denotes by $\{g(S_{k-1}), g(S_k), g(S_{k+1})\}$. We compute the average circular angle R_k for codons in the amino acid S_k based on the center codon c_k, $R_k = \frac{1}{|S_k|} \sum_{j=1}^{|S_k|} g(c_{k,1})$ and we use it as the reference value of the triplet codon. Based on the diagram of the circular angle, we can define the clockwise distance $d \, \circlearrowright$ between R_k and the circular angles of codons in S_{k-1},

$$d \, \circlearrowright (R_k, g(c_{k-1,j})) = |g(c_{k-1,j}) - R_k|, \ \forall j \in [1, |S_{k-1}|]$$

(6)

and the counterclockwise distance $d \, \circlearrowleft$ between R_k and the circular angles of codons in S_{k+1}

$$d \, \circlearrowleft \left(R_k, g(c_{k+1,j})\right) = 2\pi - |g(c_{k+1,j}) - R_k|, \ \forall j \in [1, |S_{k+1}|].$$

(7)

The codons with maximum and minimum clockwise distances in S_{k-1} will be

$$c_{k-1,max} = \arg_c \max_{c_{k-1}} d \, \circlearrowright (R_k, g(c_{k-1,j})),$$

(8)

$$c_{k-1,min} = \arg_c \min_{c_{k-1}} d \, \circlearrowright (R_k, g(c_{k-1,j}))$$

(9)

and codons with maximum and minimum counterclockwise distances in S_{k+1} will be

$$c_{k+1,max} = \arg_c \max_{c_{k+1}} d \, \circlearrowleft (R_k, g(c_{k+1,j})),$$

(10)

$$c_{k+1,min} = \arg_c \min_{c_{k+1}} d \, \circlearrowleft (R_k, g(c_{k+1,j})).$$

(11)

From Eq. (9), the maximum and minimum circular angles in S_k are

$$c_{k,max} = g(c_{k,|S_k|}),$$

(12)

$$c_{k,min} = g(c_{k,1}).$$

(13)

Finally, the watermarked triplet codons can be computed by $C_i' = \{c_{k-1}', c_k', c_{k+1}'\}$.

$$C_i' = \begin{cases} \{c_{k-1,min}, c_{k,min}, c_{k+1,min}\}, & \text{if } w_i = 0 \\ \{c_{k-1,max}, c_{k,max}, c_{k+1,max}\}, & \text{if } w_i = 1 \end{cases} \quad (14)$$

Using the above process, all bits of the watermark can be embedded into triplet codons in order.

3.3 Watermark Extracting

The process that extracts the watermark from received or pirated DNA sequence is similar to the embedding process, as shown in Fig. 2(b). During the extracting process, we generate the numerical sequence of circular angles from cDNA sequences and we detect codons at the local maxima in the DWT hard-thresholded sequences. We then align the codon index of local maxima I' to synchronize with the received index I and we obtain the embedded triplet codons, $C_i^* = \{c_{k-1}^*, c_k^*, c_{k+1}^*\}$. DNA sequences mutated by deletions or insertions are shifted the most. In this case, the alignment can synchronize I' by shifting unmatched codons while matching codons at the local maxima I' and I, codon by codon.

When extracting a watermark bit in a triplet codon C_i^*, we compute the amino acids, $f(c_{k-1}^*) = S_{k-1}^*, f(c_k^*) = S_k^*, f(c_{k+1}^*) = S_{k+1}^*$, circular angles, $g(S_{k-1}^*), g(S_k^*)$, $g(S_{k+1}^*)$, and the average angles, $\bar{g}(S_{k-1}^*), \bar{g}(S_k^*) = R_k^*, \bar{g}(S_{k+1}^*)$, for each codon.

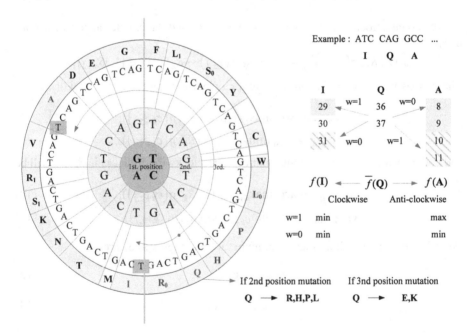

Fig. 2. Embedding example; Embed a watermark bit $w_i = 1$ into a triplet codons $C_i = \{ATC, CAG, GCC\}$.

Further, we compute the clockwise distance on R_k^* with a circular angle $g(c_{k-1}^*)$, $d \circlearrowleft (R_k^*, g(c_{k-1}^*))$ and the counterclockwise distance on R_k^* with a circular angle $g(c_{k+1}^*)$, $d \circlearrowright (R_k^*, g(c_{k+1}^*))$. The three bits of triplet codons can be computed by comparing R_k^* and the distances.

$$w_{k-1}^* = \begin{cases} 0, & \text{if } d \circlearrowleft \left(R_k^*, g(c_{k-1}^*)\right) < d \circlearrowleft \left(R_k^*, \bar{g}(S_{k-1}^*)\right) \\ 1, & \text{otherwise} \end{cases} \tag{15}$$

$$w_k^* = \begin{cases} 0, & \text{if } g\left(c_{k-1}^*\right) < R_k^* \\ 1, & \text{otherwise} \end{cases} \tag{16}$$

$$w_{k+1}^* = \begin{cases} 0, & \text{if } d \circlearrowright \left(R_k^*, g(c_{k+1}^*)\right) < d \circlearrowright \left(R_k^*, \bar{g}(S_{k+1}^*)\right) \\ 1, & \text{otherwise} \end{cases} \tag{17}$$

Therefore, a watermark bit w_i^* can be determined based on the majority rule of three bits.

$$w_i^* = floor\left(\frac{w_{k-1}^* + w_k^* + w_{k+1}^*}{3}\right) \tag{18}$$

All bits of watermark W^* can be computed using the above process. Whether a pirated DNA sequence is copied can be determined based on the similarity or bit error rate (BER) of the original watermark W and the extracted watermark W^*.

4   Experimental Results

Our in silico experiment used *Homo sapiens* CDSs (coding sequences) from the NCBI database. We evaluated the proposed method and the DNA-Crypt-based watermarking method of Heider [9, 12]. We set the pre-adjusted factor α in Eq. (7) to 0.1 and we selected the embedded triplet codons. In the DNA-Crypt method of Heider, we used an 8/4 Hamming-code for error correction and we embedded the watermark into all codons that were translated into amino acids {G,A,V,T,R,P,L,S} and which contained over four codons. Our experiment was performed using the Matlab Bioinformatics toolbox 3.

Our experiment tested intentional silent and missense mutations to evaluate the robustness. Silent mutation experiments randomly selected γ % of all codons in the CDS and substituted them with any codons for the same amino acid. A silent mutation in any of the selected codons is mutated with a probability that is the inverse of the codon cardinality of the amino acid. In our experiment we varied γ from 10 % to 100 %. Figure 3(a, b) show the mutation rate for codons and nucleotide bases in the HEXA and ANG sequences, depending on the selected codon rate γ. The mutation rate of nucleotide base was about one third of the mutation rate of codons, because one base was changed in most of the silent mutated codons. Based on this figure, we know that the mutation rate of codons was about 66–77 % when $\gamma = 100$ % and about 34 %

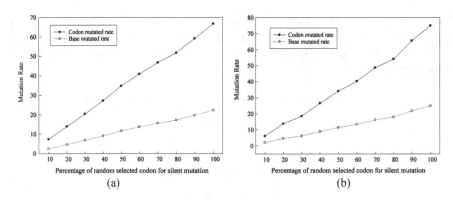

Fig. 3. Mutation rate of codon and nucleotide base of silent mutated (a) HEXA(NM_000520) and (b) ANG(NM_001145).

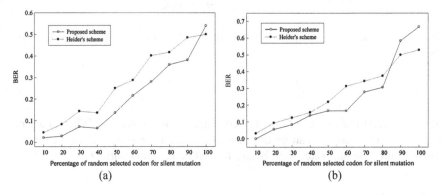

Fig. 4. BERs of watermarks extracted from silent mutated (a) HEXA and (b) ANG sequences.

when $\gamma = 50$ %. The experimental results are shown in Fig. 4(a) and (b), i.e., the BERs of the extracted watermarks. These figures show that the BER of the proposed method was less than 1.15–2.89 times that of the BER when using the Heider method. For example, when $\gamma = 50$ % (codon mutation rate of 34 %), the BER of the proposed method was about 0.1367–0.1667 whereas the BER of Heider's method was 0.21–0.25. Based on these results, we can conclude that the proposed method was more robust than Heider's method.

The missense mutation experiment randomly selected γ % of all codons and substituted them with any codons that were randomly selected from 64 codons. The probability that the selected codons were not mutated was about 1/64. One, two, or three nucleotide bases in the codons could be substituted. The codon mutation rate was the same as γ. Our experiment conducted the missense mutation by varying γ from 10 % to 50 %. The experimental results of missense mutation are shown in Fig. 5(a) and (b). When γ was 10–30 %, the BERs of both methods were less than 0.1. However, when γ was as high as 40–50 %, the BER of our method was 0.1111–0.1583 whereas the BER of Heider's method was 0.1563–0.2576, which was slightly higher

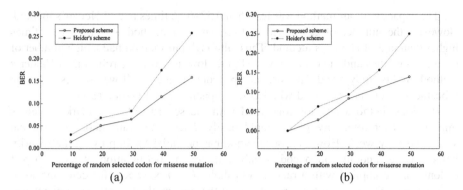

Fig. 5. BERs of watermarks extracted from missense mutated (a) HEXA and (b) ANG sequences.

than our method. These results verified that our method was more robust to missense mutations than Heider's method.

The watermark capacity data that is embedded into the CDS should be determined with consideration for amino acid preservation and robustness. Balado et al. [14–16] studied the maximum capacity that can be embedded into the non-CDS and CDS of a DAN sequence. We also investigated the capacity of our method and Heider's method, which was based on Balado's analysis.

Our method embeds a bit of the watermark into a triplet codon. If all codons except for the start and stop codons are grouped as a set of triplet codons, the maximum capacity of watermark bits, C_{max}, is $C_{max} = 1/3$ [bit/codon]. Our method uses the local maxima of DWT hard-thresholded codon sequences as the embedding targets to improve the security and detect codon singularities. The watermark capacity can be determined based on the number of local maxima $|LM|$, which depends on the factor α. Therefore, the watermark capacity of our method is C

$$C = \frac{1}{3} \times \frac{|LM|}{(|C| - 2)/3} = \frac{|LM|}{|C| - 2} \text{ [bit/codon]} \tag{19}$$

where $|C| - 2$ is the number of codons in the CDS, except for the start and end codons.

Heider's method encodes the watermark bits using an 8/4 Hamming-coding before it embeds the watermark as a unit of 2-bits into codons where the codon length is greater than four, $|S| \geq 4$. Although they used a mutation correction code of 8/4 Hamming-code and a WDH-code with a fuzzy controller, we used an 8/4 Hamming-code in this paper because it was very compact. Therefore, the watermark capacity of Heider's method is C

$$C = \frac{1}{2} \times \frac{2|Z|}{|C| - 2} = \frac{|Z|}{|C| - 2} \text{ [bit/codon]} \tag{20}$$

when $|Z|$ was the number of codons that were translated to {G,A,V,T,R,P,L,S}, where the length of codon was greater than four.

The capacity of our method was less than 0.52–0.56 times that of Heider's method. However, the number of embedded codons with our method was 1.68–1.95 times higher than that of Heider's method. Thus, although our method had a high number of embedded codons and the capacity was slightly low, it was more robust than Heider's method. In this study, we did not use an error correction code. If we use this code, the robustness would be improved whereas the capacity would be decreased.

Previous methods are not concerned with the security of watermark, which is important. Therefore, any pirate could easily detect the watermark because the security is very weak. However, our method has potential for improving the security based on the random codon coding table, where the permutation of target triplet codons can be applied with a random threshold $th = r \times m \times \alpha$ for detecting local maxima, etc. The experimental results show that our method is robust to silent and missense mutations and it has the potential for improving security. Our method can also increase the capacity by varying factor α.

5   Conclusions

The main requirements of cDNA watermarking are amino acid preservation and mutation robustness. This paper presented a cDNA watermarking method that satisfies these two requirements. The main features of the proposed method are a codon coding table, the numerical mapping of codon sequences using circular angle, singularity detection based on local maxima, and watermark embedding to ensure amino acid preservation. The codon coding table and numerical mapping was designed to be suitable for processing DNA watermarking. The watermark was robust to silent and missense mutations by embedding it into the difference of circular angles of adjacent codons. We evaluated the robustness to silent and missense mutations, the amino acid preservation, and the data capacity of the proposed method with HEXA and ANG sequences. The experimental results showed that the proposed method had superior robustness to two types of mutations and it ensured amino acid preservation. Our method has potential for improving watermark security and it also gave control over the data capacity, although it had a low capacity. In the future, we will quantitatively evaluate DNA watermark security, and we will generate a watermarked organism using our method.

Acknowledgement. This work was supported by the Korea Research Foundation Grant funded by the Korean Government (MEST)(KRF-2011- 0023118 and KRF-2011-0010902) and Tongmyong University Research Grants 2012 (2012-A001).

References

1. Sankar, P.: Genetic privacy. Annu. Rev. Med. **54**, 393–407 (2003)
2. Springer, J.A., Beever, J., Morar, N., Sprague, J.E., Kane, M.D.: Ethics, Privacy, and the Future of Genetic Information in Healthcare Information Assurance and Security. Information Assurance and Security Ethics in Complex Systems: Interdisciplinary Perspectives, 1st edn, pp. 186-205. IGI Global, Hershey (2010)

3. National Conference of State Legislatures: Genetic Privacy Laws, Genetic Information: Legal Issues Relating to Discrimination and Privacy (2008). http://www.ncsl.org
4. Clelland, C.T., Risca, V., Bancroft, C.: Hiding messages in DNA microdots. Nature **399**, 533–534 (1999)
5. Leier, A., Richter, C., Banzhaf, W., Rauhe, H.: Cryptography with DNA binary strands. Biosystems **57**, 13–22 (2000)
6. Anam, B., Sakib, K., Hossain, M.A., Dahal, K.: Review on the advancements of DNA cryptography. In: 4th International Conference on Software, Knowledge, Information Management and Applications (2010)
7. Yachie, N., Sekiyama, K., Sugahara, J., Ohashi, Y., Tomita, M.: Alignment-based approach for durable data storage into living organisms. Biotechnol. Prog. **23**, 501–505 (2007)
8. Heider, D., Barnekow, A.: DNA watermarks in non-coding regulatory sequences. BMC Bioinform. **2** (2009)
9. Heider, D., Barnekow, A.: DNA-based watermarks using the DNA-Crypt algorithm. BMC Bioinform. **8** (2007)
10. Shimanovsky, B., Feng, J., Potkonjak, M.: Hiding data in DNA. In: Proceedings of the 5th International Workshop in Information Hiding, pp. 373–386 (2002)
11. Arita, M., Ohashi, Y.: Secret signatures inside genomic DNA. Biotechnol. Prog. **20**, 1605–1607 (2004)
12. Heider, D., Barnekow, A.: DNA watermarks - a proof of concept. BMC Bioinform. **9** (2008)
13. Shuhong, J., Goutte, R.: Code for encryption hiding data into genomic DNA of living organisms. In: 9th International Conference on Signal Processing (ICSP), pp. 2166–2169 (2008)
14. Balado, F.: On the Shannon capacity Of DNA data embedding. In: IEEE International Conference on Acoustics Speech and Signal Processing (ICASSP), pp. 1766–1769 (2010)
15. Haughton, D., Balado, F.: Performance of DNA data embedding algorithms under substitution mutations. In: IEEE International Conference on Bioinformatics and Biomedicine Workshops (BIBMW), pp. 201–206 (2010)
16. Balado, F.: On the embedding capacity of DNA strands under insertion, deletion and substitution mutations. In: Proceedings of SPIE 7541, Media Forensics and Security XII, San José, USA (2010)

Optimization

Ant Local Search for Combinatorial Optimization

Nicolas Zufferey[(⊠)]

GSEM - University of Geneva, 1211 Geneva 4, Switzerland
n.zufferey@unige.ch

Abstract. In ant algorithms, each individual ant makes decisions according to the *greedy force* (short term profit) and the *trail system* based on the history of the search (information provided by other ants). Usually, each ant is a *constructive* process, which starts from scratch and builds step by step a complete solution of the considered problem. In contrast, in *Ant Local Search* (ALS), each ant is a local search, which starts from an initial solution and tries to improve it iteratively. In this paper are presented and discussed successful adaptations of ALS to different combinatorial optimization problems: graph coloring, a refueling problem in a railway network, and a job scheduling problem.

Keywords: Ant algorithms · Local search · Combinatorial optimization

1 Introduction

As exposed in [15], most ant algorithms are population based methods where at each *generation*, a set of ants provide solutions, and at the end of each generation, a central memory (the trail system) is updated. The role of each ant is to build a solution step by step from scratch. At each step, an ant adds an element to the current partial solution. Each *decision* or *move* m is based on two ingredients: the *greedy force* $GF(m)$ (short term profit for the considered ant, also called the *heuristic information*) and the *trail* $Tr(m)$ (information obtained from other ants). Let M be the set of all possible decisions. The probability $p_i(m)$ that ant i chooses decision m is given by

$$p_i(m) = \frac{GF(m)^\alpha \cdot Tr(m)^\beta}{\sum\limits_{m' \in M_i(adm)} GF(m')^\alpha \cdot Tr(m')^\beta} \tag{1}$$

where α and β are parameters, and $M_i(adm)$ is the set of admissible decisions that ant i can make. When each ant of the population has built a solution, the trails are generally updated as follows: $Tr(m) = \rho \cdot Tr(m) + \Delta Tr(m)$, $\forall m \in M$, where $\rho \in]0,1[$ is a parameter representing the evaporation of the trails (often close to 0.9), and $\Delta Tr(m)$ is a term reinforcing the trails left on decision m by the ant population of the current generation. That quantity is usually proportional

G.A. Di Caro and G. Theraulaz (Eds.): BIONETICS 2012, LNICST 134, pp. 223–236, 2014.
DOI: 10.1007/978-3-319-06944-9_16, © Institute for Computer Sciences, Social Informatics and Telecommunications Engineering 2014

to the number of times the ants selected decision m, and to the quality of the obtained solutions when decision m was made. More precisely, let N be the number of ants, then $\Delta Tr(m) = \sum_{i=1}^{N} \Delta Tr_i(m)$, where $\Delta Tr_i(m)$ is proportional to the quality of the solution provided by ant i if it has selected decision m. Overviews of ant algorithms (including *ant colony optimization*) are [3,5].

Often, in order to get competitive results, it is unavoidable to apply a local search method to the solutions provided by such *constructive* ants [6]. In contrast, as proposed in [15], a more important role can be given to each ant by considering each of them as a *local search*, where at each step, as in every ant algorithm, the considered ant makes a decision (i.e. performs a move) according to the greedy force and the trail. The resulting method is called *Ant Local Search* (ALS).

The paper is organized as follows. In Sect. 2 are briefly described the main elements of a local search and the ALS methodology. Then are presented successful adaptations of ALS to three combinatorial optimization problems, namely the graph coloring problem (Sect. 3), a refueling problem in a railway network (Sect. 4), and a job scheduling problem with setup, tardiness and abandon issues (Sect. 5). The paper ends up with a conclusion in Sect. 6.

The contribution of this paper is the following: some advantages of the ALS approach are accurately highlighted; it is showed that ALS is a flexible method, as it can be easily adapted to very different combinatorial optimization problems; a new ALS algorithm is proposed for the job scheduling problem (Sect. 5); guidelines are given to efficiently design a trail system within an ALS framework.

2 Ant Local Search (ALS)

A *local search* can be described as follows. Let f be an objective function which has to be minimized. At each step, a *neighbor* solution s' is generated from the current solution s by performing a specific modification on s, called a *move*. Let $N(s)$ denote the set of neighbor solutions of s. First, a local search needs an initial solution s_0 as input. Then, the algorithm generates a sequence of solutions $s_1, s_2, \ldots$ in the search space such that $s_{r+1} \in N(s_r)$. The process is stopped for example when an optimal solution is found (if it is known), or when a fixed number of iterations have been performed. Some famous local search algorithms are: the descent method (where at each step the best move is performed, and the process stops when a local optimum is reached), simulated annealing, variable neighborhood search, and tabu search. In tabu search, when a move is performed from a current solution s_r to a neighbor solution $s_{r+1} \in N(s_r)$, it is forbidden (with some exceptions) to perform the inverse of that move during tab (parameter) iterations: such forbidden moves are called $tabu$ moves. The solution s_{r+1} is computed as $s_{r+1} = \arg\min_{s \in N'(s_r)} f(s)$, where $N'(s)$ is a subset of $N(s)$ containing solutions which can be obtained from s by performing a non tabu move. Many variants and extensions of tabu search can be found for example in [7].

The ALS method is summarized in Algorithm 1, where N is the number of used ants. A *generation* consists in performing steps (1) and (2).

Algorithm 1. ALS

While a time limit is not reached, do

1. **for** $i = 1$ **to** N: apply the local search associated with ant i, and let s_i be the resulting solution;
2. update the trails by the use of a subset of $\{s_1, \ldots, s_N\}$.

In most ant algorithms, it is very time consuming to make a single decision according to Eq. (1). For this reason, a quick way to *select* a move, based on the greedy forces and the trails, is proposed in [15] and described below (the advantages of such a selection process are more deeply discussed here). At each iteration of the local search associated with the considered ant, let A be the set of moves with the largest greedy force (resp. trail) values. Then, the selected move is the one in A with the largest trail (resp. greedy force) value (ties are broken randomly). Of course, this process is only interesting if $|A| > 1$, otherwise the trails (resp. greedy forces) will have no impact on the search. Such a way of selecting each move at each iteration leads to several advantages over most classical ant algorithms. More precisely, it is not required anymore to:

- compute the trails (resp. greedy forces) of *all* possible moves, as only $|A|$ computations are required (note that even if the trails can be stored in a matrix which is only updated at the end of each generation, the computation of the trail of a move often needs additional specific computation, as illustrated in the three next sections);
- *normalize* the greedy forces and the trails of the possible moves (without normalization, during the search, the range of the trail values might become much larger than the range of the greedy force values, which makes the search difficult to control);
- compute the probability $p_i(m)$ associated with each possible move m;
- consider and tune the parameters α and β (as the use of Eq. (1) is avoided).

In other words, in contrast with most other ant algorithms, the greedy forces and the trails are *successively* used to make a decision (instead of *jointly*). Therefore, a significant amount of computing time is saved, and the tuning phase of the algorithm is reduced.

3   ALS for Graph Coloring

3.1   Presentation of the Problem

Given a graph $G = (V, E)$ with vertex set $V = \{1, 2, \ldots, n\}$ and edge set E, the *graph coloring problem* (GCP) [21] consists in assigning an integer (called *color*) in $\{1, 2, 3, \ldots, n\}$ to every vertex such that two adjacent vertices have different colors, while minimizing the number of used colors. The *k-coloring problem* (*k*-GCP) consists in assigning a color in $\{1, \ldots, k\}$ to every vertex such that two adjacent vertices have different colors. Thus, the GCP consists in finding a k-coloring with the smallest possible k. The GCP is usually tackled by solving a series of k-GCP's, starting with a large value of k (which is at most n) and decreasing k by one unit each time a k-coloring is found. In such a case, a solution is often represented by a partition of the vertices into k color classes, *conflicts* are allowed (i.e. adjacent vertices can have the same color), and the goal consists in minimizing the number of conflicts (if it reaches zero, a k-coloring is found and the algorithm stops). Many (meta) heuristics were proposed to solve the GCP and the k-GCP. For a recent survey, the reader is referred to [12]. As discussed in [21], there mainly exists three types of ant algorithms for the GCP. First and as in most of the cases, an ant can be a *constructive* heuristic [4]. Second, an ant can be a very simple *agent* which helps to make a minor decision [8]. Third, an ant can be a refined *local search* such as tabu search [15]. Even if the role of an ant can be defined in various ways, each decision is always based on the *greedy force*, which is associated with the self-adaptation of each ant, and the *trail system*, which represents the collaboration between the ants.

3.2 Adaptation of ALS

The ALS coloring method proposed in [15] is derived from *PartialCol* [2], an efficient tabu search algorithm for the k-GCP, where *partial legal k-colorings* are considered, which are defined as conflict-free k-colorings of a subset of vertices of G. Such colorings are represented by a partition of the vertex set into $k + 1$ subsets $V_1, \ldots, V_{k+1}$, where $V_1, \ldots, V_k$ are k disjoint color classes without any conflict, and V_{k+1} is the set of non colored vertices. V_c (with $c \leq k$) actually represents the set of vertices with color c. The objective is to minimize $|V_{k+1}|$ (if it reaches zero, a k-coloring is found and the algorithms stops). A neighbor solution can be obtained from the current solution by moving a vertex v from V_{k+1} to a color class V_c (with $c \leq k$, which means that vertex v gets color c), and by moving to V_{k+1} each vertex in V_c that is in conflict with v (such vertices are thus uncolored). Such a move m is denoted $m = (v \rightarrow V_c)$. When it is performed, it is then tabu to move v back to V_{k+1} (i.e. to remove the color c from vertex v) for a few iterations.

In ALS for the k-GCP, an ant is a tabu search procedure derived from *PartialCol*. The greedy force $GF(m)$ of a move $m = (v \rightarrow V_c)$ is defined as the inverse of the number of adjacent vertices to v that are in color class V_c (if it is zero, $GF(m)$ is set to an arbitrary large number, because there is then no need to remove the color of other vertices). The trail value $Tr(m)$ associated with move m is defined as follows. Let x and y be two vertices, and let $s_i = (V_1, \ldots, V_k; V_{k+1})$ be a solution provided by a single ant i of the population at a specific generation. If ant i gives the same color c to x and y in solution s_i (i.e. $x, y \in V_c \neq V_{k+1}$), such an information should be transmitted to the ants of the next generations, and this information should be more important if x and y are in a large color class. During the search, a non colored vertex $v \in V_{k+1}$ is likely to move to a color class V_c containing vertices with which v is used to have the same color.

Formally, let

$$\Delta Tr_i(x, y) = \begin{cases} |V_c|^2 & \text{if } x \text{ and } y \text{ have the same colour } c \text{ in } s_i; \\ 0 & \text{if } x \text{ and } y \text{ have different colours in } s_i. \end{cases}$$

At the end of each generation and as in many classical ant algorithm, the trails are globally updated as follows: $Tr(x, y) = 0.9 \cdot Tr(x, y) + \Delta Tr(x, y)$, where $\Delta Tr(x, y) = \sum_{i=1}^{N} \Delta Tr_i(x, y)$. Finally, the trail of a single move $m = (v \rightarrow V_c)$ is defined as $Tr(v \rightarrow V_c) = \sum_{x \in V_c} Tr(v, x)$.

3.3 Results

Considering a set of 14 well-known and difficult benchmark instances (see http://www.info.univ-angers.fr/pub/porumbel/graphs/), below is a representative numerical comparison for the following coloring algorithms: CAS [4], a *Constructive Ant System* where each ant is a constructive procedure (as in most classical ant algorithms); ADS [8], an *Ant Decision System* where each ant can help to color a vertex; ALS [15], the already discussed *Ant Local Search*; *PartialCol* [2], the tabu search from which ALS was derived; *Mem* [11], a memetic algorithm which can be considered as the best coloring method, as it provides the best average results on each of the benchmark instances.

A time limit of one hour and the same computer were used for *PartialCol* and the three ant coloring algorithms (i.e. CAS, ADS and ALS). The experimental conditions of *Mem* were different (e.g., a time limit of five hours [11]). The reader interested in accurate and detailed comparisons of coloring methods is referred to [21] for ant algorithms, and to [11] for other state-of-the-art algorithms. The *density* $d \in [0, 1]$ of a graph is the average number of edges between two vertices. The results are summarized in Table 1. For each graph (first column) is mentioned its number n of vertices (second column) and its density d (third column). In the next columns are given the minimum number of colors used by each method to generate conflict-free colorings. Obviously, ALS performs much better than the other ant coloring methods. Secondly, it is clear that the ingredients added to *PartialCol* to derive ALS are useful. Finally, in contrast with ALS, CAS and ADS are not competitive with the best coloring methods. The main reasons are probably the following: in CAS and ADS, too many ingredients are *simultaneously* handled in order to make the decision of changing the color of a single vertex, and these ingredients are of different *natures* and should not be mixed together. On the contrary, ALS only manipulates, *successively*, a few ingredients.

4 ALS for a Refueling Problem in a Railway Network

4.1 Presentation of the Problem

The problem proposed in the 2010 INFORMS optimization competition consists in optimizing the refueling costs of a fleet of locomotives over a railway

Table 1. Comparisons between ant and state-of-the-art coloring algorithms

Graph	n	d	CAS	ADS	ALS	*PartialCol*	*Mem*
$DSJC500.1$	500	0.1	17	15	12	12	12
$DSJC500.5$	500	0.5	68	56	48	50	48
$DSJC500.9$	500	0.9	167	135	127	127	126
$DSJC1000.1$	1000	0.1	29	25	20	21	20
$DSJC1000.5$	1000	0.5	122	104	86	89	83
$DSJC1000.9$	1000	0.9	313	255	225	226	223
$flat300\_28\_0$	300	0.48	43	36	29	28	29
$flat1000\_50\_0$	1000	0.49	120	101	50	50	50
$flat1000\_60\_0$	1000	0.49	121	102	60	60	60
$flat1000\_76\_0$	1000	0.49	120	103	85	88	82
$le450\_15c$	450	0.17	28	18	15	15	15
$le450\_15d$	450	0.17	28	18	15	15	15
$le450\_25c$	450	0.17	33	29	26	27	25
$le450\_25d$	450	0.17	33	29	26	27	25

network [9]. It is assumed that there is only one source of fuel: fueling trucks, located at yards. A solution of the problem has two important components [16]: (1) choose the number of trucks contracted at each yard, and (2) determine the refueling plan of each locomotive (i.e. the quantity of fuel that must be dispensed into each locomotive at every yard). Such components are respectively called the *truck assignment problem* (TAP) and the *fuel distribution problem* (FDP).

The constraints are the following: the capacity of the tank of each locomotive is limited, as well as the maximum amount of fuel a truck can provide the same day; a locomotive cannot be refueled at its destination yard; there is a maximum number of times (which is two) a train can stop to be refueled (excluding the origin); it is forbidden to run out of fuel. The encountered costs are the weekly operating cost of each fueling truck, the fuel price per gallon associated with each yard, and the fixed cost associated with each refueling. The problem consists in finding a feasible solution minimizing the total costs. A detailed description is provided in [9] as well as a literature review (recent papers in that field are [10, 13, 19]).

A *stop* is defined with a triplet (locomotive, yard, day). If a stop is *open*, it means that the involved locomotive can get fuel at that yard on that day (for the considered realistic instance, the same locomotive cannot stop two times at the same yard during the same day). The stop is *closed* otherwise (fuel distribution is not allowed). A *solution* of the problem can be modeled with a pair (T, S), where T and S are vectors of respective sizes equal to the number of yards and the total number of stops (among all locomotives). Component j of T is the number of contracted trucks at yard j, and component i of S indicates if stop i is open or closed. As proposed in [16], if a solution of the TAP is provided (i.e. if

T is known), the corresponding FDP can be quickly and optimally solved with a *flow* algorithm. For the flow algorithm, all the stops for which there are trucks on the associated yards are initially open. Then the provided flow solution will indicate which stop will be actually used (and the non used stops will be closed). This also means that a solution of the TAP can be evaluated by the use of the flow algorithm. Thus, only the TAP is considered below.

4.2   Adaptation of ALS

First, a descent algorithm for the TAP can be easily designed as follows. From a current solution (T, S), a move consists in adding a contracted truck to a yard (*add move*), or in removing a contracted truck from a yard (*drop move*). When a move is performed, the associated FDP is optimally solved and evaluated by the flow algorithm. During the evaluation, all the costs are considered: the refueling costs for the used gallons of fuel, the fixed refueling costs and the contracting costs of the trucks. The resulting descent method for the TAP (denoted DTAP) is proposed in Algorithm 2, with the setting $|Y^{(add)}| = |Y^{(drop)}| = 5$ (other settings were tested but did not improve the results).

Algorithm 2. DTAP: Descent for the Truck Assignment Problem

Construct an initial solution (T, S).

While a local optimum is not reached, do:

1. in solution (T, S), randomly choose a set $Y^{(drop)}$ containing yards for which drop moves are allowed; for any yard $y \in Y^{(drop)}$ and from (T, S), remove a truck from it, and apply the *flow algorithm* to evaluate such a drop candidate move;
2. in solution (T, S), randomly choose a set $Y^{(add)}$ of yards; for any yard $y \in Y^{(add)}$ and from (T, S), add a truck to it, and apply the *flow algorithm* to evaluate such an add candidate move;
3. from (T, S), perform the best move among the $| Y^{(drop)} \cup Y^{(add)} |$ above candidate moves, and rename the resulting solution as (T, S).

An ALS method for the TAP can be derived from the above descent algorithm [17], where each ant is a procedure identical to DTAP, but with a *learning process* based on a trail system, which is defined as follows. Let x and y be two yards. The trail $Tr(x, y)$ associated with yards x and y aims to indicate if it is a good idea to have trucks on both yards x and y in the same solution. At the end of the current generation, such trails are globally updated as follows: $Tr(x, y) = 0.9 \cdot Tr(x, y) + \Delta Tr(x, y)$, where $\Delta Tr(x, y)$ is the number of trucks on x and y, computed only for solutions of the current generation having trucks on both x and y. A move can be denoted by $(x \to s)$, indicating that a truck is added to or removed from yard x, in the current solution s handled by the considered ant. Let $Tr(x \to s)$ be its associated trail value. It is straightforward to set $Tr(x \rightarrow s) = \sum_{y \in s} Tr(x, y)$ if it is used as follows. If $(x \rightarrow s)$ is an add move, among the possible add moves, it is interesting to select a move with a large $Tr(x \rightarrow s)$ value (because the history of the search seems to indicate that having trucks on yard x, as well as on the yards which already contain trucks in the current solution s, is a good idea). On the contrary, if $(x \rightarrow s)$ is a drop move, among the possible drop moves, it is better to select a move with a small $Tr(x \rightarrow s)$ value.

The greedy force $GF(x \rightarrow s)$ of a move $(x \rightarrow s)$ is simply the resulting objective function value. The way that an ant selects a move at each iteration is now described, according to the greedy force and the trail system. Remember that in DTAP, the performed move is the best among the ones in the set $| Y^{(drop)} \cup Y^{(add)} |$, with $|Y^{(add)}| = |Y^{(drop)}| = 5$. For the descent algorithm associated with an ant of ALS, two sets $T^{(add)}$ and $T^{(drop)}$ of size ten are first randomly chosen (other sizes were tested but without leading to a better performance). Then, let $Y^{(add)}$ (resp. $Y^{(drop)}$) be the subset of $T^{(add)}$ (resp. $T^{(drop)}$) containing the five moves with the best trail values (which is much quicker to compute than the objective function values, as the use of the flow algorithm is avoided). The performed move among $Y^{(add)} \cup Y^{(drop)}$ is the best one according to the greedy force. Therefore, for DTAP as well as for the descent algorithm associated with an ant, the performed move has the best objective function value among a sample of ten evaluated solutions. This will allow to better measure the impact of the trail system on the search.

4.3   Results

The algorithms were tested on an *iMac* (3.06 GHz Intel Core 2 Duo processor, 4 Go 1,067 MHz RAM) with a time limit of 120 min. In order to fairly compare ALS and DTAP, the latter is restarted from scratch each time a local optimum is found (as long as the time limit is not reached). The considered realistic instance, proposed in [9], is characterized as follows: 73 yards, 213 trains, 214 locomotives, and a planning horizon of 14 days.

Table 2. Gain of ALS over DTAP (in $)

Time (min)	DTAP	Gain(ALS)
0	11,605,703	0
15	11,455,921	16,584
30	11,450,137	16,178
45	11,445,918	12,502
60	11,442,113	9,039
90	11,439,601	6,636
105	11,439,601	7,919
120	11,439,587	7,929

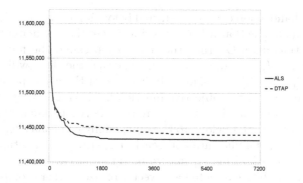

Fig. 1. Evolution of ALS and DTAP

DTAP and ALS can be compared in Table 2. In the second column is given the average value (in \$, over ten runs) of DTAP for the corresponding execution time (indicated in the first column). In the third column is indicated the average gain (in \$, over ten runs) of ALS over DTAP. In addition, Fig. 1 compares the evolution of the best encountered solution value (average over 10 runs, on the vertical axis) during 120 min (7200 s, on the horizontal axis). One can easily deduce that the learning process (i.e. the trail system) introduced to DTAP to derive ALS is relevant as it leads to non negligible savings.

The INFORMS contest involved 31 research teams. The approaches of the three best teams can be found in [9]. The winners of the contest (Kaspi and Raviv) formulated the problem as a MILP (mixed integer linear program). They found a lower bound $LB = 11,399,670.58\$$ and their best result was 0.30\$ above LB. Thus the gap between ALS and LB is approximately 0.35 %. Note that a major advantage of ALS over the MILP proposed by Kaspi and Raviv is its *flexibility*: it can be easily adapted if non linear components are added to the problem. For example, one can assume that the weekly operating cost of the fueling trucks might be concave with the number of trucks located at the same yard.

5   ALS for a Scheduling Problem with Abandon Costs

5.1   Presentation of the Problem

Consider a scheduling problem (P) where a set of n jobs have to be performed on a single machine. It is possible to abandon (reject) a job j, and in such a case, an *abandon* cost u_j is encountered (it can represent that j is allocated to an external resource). For each job j are known: its *processing* time p_j, its *release* date r_j (it is not possible to start j before that date), its *due* date d_j (the preferred completion time of job j) and its *deadline* d'_j (the latest allowed completion time of job j), such that $r_j \leq d_j \leq d'_j$. For each job, its starting time B_j (or equivalently its completion time C_j because $C_j = B_j + p_j$) has to be determined.

On the one hand, a job cannot be started before its release date (i.e. $B_j \geq r_j$), because one can assume that it is not possible to get the raw material associated with j before that date. On the other end, a job cannot be finished after its deadline (i.e. $C_j \leq d'_j$), because one can assume that the client will refuse j after that date. If $C_j \in]d_j, d'_j]$, j is said to be *late* (from the client perspective) and this will be penalized in the objective function by a component $f_j(C_j)$.

In addition, various families of jobs are considered, assuming that jobs with comparable characteristics belong to the same family or *product type*. If two jobs j and j' of different families are consecutively performed, a setup cost $c_{jj'}$ is encountered and a setup time $s_{jj'}$ has to be taken into account.

A solution s can be represented as a vector of size n where component j contains the value of B_j. If job j is unperformed, one can put a fictitious -1. Let $U(s)$ (resp. $I(s)$) be the set of unperformed (resp. performed) jobs of solution s. The goal consists in minimizing an objective function with three components: (1) the abandon costs $\sum_{j \in U(s)} u_j$; (2) the setup costs $\sum_{j \to j'} c_{jj'}$ of the consecutively performed jobs $j \to j'$ of $I(s)$; (3) the penalty costs for late completion times $\sum_{j \in I(s)} f_j(C_j)$, where $f_j(C_j)$ is a *regular* (i.e. non decreasing) function depending on C_j. In the literature, the two most popular regular objective functions are the sum of completion times (i.e. $\sum_j C_j$) and the sum of tardiness (i.e. $\sum_j T_j$, where $T_j = \max\{0; C_j - d_j\}$), as well as their weighted versions (i.e. $\sum_j w_j \cdot T_j$ and $\sum_j w_j \cdot C_j$).

Problem (P) was first proposed in [1] where the authors proposed a branch and bound algorithm able to tackle most instances with up to 30 jobs. Other relevant references in the field are [14,20]. Up to date, the only existing (meta) heuristic for (P) is a tabu search proposed in [18], denoted $Tabu(P)$, which is based on four types of moves: *reinsert* a job (i.e. remove a job $j \in I(s)$ from its current position in s and insert it somewhere else in s), *swap* two jobs of $I(s)$, *add* a job (i.e. move a job from $U(s)$ to $I(s)$), and *drop* a job (i.e. move a job from $I(s)$ to $U(s)$). At each iteration, a random sample (tuned to 15%) of the four types of move is generated, and the best move among the sample is performed. Note that if a moves leads to a non feasible solution (e.g., if the processing of some jobs overlap in time, or if a job is finished after its deadline), a repairing process is used, which allows to shift or drop jobs (see [18] for details on the repairing process).

5.2   Adaptation of ALS

The ALS method for (P) is denoted $ALS(P)$ and is summarized in Algorithm 3, where $N = 20$ and $I = 500$ were used in the experiments (other settings were tested but did not lead to better results). The local search operator of the initialization phase is $Tabu(P)$. The specificities of Algorithm 3, when compared to Algorithm 1, are the following: the trail matrix Tr is initialized before the main loop, Tr also appears in the construction operator, and Tr is updated as soon as an ant provides a solution (denoted s' in the pseudo-code).

The *greedy force GF* of a move is simply its associated objective function value. The *construction operator* starts from an empty solution s. Then, at each step, a non considered job j is scheduled within s (or rejected if it is better). More precisely, let $(j \to p, s)$ be the move consisting in inserting job j at position p of the jobs sequence of solution s (when inserting a job, it is allowed to shift, or to reject, some already scheduled jobs if it can make the solution feasible or less costly). At each step, among the q (parameter tuned to 25) less costly insertions $(j \to p, s)$, choose the one with the largest associated trail value $Tr(j \to p, s)$, which is defined as

$$Tr(j \to p, s) = \sum_{x \text{ before } p \text{ in } s} Tr(x, j) + \sum_{x \text{ after } p \text{ in } s} Tr(j, x)$$

The *trail* matrix $Tr(x, y)$ associated with the scheduling of job x before job y is based on the number of times x was positioned before y during the search (and on the quality of the associated solutions). More precisely, at the end of each generation, the trail matrix is updated as follows: $Tr(x, y) = 0.9 \cdot Tr(x, y) + 1/\hat{f}$, where $\hat{f}$ is the average value of the solutions of the current generation for which job x is scheduled before job y.

The local search operator of the main loop is exactly $Tabu(P)$, with the following modification. Every ten iterations, it is imposed to perform an add move based on Tr and GF as follows. Among the q (parameter tuned to 25) moves of type $(j \to p, s)$ with the largest trail value $Tr(j \to p, s)$, perform the less costly insertion (i.e. the one with the largest GF value).

Algorithm 3. Algorithm $ALS(P)$

Initialization of the trail matrix Tr

1. generate randomly N solutions and improve each of them with a local search operator during I iterations;
2. initialize the trail matrix Tr with the N improved solutions.

While a time limit is not reached, do:

1. construction operator: build a solution s based on the trail matrix Tr and the greedy force GF;
2. improve s by the use of a local search operator during I iterations (based on Tr and GF) and let s' be the resulting solution;
3. use s' to update the trail matrix Tr.

5.3 Results

$Tabu(P)$ and $ALS(P)$ were tested on a computer with processor Intel i7 Quand-core (2.93 GHz RAM 8 Go DDR3) during $30 \cdot n$ seconds (i.e. about four hours if $n = 500$ jobs). Note that $Tabu(P)$ is restarted every $I = 500$ iterations as long as the time limit is not reached, so that it can be fairly compared to $ALS(P)$, which uses each ant during $I = 500$ iterations. The instances are the same as the ones

Table 3. Comparison of $Tabu(P)$, $ALS(P)$ and an upper bound $UB(P)$

Instance	n	UB	$Gain(TS)$	$Gain(ALS)$
$STC_NCOS_0 1.csv$	8	920	220	220
$STC_NCOS_0 1a.csv$	8	1,010	400	400
$STC_NCOS_1 5.csv$	30	22,321	4,710	4,710
$STC_NCOS_1 5a.csv$	30	6,449	865	865
$STC_NCOS_3 1.csv$	75	6,615	−250	−250
$STC_NCOS_3 1a.csv$	75	7,590	−250	−250
$STC_NCOS_3 2.csv$	75	25,774	1,554	1795
$STC_NCOS_3 2a.csv$	75	16,908	110	110
$STC_NCOS_4 1.csv$	90	85,378	40,740	42,175
$STC_NCOS_4 1a.csv$	90	26,828	7,731	8,246
$STC_NCOS_5 1.csv$	200	308,770	169,095	164,559
$STC_NCOS_5 1a.csv$	200	318,740	107,770	170,510
$STC_NCOS_6 1.csv$	500	1,495,045	0	−463
$STC_NCOS_6 1a.csv$	500	1,821,085	6,480	6,480
Averages		**295,960**	**24,227**	**28,508**

with setups described in [1] and the cost component $f_j(C_j)$ is $w_j \cdot T_j$. In Table 3 are compared the upper bounds $UB(P)$ provided by the branch and bound algorithm of [1], and the average gain (over 10 runs) of $Tabu(P)$ and $ALS(P)$ over $UB(P)$. One can remark that the average gain of $ALS(P)$ is more than 4,000 above the average gain of $Tabu(P)$: it is thus worthy to use the proposed trail system.

6   Discussion and Conclusion

Within the ant algorithms field, paper [21] was a first try to answer the question: *"What should be the role of a single ant?"*. In most ant algorithms, an ant is a constructive heuristic. In contrast, an ant is a local search in ALS. Even if the role of an ant can be defined in various ways, each decision is always based on the greedy force (representing the self-adaptation of each ant), and the trail system (modeling the collaboration between the ants). This paper shows that ALS is a promising algorithm for combinatorial problems: it obtained competitive results for three very different combinatorial optimization problems (graph coloring, a refueling problem, and a job scheduling problem). For the graph coloring problem, it was numerically showed that ALS performs much better than a standard ant algorithm. Therefore, a straightforward avenue of research would be to adapt ALS to other combinatorial problems.

Another important issue is indirectly tackled in this paper: *"How should be defined an efficient trail system?"*. For the three considered problems, it would not be relevant to respectively transmit the following *myopic* (or very local) information to the ants of the next generations: (1) a pair (vertex x, color c), as two colorings can be equivalent if the color indexes are permuted; (2) a pair (yard y, truck t), as yard y might be empty if trucks are located in a yard close to y; (3) a pair (position p, job j), as the scheduling of a job strongly depends on the scheduling of the other jobs (especially if setups are considered). In contrast, the above proposed trail systems are respectively based on: (1) the assignment of the *same color* to some vertices; (2) the assignment of trucks to a *same set* of yards; (3) the *relative order* in which jobs appear. In other words, a trail system should *globally cover* specific characteristics of the considered problem.

References

1. Baptiste, P., Le Pape, C.: Scheduling a single machine to minimize a regular objective function under setup constraints. Discrete Optim. **2**, 83–99 (2005)
2. Bloechliger, I., Zufferey, N.: A graph coloring heuristic using partial solutions and a reactive tabu scheme. Comput. Oper. Res. **35**, 960–975 (2008)
3. Blum, C.: Ant colony optimization: introduction and recent trends. Phys. Life Rev. **2**(4), 353–373 (2005)
4. Costa, D., Hertz, A.: Ants can colour graphs. J. Oper. Res. Soc. **48**, 295–305 (1997)
5. Dorigo, M., Birattari, M., Stuetzle, T.: Ant colony optimization – artificial ants as a computational intelligence technique. IEEE Comput. Intell. Mag. **1**(4), 28–39 (2006)
6. Dorigo, M., Stuetzle, T.: The ant colony optimization metaheuristic: algorithms, applications, and advances. In: Glover, F., Kochenberger, G. (eds.) Handbook of Metaheuristics, vol. 57, pp. 251–285. Kluwer Academic Publishers, Boston (2003)
7. Glover, F., Laguna, M.: Tabu Search. Kluwer Academic Publishers, Boston (1997)
8. Hertz, A., Zufferey, N.: A new ant colony algorithm for graph coloring. In: Pelta and Krasnogor (eds.) Proceedings of the Workshop on Nature Inspired Cooperative Strategies for Optimization, NICSO 2006, pp. 51–60, Granada, Spain, 29–30 June 2006
9. INFORMS RAS Competition (2010). http://www.informs.org/community/ras/problem-solving-competition/2010-ras-competition
10. Kuby, M., Lim, S.: The flow-refueling location problem for alternative-fuel vehicles. Socio Econ. Plann.Sci. **39**(2), 125–145 (2005)
11. Lu, Z., Hao, J.-K.: A memetic algorithm for graph coloring. Eur. J. Oper. Res. **203**, 241–250 (2010)
12. Malaguti, E., Toth, P.: A survey on vertex coloring problems. Int. Trans. Oper. Res. **17**(1), 1–34 (2010)
13. Nourbakhsh, S.M., Ouyang, Y.: Optimal fueling strategies for locomotive fleets in railroad networks. Transp. Res. Part B **44**(8–9), 1104–1114 (2010)
14. Oguz, C., Salman, F.S., Yalcin, Z.B.: Order acceptance and scheduling decisions in make-to-order systems. Int. J. Prod. Econ. **125**, 200–2011 (2010)
15. Plumettaz, M., Schindl, D., Zufferey, N.: Ant local search and its efficient adaptation to graph colouring. J. Oper. Res. Soc. **61**, 819–826 (2010)
16. Schindl, D., Zufferey, N.: A local search for refueling locomotives. In: Proceedings of the 54th Annual Conference of the Administrative Science Association of Canada - Production and Operations Management Division (ASAC 2011), pp. 53–61, Montreal, Canada, 2–5 July 2011

17. Schindl, D., Zufferey, N.: Ant local search for fuel supply of trains in America. In: Proceedings of the 1st International Conference on Logistics Operations Management, Le Havre, France, 17–19 October 2012
18. Thevenin, S., Zufferey, N., Widmer, M.: Tabu search to minimize regular objective functions for a single machine scheduling problem with rejected jobs, setups and time windows. In: Proceedings of the 9th International Conference on Modeling, Optimization and Simulation (MOSIM 2012), Bordeaux, France, 6–8 June 2012
19. Vaidyanathan, B., Ahuja, R.K., Liu, J., Shughart, L.A.: Real-life locomotive planning: new formulations and computational results. Transp. Res. Part B **42**(2), 147–168 (2008)
20. Yang, B., Geunes, J.: A single resource scheduling problem with job-selection flexibility, tardiness costs and controllable processing times. Comput. Ind. Eng. **53**, 420–432 (2007)
21. Zufferey, N.: Optimization by ant algorithms: possible roles for an individual ant. Optim. Lett. **6**(5), 963–973 (2012)

An Enhanced Ant Colony System
for the Probabilistic Traveling Salesman Problem

Dennis Weyland[✉], Roberto Montemanni, and Luca Maria Gambardella

IDSIA - Dalle Molle Institute for Artificial Intelligence/USI/SUPSI,
Galleria 2, 6928 Manno, Switzerland
{dennis,roberto,luca}@idsia.ch

Abstract. In this work we present an Enhanced Ant Colony System algorithm for the Probabilistic Traveling Salesman Problem. More in detail, we identify drawbacks of the well-known Ant Colony System metaheuristic when applied to the Probabilistic Traveling Salesman Problem. We then propose enhancements to overcome those drawbacks. Comprehensive computational studies on common benchmark instances reveal the efficiency of this novel approach. The Enhanced Ant Colony System algorithm clearly outperforms the original Ant Colony System metaheuristic. Additionally, improvements over best-known results for the Probabilistic Traveling Salesman Problem could be obtained for many instances.

Keywords: Ant colony system · Stochastic vehicle routing · Probabilistic traveling salesman problem · Monte carlo sampling

1   Introduction

The Probabilistic Traveling Salesman Problem (PTSP, [15]) is a well-known Stochastic Vehicle Routing Problem [14]. It is a generalization of the famous Traveling Salesman Problem [16] and therefore a NP-hard optimization problem. Many different algorithms have been proposed for this problem in recent years [1–8,20] and computing good solutions is a great challenge. For many Vehicle Routing Problems the Ant Colony System metaheuristic [10] is able to obtain good solutions. Although this is also the case for the PTSP [3], there also exist competitive approaches [2,20]. Therefore, in this work we critically analyze the original Ant Colony System metaheuristic in the context of the PTSP. We identify some weaknesses of this method and propose appropriate changes to the original paradigm. The resulting approach is called the Enhanced Ant Colony System algorithm. We perform comprehensive computational studies of this new approach on the PTSP. The results show the effectiveness of the enhancements introduced. A more general paper discussing similar aspects in a wider context has been published recently [13].

The remaining part of this paper is organized in the following way. In Sect. 2 we introduce the Probabilistic Traveling Salesman Problem. We give a formal

G.A. Di Caro and G. Theraulaz (Eds.): BIONETICS 2012, LNICST 134, pp. 237–249, 2014.
DOI: 10.1007/978-3-319-06944-9_17, © Institute for Computer Sciences, Social Informatics
and Telecommunications Engineering 2014

definition of this problem and we discuss the relevant literature. Then we give an overview about the original Ant Colony System metaheuristic in Sect. 3. Section 4 starts with identifying drawbacks of the original Ant Colony System approach. Enhancements are introduced to overcome those drawbacks and the resulting Enhanced Ant Colony System approach is formally defined. In Sect. 5 we continue with comprehensive computational studies of the Enhanced Ant Colony System approach, applied to the Probabilistic Traveling Salesman Problem. Finally, we finish the paper with a short discussion and conclusions in Sect. 6.

2   The Probabilistic Traveling Salesman Problem

Stochastic Vehicle Routing Problems [14] have received increasing attention in recent years. Those problems make use of stochastic input to obtain more realistic models of real world problems. One of the best-known Stochastic Vehicle Routing Problems is the Probabilistic Traveling Salesman Problem (PTSP, [15]). This problem is a generalization of the famous Traveling Salesman Problem [16]. Therefore this problem is NP-hard and computing good solutions is a great challenge. The problem was first introduced in [15] and many different solution approaches have been proposed. In [4] homogeneous instances of the PTSP are optimized using an Ant Colony Optimization approach. This approach has been extended in [5] for the optimization of heterogeneous instances. Another Ant Colony Optimization approach has been introduced in [7]. A method using an approximation of the objective function based on temporal aggregation is presented in [8]. A Local Search algorithm using the technique of delta evaluation is shown in [6] and improvements using adaptive sample sizes and importance sampling are presented in [1]. The same authors also propose an Ant Colony Optimization approach hybridized with a Local Search algorithm [3]. Local Search Algorithms based on efficient approximations of the objective function are introduced in [19]. Those Local Search algorithms are then used within heuristics leading to the current state-of-the-art methods for the PTSP [20].

In contrast to the Traveling Salesman Problem, the customers are modeled in a stochastic way for the PTSP. Each customer has assigned a certain probability which indicates how likely it is that the customer requires to be visited. The event that a specific customer requires a visit is independent of the requests of other customers. A solution for the PTSP is like for the Traveling Salesman Problem a tour visiting all customers exactly once. Such a solution is called a-priori solution in the context of the PTSP. For a specific realization of the random events, the a-priori solution is used to derive a so called a-posteriori solution. Here the customers that require to be visited are processed in the same order as defined by the a-priori solution, while customers that do not require to be visited are just skipped. The cost for such an a-posteriori solution is the total travel time. The optimization goal is now to find an a-priori solution with minimum expected costs over the a-posteriori solutions with respect to the given probabilities.

More formally, we can define the PTSP over a complete undirected edge- and node-weighted graph $G = (V, c, p)$. $V = \{1, 2, \ldots, n\}$ is the set of nodes, which represent the customers, $p : V \longrightarrow [0, 1]$ is the probability function, that assigns to each node the probability that the node requires a visit and $c : V \times V \longrightarrow \mathbb{R}^+$ is the symmetric cost function, that represents the non-negative travel costs between any two nodes. The goal is to find a permutation $\phi : V \to V$, the a-priori solution, which minimizes the expected cost over the a-posteriori solutions.

Given an a-priori solution ϕ, the total expected cost of ϕ is the sum of the expected cost for each edge in an a-posteriori solution. According to [15] this can be mathematically described in the following way.

$$
f_{\text{ptsp}}(\phi) = \sum_{i=1}^{n} \sum_{j=i+1}^{n} c(\phi_i, \phi_j)\, p(\phi_i)\, p(\phi_j) \prod_{k=i+1}^{j-1} (1 - p(\phi_k))
$$
$$
+ \sum_{i=1}^{n} \sum_{j=1}^{i-1} c(\phi_i, \phi_j)\, p(\phi_i)\, p(\phi_j) \prod_{k=i+1}^{n} (1 - p(\phi_k)) \prod_{k=1}^{j-1} (1 - p(\phi_k))
$$

A certain edge (v, w), $v, w \in V$, occurs in the a-posteriori solution, if the customers corresponding to v and w require a visit and all customers in between v and w do not require a visit. This formula can be evaluated in a computational time of $\mathcal{O}(n^2)$ by adding the summands in a specific order. For more details we refer to [15].

3   The Original Ant Colony System Algorithm

The Ant Colony System metaheuristic (ACS) is part of the algorithm family called Ant Colony Optimization (ACO). The first Ant Colony Optimization algorithm, Ant System (AS), has been proposed in [9,11]. This computational paradigm is inspired by the behavior of real ant colonies. The main idea adapted from the behavior of ants is to parallelize the search over multiple constructive computational threads. The construction process of each thread is guided implicitly by a dynamic memory structure, which holds information on the effectiveness of previously created solutions. The behavior of each single thread is inspired by the behavior of real ants.

The main elements of this approach are *ants*. They are simple computational agents which individually and iteratively construct solutions for the problem. The solutions are constructed step by step, starting from an empty solution. The partial solutions that occur during the construction process are called *states*. While constructing new solutions, ants move from one state to another state, corresponding to a more complete partial solution. This process terminates if a complete solution is obtained. At each constructive step, every ant computes a set of feasible states which can be reached from the current state. Then probabilistically the ant moves to one of these states. The following two values are used for computing the probability p_{ab}^k that ant k performs a move from state a to state b:

1. The attractiveness μ_{ab} of the move, indicating the a-priori desirability of that move.
2. The trail level τ_{ab} of the move, indicating how beneficial it has been in the past to make that particular move. It is therefore an a-posteriori indication of the desirability of that move.

In each iteration the values for the trail levels are updated, increasing the values of those moves that were part of good solutions, while decreasing all the others. For all feasible states $b \in B$ that can be reached by ant k from the current state a the probabilities are defined by

$$p_{ab}^k = \frac{(\tau_{ab})\xi + (\mu_{ab})(1 - \xi)}{\sum_{c \in B}((\tau_{ac})\xi + (\mu_{ac})(1 - \xi))}. \tag{1}$$

With probability q_0 the next state is chosen as the state b that maximizes the probability p_{ab}^k, while otherwise (with a probability of $1 - q_0$) the next state is chosen according to the probabilities. Here ξ is a parameter controlling the relative importance of the trail τ_{ab} and the a-priori attractiveness μ_{ab}.

After all the m ants have completed a solution, the trail level of each move (a, b) of the best solution seen so far is increased. In this way future ants will use this information to generate new solutions in a neighborhood of this preferred solution. The best solution seen so far is denoted by BestSol and its cost is denoted by BestCost. For minimization problems the update of the pheromone values can then be described mathematically as

$$\tau_{ab} = (1 - \rho)\tau_{ab} + \rho/\text{BestCost}. \tag{2}$$

Additionally, the pheromone values are also updated during the constructive phase. Here the pheromone values for moves that are chosen and performed are decreased. In case a move (a, b) is performed, the following rule is applied.

$$\tau_{ab} = (1 - \psi)\tau_{ab} + \psi \cdot \tau_{\text{init}} \tag{3}$$

Here ψ is a parameter regulating the evaporation of the pheromone trace over time and τ_{init} is the initial value of trails. In this way a certain variety among the generated solutions is maintained.

Finally, after a solution has been constructed and prior to the evaluation of the solution, a Local Search algorithm is applied to improve the solution. Algorithm 1 summarizes the whole process in pseudo-code.

3.1 ACS for the PTSP

Before we discuss the drawbacks of the Ant Colony System metaheuristic and overcome them by introducing several enhancements, we want to show how the original ACS metaheuristic can be used for the optimization of the PTSP. Following Algorithm 1, the Ant Colony System paradigm can be easily used for solving the PTSP. In the constructive phase the ants start with a customer chosen uniformly at random. Then step by step the intermediate solutions are

Algorithm 1. Pseudo-code of the ACS metaheuristic.

1: BestCost := ∞
2: **for** each move (a, b) **do**
3:     $\tau_{ab} := \tau_{\text{init}}$
4: **end for**
5: **while** termination criteria not met **do**
6:    **for** $k := 1$ to m **do**
7:       **while** ant k has not completed its solution **do**
8:          **for** every feasible move (a, b) from the current state a **do**
9:             Compute p_{ab}^k according to equation (1)
10:          **end for**
11:          **if** uniform random number in $[0, 1] > q_0$ **then**
12:             Choose the move randomly according to the probabilities
13:          **else**
14:             Choose the move with the maximum probability
15:          **end if**
16:          Update the trail level τ_{ab} by means of (3)
17:       **end while**
18:       Apply a Local Search to improve the current solution
19:       Cost := Cost of the improved current solution
20:       **if** Cost < BestCost **then**
21:          BestCost := Cost
22:          BestSol := current solution
23:       **end if**
24:    **end for**
25:    **for** each move (a, b) **do**
26:       Update the trail level τ_{ab} by means of (2)
27:    **end for**
28: **end while**
29: **return** BestSol

completed by inserting customers, which have not been visited yet, at the end
of the partial tours. For the Local Search algorithm used within the ACS we use
state-of-the-art methods introduced in [19,20]. In particular, we use the *2.5-opt
combined* Local Search algorithm. For the evaluation of the objective function
within the local search, two different approximations are used. The first one is
based on Monte Carlo Sampling (sampling approximation) and the second one
is an analytical approximation (depth approximation) truncating the sums in
the formula for the exact evaluation at a certain depth. The *2.5-opt combined*
Local Search algorithm starts with a given solution and performs a local search
using the 2.5-opt neighborhood [16] and the sampling approximation. The final
solution is then used as the starting point for a Local Search using again the
2.5-opt neighborhood, but now with the depth approximation. This process can
also be iterated for a few times. It has been shown in [20] that this approach is
extremely efficient and the current state-of-the-art local search method for the
PTSP.

The usage of *neighborhood lists* and *don't look bits* are an integral part of the *2.5-opt combined* Local Search algorithm. For each customer a *neighborhood list* with promising tour neighbors, based on a-priori desirability, is used. During the Local Search only those moves are considered that lead to at least one new connection between some customer and a customer in its *neighborhood list*. We also use a *don't look bit* for each customer. This bit is initially set to *false* and indicates that moves involving this customer should be explored. If there are no improving moves involving a specific customer during the run of the algorithm, the corresponding *don't look bit* is set to *true* and moves involving this customer are not considered anymore. If an improving move is performed, changing the neighbors of a customer with active *don't look bit*, or the position of that customer, the corresponding *don't look bit* is set to *false* and moves involving this customer are again considered in the Local Search.

4   An Enhanced Ant Colony System

In this section we critically analyze the original ACS metaheuristic in the context of the Probabilistic Traveling Salesman Problem. We identify weaknesses of this approach when applied to the PTSP and overcome them by introducing certain enhancements.

One known drawback of the ACS metaheuristic is the large computational time that is required to construct new solutions. Let n be the number of steps that are necessary to construct a new solutions. Usually each of the n steps requires a computational time of $\mathcal{O}(n)$, leading to a total computational time of $\mathcal{O}(n^2)$ for the construction of one solution. This is acceptable in case of small problems, but it is no longer feasible if larger problems are considered. In fact, for some problems the good results reported for the ACS metaheuristic cannot be replicated on larger instances. We therefore propose an improved constructive phase that requires less computational effort to construct new solutions. We discuss the details of this improved constructive phase later in this section.

The second drawback we identified for the original ACS metaheuristic is the following. The Local Search algorithm used to improve the solutions that have been constructed by the ants is usually treated as a black box. That means no information are exchanged between the ACS and the Local Search used within the ACS. Solutions constructed by the ACS do not necessarily differ a lot from the best solution seen so far. This holds in particular for solutions that are generated by the improved construction phase, which will be discussed later in this section. Such additional information could be used to increase the efficiency of the Local Search algorithm. We therefore propose a better integration between the construction of new solutions and the Local Search algorithm. In the next section we show how this can be achieved in the context of the PTSP.

4.1   An Enhanced Ant Colony System for the PTSP

The idea for improving the construction phase is to directly consider the best solution seen so far during that phase. In the original ACS metaheuristic an ant selects the next state according to a probabilistic criterion. With a probability of q_0 the state with the best weighted value between the pheromone trail and the a-priori desirability is chosen, while with a probability of $1 - q_0$ the state is selected probabilistically (with probabilities proportional to the weighted values mentioned before). In our new approach, the state selected with a probability of q_0 is the same state as in the best solution seen so far (in case this state is not feasible, the classic mechanism is applied). Otherwise, that means with a probability of $1 - q_0$ the next state is selected probabilistically. Since the probability q_0 is usually very close to 1.0, the computational time is reduced a lot by the new approach.

By using the new construction approach, solutions in the neighborhood of the best solution seen so far are obtained in most of the cases. Such solutions have many parts in common with the best solution seen so far and it does not make sense to apply the Local Search algorithm on all the components of the solution. Instead of that, we propose to initially apply the Local Search algorithm only on those parts of the new solution that have changed with respect to the best solution seen so far. By initializing the *don't look bit* data structure of the underlying Local Search algorithm appropriately, we can efficiently achieve the desired behavior. More in detail, we set the *don't look bit* to *false* only for those customers that are assigned to a different position or whose predecessor/successor have changed compared to the best solution seen so far. In this way the Local Search algorithm is initially only applied on those parts of the solution that differ with respect to the best solution seen so far. Depending on the Local Search moves that are performed, the exploration of other parts of the solution may be re-enabled as described in the previous section. Overall, in this way the computational time can be further decreased, especially if q_0 is chosen close to 1.0.

We call the resulting algorithm incorporating both enhancements the Enhanced Ant Colony System algorithm (EACS). Algorithm 2 summarizes the EACS using pseudo-code: The main differences with respect to the ACS metaheuristic are the constructive phase depicted in lines 8–25 and the integration with the local search in line 28.

5   Computational Experiments

In this section we perform comprehensive computational studies on common benchmark instances of the PTSP. The main goal is to perform a comparison between the original Ant Colony System metaheuristic and the Enhanced Ant Colony System algorithm. In this way we are able to assess the enhancements introduced in Sect. 4. Furthermore, we also want to compare the Enhanced Ant Colony System algorithm with state-of-the-art approaches. We start with describing the benchmark instances used for our experiments. Then we discuss the experimental setup in detail. Finally, we present the results and we finish with a detailed analysis of those results.

Algorithm 2. Pseudo-code of the EACS approach.

```
 1: BestCost := ∞
 2: for each move (a, b) do
 3:     τab := τinit
 4: end for
 5: while termination criteria not met do
 6:     for k := 1 to m do
 7:         while ant k has not completed its solution do
 8:             if uniform random number in [0, 1] > q0 then
 9:                 // This part is executed with low probability
10:                 for every feasible move (a, b) from the current state a do
11:                     Compute p^k_ab according to equation (1)
12:                 end for
13:                 Choose the move randomly according to the probabilities
14:             else
15:                 b* := state such that move(a, b*)is part of BestSol
16:                 if (a, b*) is a feasible move then
17:                     Chose the move (a, b*)
18:                 else
19:                     // This part is executed with low probability
20:                     for every feasible move (a, b) from the current state a do
21:                         Compute p^k_ab according to equation (1)
22:                     end for
23:                     Choose the move with the maximum probability
24:                 end if
25:             end if
26:             Update the trail level τab by means of (3)
27:         end while
28:         Apply a Local Search to improve the current solution (considering initially
             only components of the new solution that differ from BestSol)
29:         Cost := Cost of the improved current solution
30:         if Cost < BestCost then
31:             BestCost := Cost
32:             BestSol := current solution
33:         end if
34:     end for
35:     for each move (a, b) do
36:         Update the trail level τab by means of (2)
37:     end for
38: end while
39: return BestSol
```

5.1 Benchmark Instances

For our experiments we use common benchmark instances for the Probabilistic Traveling Salesman Problem. Currently there are three important classes of benchmark instances used. The first class consists of instances for the Traveling Salesman Problem from the TSPLIB benchmark [18] supplemented with the probabilities for the customers. In this way we are able to use data from real world routing problems in our experiments. We refer to those instances as *tsplib* instances. The second class consists of Euclidean instances in which customers are distributed uniformly at random in a square. Those instances are called *uniform* instances. They are widely used to assess the performance of algorithms for the PTSP. Nonetheless, those instances are more artificial and not directly related to real world scenarios. This changes for the last class of instances. This class contains Euclidean instances in which customers are distributed around a certain number of centers which themselves are distributed uniformly at random in a square. Those instances represent the situation in which customers are located in different cities which are highly populated around their centers and less populated in suburban areas. We refer to those instances as *clustered* instances. For the customers' probabilities we use homogeneous values of $0.05, 0.1$ and 0.2, which are a common choice for benchmark instances for the PTSP. For more details about the benchmark instances we refer the interested reader to [6].

5.2 Experimental Setup

The algorithms have been coded in ANSI C and the experiments have been performed on a Quad-Core AMD Opteron system running at 2 GHz. For the underlying Local Search algorithm we have decided to use the combined local search with 300 samples, a threshold value of 0.001 and 3 successive iterations (cf. [20]). This method is currently the state-of-the-art Local Search algorithm for the Probabilistic Traveling Salesman Problem. To obtain reasonable parameters for the Ant Colony System metaheuristic and the Enhanced Ant Colony System algorithm we have performed several preliminary experiments. Here we have performed experiments using all the possible combinations of different values for the parameters on a certain set of benchmark instances (different from the instances used later). Among all those parameter settings we have chosen the one which achieved the best performance. For both approaches $m = 10$ ants are used for the construction of new solutions. The other parameters are $\tau_{init} = 1$, $\xi = 0.005$, $\psi = 0$, $\rho = 0.8$ and $q_0 = 0.99$. It is interesting to note that the best value for ψ is 0, which means that the diversification introduced by this parameter does not seem to play a crucial role in the context of the Probabilistic Traveling Salesman Problem. Nonetheless, this observation requires some further investigations. Due to the inherent stochasticity of our approaches, we perform for each algorithm and for each benchmark instance a total of 20 independent runs. To allow a fair comparison with state-of-the-art approaches for the PTSPD, we additionally limit the total computational time for each run to one hour. During the execution of the algorithm we monitor the costs of the best solutions seen at different points in time.

5.3 Results

The results of our experiments are summarized in Table 1. The average solution costs for the Ant Colony System metaheuristic and the Enhanced Ant Colony

System over 20 runs are reported. Additionally, the best-known (average) results for those instances are given. Those are also average results over multiple runs using algorithms from [20] and they are reported in the supplementary material of that publication. Note that the best-known (average) results are not necessarily obtained by the same algorithm for all the different instances. That means, for each instance we perform a comparison between the ACS, the EACS and the best approach for that particular instance. For each instance we have emphasized the best result using bold font.

We can see that the Enhanced Ant Colony System algorithm clearly outperforms the original Ant Colony System metaheuristic. The EACS was able to obtain better final solutions than the ACS for all of the benchmark instances used in our experiments. All those results show a statistical significance due to a paired student's t-test with a significance level of 97.5 %. As we will see at the end of this section, the EACS is also able to obtain such solutions much faster than the ACS. Furthermore, for 75 % of the benchmark instances the EACS could even improve over previous best-known results. Unfortunately, for this comparison we could not apply a paired statistical test, since different randomly created instances (belonging to the same benchmark set) were used. Due to the variance within the different instances of each benchmark class no statistical significance can be reported. Still, those results are very impressive, because for each instance we compare the EACS with the best-known approach for that particular instance. All in all, the results clearly confirm the effectiveness of the EACS.

Additionally, Fig. 1 illustrates the temporal behavior of the Ant Colony System metaheuristic and the Enhanced Ant Colony System algorithm on instances with probabilities of 0.1. Here the x-axis corresponds to the computational time in seconds and the y-axis corresponds to the solution costs. We can see that the EACS is much more efficient than the original ACS during the whole execution of the algorithms. The final solutions obtained by the EACS are always of better quality than those obtained by the ACS. Moreover, the solutions obtained by the EACS after only a small fraction of the total computational time are already better than the final solutions obtained by the ACS. Those results further confirm the effectiveness of the EACS.

6   Discussion and Conclusions

In this work we have presented an Enhanced Ant Colony System algorithm for the Probabilistic Traveling Salesman Problem. The new algorithm is based on the well-known Ant Colony System metaheuristic. We have identified drawbacks of this metaheuristic when applied to the Probabilistic Traveling Salesman Problem and proposed enhancements on top of the original approach to overcome those drawbacks. The resulting approach was then applied to the Probabilistic Traveling Salesman Problem. Comprehensive computational studies on common benchmark instances for this problem reveal the efficiency of the new approach. A significant improvement over the original Ant Colony System algorithm could

Table 1. A comparison between the average solution costs for the EACS, the ACS and best known results on different instances for the PTSP.

Instance	Best known [20]	ACS	EACS
uniform, n = 1000, p = 0.05	6414562.00	6433861.00	**6414158.00**
uniform, n = 1000, p = 0.10	8848432.00	8873150.00	**8844323.00**
uniform, n = 1000, p = 0.20	12177496.00	12194659.00	**12150359.00**
clustered, n = 1000, p = 0.05	**3855327.00**	3857363.00	3855981.00
clustered, n = 1000, p = 0.10	4880419.00	4884078.00	**4878900.00**
clustered, n = 1000, p = 0.20	6285745.00	6288443.00	**6273528.00**
tsplib, att532, n = 532, p = 0.05	**25483.53**	25483.87	25483.81
tsplib, att532, n = 532, p = 0.10	33664.74	33665.83	**33663.22**
tsplib, att532, n = 532, p = 0.20	44667.65	44673.06	**44653.37**
tsplib, rat783, n = 783, p = 0.05	**2305.18**	2307.70	2305.36
tsplib, rat783, n = 783, p = 0.10	3237.96	3239.52	**3235.55**
tsplib, rat783, n = 783, p = 0.20	4545.20	4548.93	**4534.06**

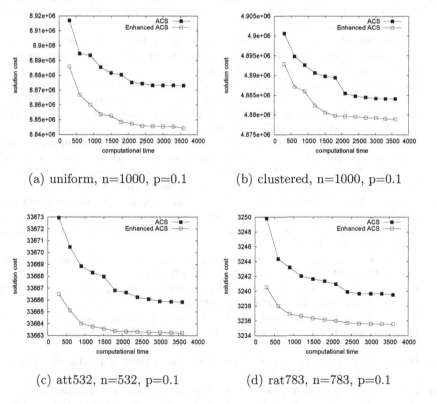

(a) uniform, n=1000, p=0.1

(b) clustered, n=1000, p=0.1

(c) att532, n=532, p=0.1

(d) rat783, n=783, p=0.1

Fig. 1. An illustration of the temporal behavior of the ACS algorithm and the EACS algorithm on benchmark instances with probabilities of 0.1.

be observed. Additionally, improvements over best-known results could be obtained for many instances.

Until now the efficiency of the Enhanced Ant Colony System algorithm has been demonstrated for the Sequential Ordering Problem [12], the Team Orienteering Problem with Time Windows [17] and the Probabilistic Traveling Salesman Problem. Due to the efficiency and the generality of the new approach, applications to other Combinatorial Optimization Problems seem very promising.

Acknowledgments. The first author's research has been supported by the Swiss National Science Foundation, grants 200021-120039/1 and 200020-134675/1.

References

1. Balaprakash, P., Birattari, M., Stützle, T., Dorigo, M.: Adaptive sample size and importance sampling in estimation-based local search for the probabilistic traveling salesman problem. Eur. J. Oper. Res. **199**, 98–110 (2009)
2. Balaprakash, P., Birattari, M., Stützle, T., Dorigo, M.: Estimation-based metaheuristics for the probabilistic traveling salesman problem. Comput. Oper. Res. **37**(11), 1939–1951 (2010)
3. Balaprakash, P., Birattari, M., Stützle, T., Yuan, Z., Dorigo, M.: Estimation-based ant colony optimization and local search for the probabilistic traveling salesman problem. Swarm Intell. **3**(3), 223–242 (2009)
4. Bianchi, L., Gambardella, L.M., Dorigo, M.: Solving the homogeneous probabilistic traveling salesman problem by the ACO metaheuristic. In: Dorigo, M., Di Caro, G.A., Sampels, M. (eds.) Ant Algorithms 2002. LNCS, vol. 2463, pp. 176–187. Springer, Heidelberg (2002)
5. Bianchi, L., Gambardella, L.M., Dorigo, M.: An ant colony optimization approach to the probabilistic traveling salesman problem. In: Guervós, J.J.M., Adamidis, P.A., Beyer, H.-G., Fernández-Villacañas, J.-L., Schwefel, H.-P. (eds.) PPSN 2002. LNCS, vol. 2439, pp. 883–892. Springer, Heidelberg (2002)
6. Birattari, M., Balaprakash, P., Stützle, T., Dorigo, M.: Estimation-based local search for stochastic combinatorial optimization using delta evaluations: a case study on the probabilistic traveling salesman problem. INFORMS J. Comput. **20**(4), 644–658 (2008)
7. Branke, J., Guntsch, M.: Solving the probabilistic TSP with ant colony optimization. J. Math. Model. Algorithms **3**(4), 403–425 (2004)
8. Campbell, A.M.: Aggregation for the probabilistic traveling salesman problem. Comput. Oper. Res. **33**(9), 2703–2724 (2006)
9. Colorni, A., Dorigo, M., Maniezzo, V.: Distributed optimization by ant colonies. In: Proceeding of ECAL - European Conference on Artificial Life, pp. 134–142 (1991)
10. Dorigo, M., Di Caro, G., Gambardella, L.M.: Ant algorithms for discrete optimization. Artif. Life **5**, 137–172 (1999)
11. Dorigo, M., Maniezzo, V., Colorni, A.: The ant system: optimization by a colony of cooperating agents. IEEE Trans. Syst. Man Cybern. - Part B: Cybern. **26**(1), 29–41 (1996)

12. Gambardella, L.M., Montemanni, R., Weyland, D.: An enhanced ant colony system for the sequential ordering problem. In: Proceedings of OR 2011 - International Conference on Operations Research, Zurich, Switzerland (2011)
13. Gambardella, L.M., Montemanni, R., Weyland, D.: Coupling ant colony systems with strong local searches. Eur. J. Oper. Res. **220**(3), 831–843 (2012)
14. Gendreau, M., Laporte, G., Seguin, R.: Stochastic vehicle routing. Eur. J. Oper. Res. **88**(1), 3–12 (1996)
15. Jaillet, P.: Probabilistic traveling salesman problems. Ph.D. thesis, M. I. T., Dept. of Civil Engineering (1985)
16. Johnson, D.S., McGeoch, L.A.: The traveling salesman problem: a case study in local optimization. In: Local Search in Combinatorial Optimization, pp. 215–310 (1997)
17. Montemanni, R., Weyland, D., Gambardella, L.M.: An enhanced ant colony system for the team orienteering problem with time windows. In: ISCCS 2011 - The 2011 International Symposium on Computer Science and Society, pp. 381–384. IEEE (2011)
18. Tsplib. http://www.iwr.uni-heidelberg.de/groups/comopt/software/TSPLIB95/
19. Weyland, D., Bianchi, L., Gambardella, L.M.: New approximation-based local search algorithms for the probabilistic traveling salesman problem. In: Moreno-Díaz, R., Pichler, F., Quesada-Arencibia, A. (eds.) EUROCAST 2009. LNCS, vol. 5717, pp. 681–688. Springer, Heidelberg (2009)
20. Weyland, D., Bianchi, L., Gambardella, L.M.: New heuristics for the probabilistic traveling salesman problem. In: Proceedings of the VIII Metaheuristic International Conference (MIC 2009) (2009)

Exploiting Synergies Between Exact and Heuristic Methods in Optimization: An Application to the Relay Placement Problem in Wireless Sensor Networks

Eduardo Feo Flushing[✉] and Gianni A. Di Caro

Dalle Molle Institute for Artificial Intelligence (IDSIA), Manno-Lugano, Switzerland
{eduardo,gianni}@idsia.ch

Abstract. Exact and heuristic methods for solving difficult optimization problems have been usually considered as two completely different approaches, with each of them being extensively studied by a different research community and used in different fields of application. In this work we propose a scheme for the integration and cooperation of an exact method, based on mixed-integer linear programming (MILP), and a bio-inspired metaheuristic, a genetic algorithm (GA), for the solution of a relay node placement problem in wireless sensor networks.

The integration aspect relates to the use of a MILP solver inside GA's operators. The cooperation aspect is implemented through a shared incumbent environment in which a MILP solver and the GA work in parallel and exchange relevant information in order to overcome their individual weaknesses and provide better solutions within a shorter time.

Experimental results show a significant increase of performance when integration and cooperation take place, in comparison to the performance of the MILP and the GA when used independently from each other.

Keywords: MILP · Genetic algorithm · Relay placement · Hybrid metaheuristic · Wireless sensor networks · Shared incumbent · Optimization

1 Introduction

In this work we implement a novel synergistic search scheme for solving complex optimization problems. Our approach is based on a social metaphor: a community of problem solvers, of possibly different nature and featuring different skills, communicate and cooperate with each other in order to find the best solution to a common problem. The solvers actively and concurrently exploit their mutual differences in order to produce community-level synergies and overcome individual weaknesses. More specifically, we consider the combination of exact and heuristic methods, considering a *mixed-integer linear programming (MILP)* approach and a *genetic algorithm (GA)*, a bio-inspired metaheuristic.

MILP and metaheuristics are two different paradigms for solving optimization problems. The former is characterized by the definition of a mathematical

G.A. Di Caro and G. Theraulaz (Eds.): BIONETICS 2012, LNICST 134, pp. 250–265, 2014.
DOI: 10.1007/978-3-319-06944-9_18, © Institute for Computer Sciences, Social Informatics
and Telecommunications Engineering 2014

program including linear equations and both continuous and integer variables. A solution is determined according to methods that intelligently explore the search space and provide a guarantee on the optimality of the solution found. Metaheuristics, on the other hand, aim to exploit problem-specific knowledge and to find good (possibly optimal) solutions within short time, but not providing any guarantees on the quality of the solution found. Given their structural differences, both approaches exhibit complementary properties, such that the strengths of one can help to overcome the weaknesses of the other, and vice versa. For example, the utter exploration of the solution space performed by a MILP solver can be exploited by the metaheuristic search in order to avoid fast convergence to or get trapped in local optima. Also, the MILP solver may use the local optima identified by the metaheuristic search to improve its search bounds and therefore be able to identify areas of the solution space which can be safely excluded from the search. These potential synergies suggest the combination of both approaches to get the best of two worlds. The relatively new domain of research in *matheuristics* [19] precisely addresses the study of the integration of exact and heuristic methods.

The specific optimization problem that we tackle in this paper is the *relay node placement for performance enhancement* problem (RNP-PE in short), a complex combinatorial optimization problem with several applications in Wireless Sensor Networks (WSNs). In general terms, the relay node placement optimization problem can be defined as selecting, from a given set of possible locations, a limited number of positions where some additional nodes (called relay nodes) can be deployed. The objective of the deployment is to improve the network performance metrics of interest. In the case of our RNP-PE, we define these metrics as the *end-to-end communication delays* and the overall data *throughput* obtained at the *sink nodes* of the WSN.

Realistically large instances of the RNP-PE problem present an exponential number of possible deployments. Moreover, for each single deployment, an optimal routing problem must be solved in order to evaluate its quality. Unfortunately, the optimal routing problem belongs to the family of $\mathcal{NP}$-hard network design problems [13,25]. For this reason, standard approaches fail to find optimal solutions to the RNP-PE, even for small size instances, in reasonably short time. To overcome these limitations and find good solutions to the RNP-PE within short time and with quality guarantees, we employ a social approach for problem solving which is based on cooperation and integration. The *cooperation* aspects involve the use of different strategies, namely a MILP solver and a GA heuristic, working in parallel and continuously exchanging relevant search information. The exchanged information is used by each side to improve its performance. The *integration* lies in the decomposition of the problem, and the inclusion of another MILP solver inside the GA. The decomposition approach enables the heuristic to obtain valuable information which is exploited by the genetic operators.

2   Related Work

Bio-inspired approaches, and more specifically GAs, have been used to tackle the complexity of many optimization problems in wireless networks, including clustering [2], optimal design [12], routing and link scheduling [4], network planning [23] and node placement [7,22,26,31,32]. Despite the success of GAs and other metaheuristics in solving complex optimization problems, there is a growing feeling that pure metaheuristics are reaching their limits [5]. As technology advances, new and increasingly complex optimization problems arise. These observations, together with the exhaustion of new design ideas for pure metaheuristics, is leading the researchers to explore the combination of different techniques and the proposal of hybrid methods [19].

The main motivation behind the hybridization of different algorithms is to exploit their potential synergies. Not until recently, hybrid approaches began to appear in networking optimization problems. In [3], a tabu search is embedded in the genetic operators of a conventional GA to solve a frequency assignment problem, while in [17], tabu search is combined with simulated evolution to tackle a network routing problem. In [26], local search heuristics are used to evaluate GA's individuals to solve a switch location problem in cellular networks.

Within the vast possibilities of hybrid methods for combinatorial and discrete optimization, there is a promising direction of research in combining exact mathematical programming techniques and meta-heuristic approaches. According to the structure of these different solution approaches, two main categories have been identified: *integrative and collaborative combinations*. In integrative combinations, one technique is usually embedded inside another technique, hence the latter is seen as a functional component of the former. On the other hand, collaborative combinations feature two or more methods running sequentially, intertwined or in parallel, which are not part of each other [24].

A *shared incumbent environment* is a general methodology to realize collaborative combinations of mathematical programming and metaheuristic approaches. The main idea of the shared environment is to allow both components (i.e., a MILP solver and a metaheuristic algorithm) to exchange information about their current best known solutions. This information is used by the MILP solver to improve its current incumbent and prune the branch and bound tree, and by the meta-heuristic to guide the search to more promising regions of the solution space and to prevent getting stuck in local minima [28].

Compared to previous approaches, our work exploits the synergies between different methods in two dimensions both in integration and cooperation. We consider a heterogeneous group of solvers, composed of standard mathematical solvers and a well-known bio-inspired metaheuristic.

3   The Problem

A *wireless sensor network* (WSN) consists of a set nodes that are equipped with sensing and limited processing capabilities, and that can locally communicate with each other through a wireless medium [1]. The *sensor nodes* (SNs) composing a WSN are usually inexpensive and low-powered, such as they can be deployed in large numbers to provide monitoring and sensing services for long time periods. In typical applications, the data generated by the sensor nodes need to be transmitted to and aggregated and processed at *base stations* (BSs). The general model for the forwarding of the data from SNs to BSs is based on the definition and the use of *multi-hop routing paths*.

Since a WSN can operate for relatively long times and/or it can be embedded in dynamic or hostile environments, a core issue in WSNs is the definition of effective strategies to maintain the network aoperational for long time period and/or for its adaptation to external or internal changes. In this direction, a wealth of research has considered the use of special nodes, referred to as *relay nodes* (RNs), that can be deployed and added to the WSN after the network has been put in place. RNs can be positioned by hand, or can be part of a mobile robotic unit, such that they can be deployed autonomously or on-demand. Possible roles of RNs include the provisioning of connectivity [6,9,18,20,27], *extend the network lifetime* [15,30], *energy-efficient or balanced data gathering* [10,21], and to provide *survivability and fault tolerance* [14,16,20,27,33].

The *relay node placement for performance enhancement* problem is defined as follows: given a set of possible locations where to deploy a restricted number of available RNs, we aim to select from this set the locations in which the additional nodes can be positioned to improve throughput and end-to-end packet delays for the data gathered at BSs. Although the primary objective is determining the physical locations where RNs should be placed to, the solution also specifies the way these RNs should be used. This specification comes in the form of *optimal routing paths* from SNs to BSs to forward the data flows. We assume that the initial WSN is connected, therefore the use of RNs is entirely devoted to improve the performance of the network. We present the RNP-PE as a *linear, mixed integer mathematical program* (MILP). The formulation includes a number of constraints and penalty components, aimed at closely modeling the specific characteristics of the wireless environment.

3.1   MILP Model

We model the WSN as a set of SNs and BSs located in a set of known positions $\mathcal{S}$ and $\mathcal{B}$, respectively. SNs both generate and forward data packets towards one of the BSs in multi-hop fashion (a data flow can be split over multiple paths). We assume that the characteristics of data generation characteristics for each SN are known. All nodes communicate with each other within the *communication range* **r**. A set of K RNs is also available, their role is to forward data received from other nodes. The placement of node relays is restricted to a *numerable set of candidate locations* denoted as $\mathcal{R}$. We formalize the RNP-PE by a MILP model based on a *minimum cost flow* formulation as follows.

Let $G = (V, E)$ be a connected digraph representing a WSN, where $V = \mathcal{N}$ is the set of nodes, and E is the set of communication links. $\gamma : E \mapsto \Re$ is a *link cost* function, and $\tau : \mathcal{S} \mapsto \Re$ is a data generation (*traffic load*) function, expressed in the *data per second* generated by an SN. In the following, we measure τ in terms of *flow units*, f_{unit}, expressed as bytes/sec. Data flows and relay positions define the two sets of *decision variables*. The *flow variable* f_{ij} denotes the amount of flow through link (i, j), that is, the data traffic to be sent from node n_i to node n_j located at positions i, $j \in \mathcal{N}$ respectively. f_{ij} values are expressed in *flow units*. The *binary positional variable* y_i indicates whether location $i \in \mathcal{R}$ is used to circulate flow or not. When y_i is set to 1 in a solution, an RN is to be positioned at the corresponding relay location. A full solution specifies both flows and relay positions. The SN-to-BS routes are defined in the *routing-tree* induced by the set $\{(i, j) \in E \mid f_{ij} > 0\}$. The complete MILP model is presented in Fig. 1.

$$\min \text{ RNP-PE} = \sum_{(i, j) \in E} \gamma_{ij} f_{ij} + \hat{R} \sum_{i \in \mathcal{R}} y_i + \alpha \sum_{i \in \mathcal{S}} p_i \hat{F} \,. \tag{1}$$

$$\text{subject to: } \sum_{(i, j) \in E} f_{ij} - \sum_{(j, i) \in E} f_{ji} = \begin{cases} \tau_i & \text{if } i \in \mathcal{S}, \\ 0 & \text{if } i \in \mathcal{R} \end{cases} \tag{2}$$

$$\sum_{i \in \mathcal{B}} \sum_{(j, i) \in E} f_{ji} = \sum_{k \in \mathcal{S}} \tau_k \tag{3}$$

$$y_i = 1 \iff \sum_{j \in \mathcal{N}} f_{ji} > 0 \quad \forall i \in \mathcal{R} \tag{4}$$

$$\sum_{i \in \mathcal{R}} y_i \leq K \tag{5}$$

$$\sum_{(i, j) \in E} f_{ij} + \sum_{(j, i) \in E} f_{ji} \leq L_{cap}, \quad \forall i \in \mathcal{N} \tag{6}$$

$$b_{ij} = 1 \iff f_{ij} > 0 \quad \forall i, j \in \mathcal{N} \tag{7}$$

$$\sum_{(j, i) \in E} b_{ji} \leq D \quad \forall i \in \mathcal{S} \tag{8}$$

$$p_i = 1 \iff \sum_{(i, j) \in E} \sum_{(j, k) \in E} f_{jk} \geq \bar{F}_{max} \tag{9}$$

Fig. 1. MILP formulation of the RNP-PE problem.

Constraints (2–3) correspond to the flow definition. The number of available RNs is limited to K constraints (4–5). Given that the optimal solution can correspond to a number of RNs $k < K$, we define a penalty factor in the objective (1) to favor the use of a minimal amount of RNs: any optimal solution using n relays needs to provide a minimal gain $\hat{R}$ with respect to the solution obtained using $n - 1$ relays. Parameter $\hat{R}$ can be adjusted according to the problem instance (e.g., relay node availability, economic cost).

Shared wireless channels in WSNs are necessarily *bandwidth-limited*. This condition is reflected by *link capacity* parameter L_{cap}, which is the nominal amount of data (bytes/sec) that can be transmitted by a wireless link in the network (assuming the same capacity for all links), and constraint (6).

For a node n, the *routing in-degree* is the number of n's neighbors using n to relay data. Because of shared medium and contention access, this number strongly impacts on the effective node capacity and network load balancing. Hence, node degree in the routing trees are limited by constraints (7–8).

To minimize wireless interference and produce balanced routing trees, which allow balanced energy depletion, we need to setup *minimally interfering flow paths*. We enforce this by including in the objective function a penalty component based on the *maximum local flow*, $\bar{F}_{max}$, defined as the maximum amount of flow that can circulate within a disk of radius r centered on an SN. The calculation of the flow circulating within the r-disk of an SN i, requires to sum up the outgoing flows from all i's neighbors. This notion is included in constraints (9), where p_i is a binary penalty variable that takes value 1 when the flow through the i's r-disk violates the maximum allowed amount. In order to use p for inclusion in the objective function as penalty, we derive a rough estimation, $\hat{F}$, of the optimal solution value of problem, without penalties, and we use it as a penalty score for the violation of the circulating flow limit. Using $\hat{F}$ and p, the penalty for the violation in maximum local flow is therefore included in the objective function. The parameter α weighs the penalty in the objective function. In the experiments we set $\alpha = 0.1$. We refer the interested reader to [8,11] for a full description of the parameters and an extensive evaluation of the RNP-PE model.

4   Integration: A Hybrid Genetic Algorithm

In the RNP-PE formulation, it is primarily the number of possible assignments of relay positions which affects the size of the problem. In instances where $|\mathcal{R}|$ is in terms of thousands (which is a typical scenario), even for small number of relays to select, the solution space is exponentially large.

Considering that the RNP-PE jointly solves two problems, namely the node placement and data routing, we decompose the problem in two hierarchical levels. At the top level, the GA is used to explore possible relay placements, which involves iterating over assignments of the variables y_i of the full MILP model. At the bottom level, a *simplified* MILP is used to compute the optimal routing of each one of these placements. This means that, for a certain assignment of y_i (i.e., positioning of relays), a MILP solver computes the best possible flow routing scheme achievable (based on the RNP-PE formulation). The optimization problem obtained by fixing the variables y_i is much smaller and simpler than the original one, therefore its computation time is much reduced, a property which is effectively exploited by the decomposition method.

4.1   Solution Encoding

The encoding of individuals (also known as *chromosome encoding*) is fundamental to the implementation of GAs in order to efficiently transmit the genetic information from parents to offsprings. In our case, an individual of the population represents a deployment of relay nodes. Since RNs can be placed at one of the set of candidate locations $\mathcal{R}$, the location of each RN can be conveniently specified as the index of the element in $\mathcal{R}$ to which it corresponds to. Accordingly, a population member is encoded as an array of indexes values. The array size is variable, given that the problem can admit solutions with $k \leq K$ relays.

Apart from the encoding, another fundamental aspect of a GA is the design of its genetic operators. We propose a set of application-specific genetic operators designed to achieve a good trade-off between exploration and exploitation.

4.2   Evaluation of Individuals

Let $R^* \subseteq \mathcal{R}$, be an individual of the genetic population, with $|R^*| \leq \mathcal{K}$. In order to evaluate the quality of R^*, we compute the optimal routing scheme based on the relays located at the positions in R^*. A simple way to perform this computation consists in constructing and solving a MILP problem, using the RNP-PE formulation and replacing $\mathcal{R}$ with R^*. However, in practice, this approach involves an additional overhead in terms of computation time, since a complete mathematical model (which needs to be constructing and initialized) must be built each time. The overhead is more evident after the evaluation of hundreds (or even thousands) of individuals, as usually performed by a GA.

To overcome this issue, we propose to initially build a complete model, using the whole set $\mathcal{R}$, and add and remove constraints to the original formulation every time an individual need to be evaluated. This way turns out to be both efficient and time-saving. To evaluate R^* using the original formulation presented in Sect. 3.1, the following constraints must be included:

$$y_i = 1 \quad \forall i \in R^* \tag{10}$$

$$y_i = 0 \quad \forall i \in \mathcal{R} \setminus R^*. \tag{11}$$

Additionally, since the best data routing schemes may be achieved using less than $|R^*|$ relays, the model need to be enabled to obtain solutions in which not all of the positions in R^* are used for data *flows*. However, the original formulation enforces the generation of routing solutions such that, if y_i is equal to 1, some flow variable in the set $\{f_{ji}, j \in \mathcal{N}\}$, must be positive. This implies that by adding constraints (10), we are forcing the model to produce routing path that require to use all the positions specified in R^*. To avoid this undesirable behavior, we relax the constraints (4) by replacing them with:

$$y_i = 1 \Leftarrow \sum_{j \in \mathcal{N}} f_{ji} > 0 \quad \forall i \in \mathcal{R}. \tag{12}$$

After solving the now simplified MILP, the obtained solution is checked for *unused* relays, that is, for positions $i \in R^*$ such that $\sum_{j \in \mathcal{N}} f_{ji} = 0$. For these positions, the objective function value is recomputed assuming $y_i = 0$, hence obtaining an accurate assessment of the optimal routing paths.

4.3   Routing-Aware Placement (RAP) Crossover

We propose an application-specific crossover operator, referred to as *Routing-aware placement* (RAP) crossover, aimed to boost the performance of the GA search. The main ideas behind our operator are the design of a multi-parent crossover operation and the inclusion of the extracted information from the routing trees obtained by the evaluation of previous individuals.

During the progress of the GA, the routing trees obtained after the evaluation of each solution are examined and analyzed to extract useful information about the shape of the generated sub-optimal solutions. In these routing trees, some relay nodes (and hence their corresponding positions) are more important than others, in the sense that they are used to carry more data. Therefore, we measure the *degree of importance* of relay positions by looking at the amount of flow circulating through the positioned relay nodes each time we evaluate an individual. The most important relay positions form a set which we refer to as **preferential relay set**, hereafter denoted as $PS \subseteq \mathcal{R}$.

Since close-by relay positions are likely to be used in a similar manner, the set PS might become a group of *clustered relay positions*. To promote diversity in PS, we propose a strategy which consists on partitioning the area into a number of regions, where to each region is assigned a quota in PS, corresponding to the maximum number of relay positions located in that region. The maximum size of PS is limited by a user-defined parameter (set to $2K$ in this paper).

Another structural property that we observe in the routing trees, is the presence of **relay chains**. That is, the combined use of relays forming chains to link different regions of the network. A pair of *chained relay positions* appears in a solution whenever RNs located at these positions are linked to each other, meaning that data flows directly from one relay to the other. After every solution evaluation, we extract all pairs of chained relay positions and store them inside a map structure, $CR : \mathcal{R} \mapsto 2^{\mathcal{R}}$, where the set $CR(i) \subseteq \mathcal{R}$ contains all the positions that have formed a chain with i.

Finally, we also identify possible **conflicts** between relay positions. A conflict is identified whenever a relay position is included as an individual but is not used to carry data in the routing solution. This situation suggests that the potential benefits of using the ignored relay position have been achieved by other RNs, placed in different positions. Hence, we say that the ignored relay positions are *in conflict* with the currently utilized ones, and vice versa. To this end, we keep a conflict map $CM : \mathcal{R} \mapsto 2^{\mathcal{R}}$ that associates to each RN position a set of positions that appear to be in conflict with any of the routing tree solutions. Figure 2 illustrates the analysis of the routing solutions performed after the evaluation of an individual. In the example, one pair of chained relays, one preferential relay position (carrying data from four static nodes), and one conflict (unused position) are detected in the routing tree. The description of the algorithmic procedure of the RAP crossover is presented in Fig. 1. The parameter $pChainedRelay$ regulates the use of chained relay positions. We denoted as Mom and Dad the genetic parents. The size of the new child individual is taken randomly in the interval defined by the size of both parents involved in the

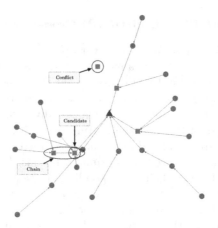

Fig. 2. Elements extracted from MILP solution

operation. The operator first tries to construct a conflict-free set of RN positions by randomly taking genes from both parents. Each time a gene g is selected, it also takes a chained relay from $CR(g)$ with probability $pChainedRelay$. After this attempt to construct a child, the number of selected genes might not be enough to achieve the target child size. Therefore, a second step is performed in which the remaining RN positions are randomly selected from the preferential set, and in case there are not enough preferential relays, the remaining genes are selected taken randomly from $\mathcal{R}$. In this way, the operator produces a new chromosome which shares genetic material from both parents and from the set PS. Moreover, the RAP crossover operator reinforces the generation of better individuals by excluding conflicts previously seen, replacing those conflicts by potentially useful relays (i.e., from PS), and by introducing chained relays positions, which are usually very beneficial, but not likely to appear frequently when using a completely random procedure.

4.4 Mutation

The proposed mutation operator allows a controlled exploration of new regions of the solution space by inducing small perturbations to existing individuals. Mutation is implemented by displacing the relay positions within a circular area of size $2r$. Apart from changing the positions of the relays, the mutation may also modify the size of the solution, that is, the number of positions of relays. With a given probability $pSizeChange$, the operator varies the size, removing existing positions or adding new ones.

5   Cooperation: A Shared Incumbent Environment

At this point, we are able to obtain solutions to the RNP-PE in two different ways: (a) using a standard mathematical solver to solve the MILP model presented in Sect. 3.1, and/or (b) using the metaheuristic procedure introduced in

Sect. 4. Both approaches offer different advantages and disadvantages. Approach (a) offers the possibility (if we are lucky enough) to find an optimal solution. However, depending on the problem instance, we might run out of computational resources before obtaining a good (or even at least any) feasible solution. Approach (b) offers solutions whose quality depends on the amount of time we provide to the GA procedure. In some cases, the evolution process can get stuck and no improvements can be made to the current solution. Moreover, even though we can get optimality bounds for the provided solutions, these bounds might not be tight. Hence, we won't be able to have a strong assessment about the quality of the obtained solution (even if the solution is in fact optimal).

Input: Parents: $Mom, Dad \subseteq \mathcal{R}, |Dad| \leq |Mom| \leq K$
Input: Preferential relay set: $PS \subseteq \mathcal{R}, |PS| \leq 2K$
Input: Chained relays: $CR : \mathcal{R} \mapsto 2^{\mathcal{R}}$, Conflict map: $CM : \mathcal{R} \mapsto 2^{\mathcal{R}}$
Result: $Child \subseteq \mathcal{R}$
if $Mom = Dad$ **then**
| $Child \longleftarrow RandomChild()$ **return** Child
end
$childSize = RandInteger(|Dad|, |Mom|)$
$genePool = Mom \cup Dad$
$Child = \emptyset$
while $|Child| < childSize$ **and** $genePool \neq \emptyset$ **do**
| pick random relay $g \in genePool$ **if** $\exists \hat{g} \in Child : g \in CM(\hat{g})$ **then**
| | $genePool \longleftarrow genePool \setminus \{g\}$
| **else**
| | $Child \longleftarrow Child \cup \{g\}$ $genePool \longleftarrow genePool \setminus \{g\}$ **if** $Rand(0, 1) \leq pChainedRelay$
| | **and** $|Child| < childSize$ **then**
| | | pick random relay $h \in CR(g)$ $Child \longleftarrow Child \cup \{h\}$ $genePool \longleftarrow genePool \setminus \{h\}$
| | **end**
| **end**
end
if $(childSize - |Child|) \leq |PS|$ **then**
| **while** $|Child| < childSize$ **do**
| | pick random relay $g \in PS$ $Child \longleftarrow Child \cup \{g\}$
| **end**
else
| $Child \longleftarrow Child \cup PS$ **while** $|Child| < childSize$ **do**
| | pick random relay $g \in \mathcal{R}$ $Child \longleftarrow Child \cup \{g\}$
| **end**
end
return $Child$

Algorithm 1: RAP crossover operator

A *cooperative environment* is implemented by the execution of both approaches as two independent processes able to *communicate* with each other. We use a shared incumbent environment as the cooperation scheme between both solvers, which consists in letting both methods continuously exchange their best found solutions so far. In the MILP solver, this corresponds to the best upper bound (also known as the *incumbent solution*). In the GA, it is simply represented by the best individual that has been evaluated so far.

The hybrid nature of the GA and its MILP functional component used in the evaluation operator, facilitates the implementation of the shared incumbent environment. Each time an individual is evaluated by the GA, we can easily obtain, from the reduced MILP, the values of the variables corresponding to the optimal routing solution. These values, together with those corresponding to the positions (which are available from the encoding), allows to compose

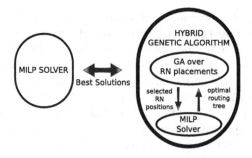

Fig. 3. Cooperative environment based on the shared incumbent.

the solution vector of the complete MILP model. Due to the potentially large number of variables, we use a data compression scheme for the communication of the solution vectors. A graphical representation of the proposed bio-inspired shared environment is presented in Fig. 3.

6   Evaluation

To evaluate the proposed method, we considered a number of randomly generated network instances representing complex scenarios of the RNP-PE. In total, we considered 20 network instances generated with different topological characteristics, i.e. *uniform, clustered,* and *small world.* Networks were embedded in an area of size $150 \, \text{m} \times 150 \, \text{m}$, and the set of possible relay positions was determined using a uniform grid, with the grid points separated by $\Delta = 2 \, \text{m}$ of distance. Considering the grid resolution and the size of the area, the number of possible relay positions becomes particularly large (up to 5000). We have considered two group of experiments varying the value of $\mathcal{K}$ since this parameter determines the number of deployments, and has an impact on the performance of the GA. Table 1a indicates the parameters of network topologies considered. Figure 4 depicts one of the topology instances. The evaluation consisted on three steps.

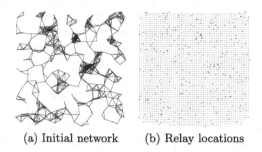

(a) Initial network   (b) Relay locations

Fig. 4. Example of network topology instance

First, all instances are exhaustively solved using the MILP model presented in Sect. 3.1. To solve the MILP we used the CPLEX® solver under its default parameters. In this step, the mathematical solver was given a larger amount of computational resources, in terms of both CPU time (10 h) and physical memory (8 GB RAM), in comparison to the following experiments. In this way, we aimed at obtaining the optimal or best sub-optimal solutions to each of the problem instances, through an intensive and exhaustive solving process.

In the second step, we solved the same instances using the hybrid Genetic Algorithm presented in Sect. 4. To evaluate the efficacy of the genetic operators, we also use a simplified version of the GA, in which the crossover and mutation operators were replaced by a one-point random crossover and a uniform random mutation operators. In the implementation, we use a steady state genetic algorithm, and the *GAlib library* [29]. The main parameters of the GA are listed in Table 1b. Finally, in the last step, we solve the instances using the cooperative environment as described in Sect. 5. Both components (i.e., the mathematical solver and the GA) were executed on similar CPUs, and the communication was implemented using sockets. All methods, except the initial step, were given a maximum CPU time of 2 h and 2 GB of physical memory.

6.1   Experimental Results

After performing the first step of the evaluation, we obtained solutions to each of the instances considered. Table 2 shows objective function values, optimality gap, and solving time for some of the instances considered. The mathematical solver was not able to prove the optimality of any of the considered instances. For some instances, the solver ran out of memory and finished before consuming the available CPU time, as noted in the table. In the second step, we analyze performance of the solution approaches, namely the GA with one-point crossover and uniform mutation (GA-one point), the GA with the RAP crossover and the proposed mutation operators (GA-RAP), and the cooperative environment featuring a MILP solver and the GA within a shared incumbent environment (MILP+GA). To measure the performance of each approach, we consider the ratio between the objective function value of their best solutions found and the solutions obtained in the first step. We refer to this value as the *performance ratio*. For each instance, we performed 20 independent runs (to account random-

Table 1. Experimental setup.

(a) Instances parameters

Area	$150 \times 150 \text{ m}^2$
$|\mathcal{S}|$	200
$|\mathcal{B}|$	5
TX range (r)	10m

(b) GA parameters

Population size	100
Scaling scheme	Linear
Selection scheme	Tournament
Crossover probability	0.9
Mutation probability	0.1

Table 2. Solutions obtained by the MILP.

Obj. value	Gap (%)	CPU time (s)
1576.79	11.79	30290.7
2111.38	13.87	36002
1403.98	16.52	19397.2
1318.45	9.53	16725.7

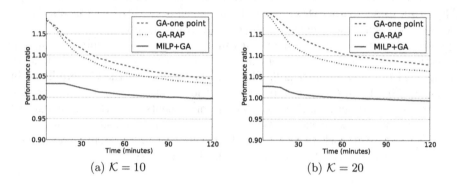

(a) $\mathcal{K} = 10$ (b) $\mathcal{K} = 20$

Fig. 5. Comparison of the proposed solution methods

ness in the GA procedure), and we took the median value of the performance ratio at different points of time.

Figure 5 shows the average performance ratio over all the instances considered for each value of $\mathcal{K}$. We can observe a variation of performance depending on the choice of the genetic operators. The GA-RAP method is more effective than its naive counterpart and provides better solutions. However, both tend to converge fast and have difficulties in improving their best solution as time advances. On the other hand, the performance of the MILP+GA cooperative scheme is considerably better, and in some of the instances is also able to provide solutions better than those obtained with the first step.

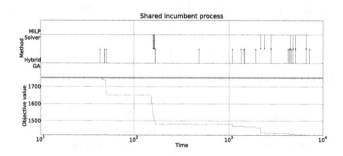

Fig. 6. Illustration of cooperation process.

To illustrate the interactions between both methods within the cooperative environment, we selected one execution of the MILP+GA method. Figure 6 shows the interactions and the evolution of the best solution over time (logarithmic scale). We can observe two clear behaviors: the reactive interactions (sequences of solution exchanges separated by short intervals, and stand-alone interactions. We conclude that none of the methods strictly dominates the others, and the performance obtained by the combination is much superior than the performance achieved by each method individually.

7 Conclusions

We considered an integrative and cooperative environment for solving a complex mixed-integer optimization problem in WSNs. The approach features the use of two problems solvers, of different nature and with orthogonal properties, which interact and cooperate in order to improve their joint performance. Specifically, we presented the case of a bio-inspired metaheuristic (i.e., a genetic algorithm) and a traditional MILP solver for the relay node placement problem in WSNs.

The contributions of this work lie along three lines/ First, we propose a GA which makes use of a decomposition strategy and a MILP solver to evaluate the individuals of the population. Secondly, we designed a cooperative environment in which a MILP solver and the proposed GA work, solve the same problem in parallel, interacting and continuously exchanging relevant information. The objective is to help ach other to overcome their weaknesses and speed-up the solving process. Finally, we demonstrate through extensive experiments the effectiveness of the proposed strategy and show a significant performance improvement when the solvers are combined, compared to their use as independent solvers.

Future work involves considering larger groups (three or more) of solvers, including other types of metaheuristics and mathematical solvers, and studying and assessing the collaboration level achieved according to the size and nature of the group of solvers. Together with this, we will also consider the application of this strategy to other, possibily related, optimization problems.

References

1. Akyildiz, I.: Wireless sensor networks: a survey. Comput. Netw. **38**(4), 393–422 (2002)
2. Al-Obaidy, M., Ayesh, A., Sheta, A.F.: Optimizing the communication distance of an ad hoc wireless sensor networks by genetic algorithms. Artif. Intell. Rev. **29**(3–4), 183–194 (2009)
3. Alabau, M., Idoumghar, L., Schott, R.: New hybrid genetic algorithms for the frequency assignment problem. IEEE Trans. Broadcast. **48**(1), 27–34 (2002)
4. Badia, L., Botta, A., Lenzini, L.: A genetic approach to joint routing and link scheduling for wireless mesh networks. Ad Hoc Netw. **7**(4), 654–664 (2009)
5. Blum, C., Puchinger, J., Raidl, G.R., Roli, A.: Hybrid metaheuristics in combinatorial optimization: a survey. Appl. Soft Comput. **11**(6), 4135–4151 (2011)

6. Cheng, X., Du, D.Z., Wang, L., Xu, B.: Relay sensor placement in wireless sensor networks. Wirel. Netw. **14**(3), 347–355 (2007)
7. Chu, S., Wei, P., Zhong, X.: Deployment of a connected reinforced backbone network with a limited number of backbone nodes. IEEE Trans. Mob. Comput. 1–14 (2012)
8. Di Caro, G.A., Feo, E.: Optimal relay node placement for throughput enhancement in wireless sensor networks. In: Proceedings of the 50th FITCE International Congress (2011)
9. Efrat, A., Fekete, S., Gaddehosur, P., Mitchell, J., Polishchuk, V.: Improved approximation algorithms for relay placement. In: Algorithms-ESA (2008)
10. Ergen, S., Varaiya, P.: Optimal placement of relay nodes for energy efficiency in sensor networks. In: Proceedings of IEEE ICC, vol. 8, pp. 3473–3479 (2006)
11. Feo, E.: Optimal relay node placement for throughput enhancement in wireless sensor networks. Master's thesis, Trento, Italy, December 2010
12. Ferentinos, K., Tsiligiridis, T.: Adaptive design optimization of wireless sensor networks using genetic algorithms. Comput. Netw. **51**(4), 1031–1051 (2007)
13. Garey, M.R., Johnson, D.S.: Computers and Intractability; A Guide to the Theory of NP-Completeness. W. H. Freeman & Co., New York (1990)
14. Han, X., Cao, X., Lloyd, E., Shen, C.: Fault-tolerant relay node placement in heterogeneous wireless sensor networks. IEEE Trans. Mob. Comput. 1 (2009)
15. Hou, Y., Shi, Y., Sherali, H., Midkiff, S.: Prolonging sensor network lifetime with energy provisioning and relay node placement. In: Proceedings of IEEE SECON'05, pp. 295–304 (2005)
16. Kashyap, A., Khuller, S., Shayman, M.: Relay placement for higher order connectivity in wireless sensor networks. In: Proceedings of IEEE INFOCOM, pp. 1–12 (2006)
17. Khan, S.A., Baig, Z.A.: A Simulated Evolution-Tabu search hybrid metaheuristic for routing in computer networks. In: Proceedings of IEEE CEC, pp. 3818–3823 (2007)
18. Lloyd, E., Xue, G.: Relay node placement in wireless sensor networks. IEEE Trans. Comput. **56**(1), 134–138 (2007)
19. Maniezzo, V., Stützle, T., Voß, S. (eds.): Hybridizing Metaheuristics and Mathematical Programming. Annals of Information Systems. Springer, Heidelberg (2009)
20. Misra, S., Hong, S.D., Xue, G., Tang, J.: Constrained relay node placement in wireless sensor networks: formulation and approximations. IEEE/ACM Trans. Netw. **18**(2), 434–447 (2010)
21. Patel, M., Chandrasekaran, R., Venkatesan, S.: Energy efficient sensor, relay and base station placements for coverage, connectivity and routing. In: Proceedings of IEEE IPCCC, pp. 581–586 (2005)
22. Perez, A., Labrador, M., Wightman, P.: A multiobjective approach to the relay placement problem in WSNs. In: Proceedings of IEEE WCNC, pp. 475–480 (2011)
23. Pries, R., Staehle, B., Staehle, D., Wendel, V.: Genetic algorithms for wireless mesh network planning. In: Proceedings of the 13th ACM MSWIM, p. 226 (2010)
24. Raidl, G., Puchinger, J.: Combining (integer) linear programming techniques and metaheuristics for combinatorial optimization. In: Blum, C., Aguilera, M.J.B., Roli, A., Sampels, M. (eds.) Hybrid Metaheuristics, Studies in Computational Intelligence, pp. 31–62. Springer, Heidelberg (2008)
25. Ravi, R., et al.: Approximation algorithms for degree-constrained minimum-cost network-design problems. Algorithmica **31**(1), 58–78 (2001)
26. Salcedo-Sanz, S., et al.: Optimal switch location in mobile communication networks using hybrid genetic algorithms. Appl. Soft Comput. **8**(4), 1486–1497 (2008)

27. Tang, J., Hao, B., Sen, A.: Relay node placement in large scale wireless sensor networks. Comput. Commun. **29**(4), 490–501 (2006)
28. Toklu, N.E., Montemanni, R., Di Caro, G.A., Gambardella, L.M.: A shared incumbent environment for the minimum power broadcasting problem in wireless networks. In: Proceedings of the ICICN 2012. IACSIT Press, Singapore (2012)
29. Wall, M.: GAlib: A C++ library of genetic algorithm components. Mechanical Engineering Department, Massachusetts Institute of Technology (1996)
30. Wang, G., Huang, L., Xu, H., Li, J.: Relay node placement for maximizing network lifetime in wireless sensor networks. In: Proceedings of IEEE WiCOM, pp. 1–5 (2008)
31. Xhafa, F., Sánchez, C., Barolli, L.: Genetic algorithms for efficient placement of router nodes in wireless mesh networks. In: Proceedings of the IEEE AINA, pp. 465–472 (2010)
32. Youssef, W., Younis, M.: Optimized asset planning for minimizing latency in wireless sensor networks. Wirel. Netw. **16**(1), 65–78 (2008)
33. Zhang, W., Xue, G., Misra, S.: Fault-tolerant relay node placement in wireless sensor networks: problems and algorithms. In: Proceedings of IEEE INFOCOM'07, pp. 1649–1657 (2007)

Bio-Inspired Modeling

Extracting Communities from Citation Networks of Patents: Application of the Brain-Inspired Mechanism of Information Retrieval

Hiroshi Okamoto[1,2(✉)]

[1] Research & Development Group, Fuji Xerox Co., Ltd, 6-1 Minatomirai,
Nishi-ku, Yokohama-shi, Kanagawa 220-8668, Japan
hiroshi.okamoto@fujixerox.co.jp
[2] RIKEN Brain Science Institute, 2-1 Hirosawa, Wako,
Saitama 351-0198, Japan

Abstract. We consider a citation network of technological documents such as patents and define a 'community' as a group of patents that are densely connected within this group but are less connected with patents outside this group; the citation network is supposed to be covered by several communities, with each corresponding to a specific topic of technology. We propose a computational method of extracting a relevant community from the citation network in a manner reflecting user's specific interest in some technological topic. By use of this method, the user readily gets a list of patents to read in order of priority. The algorithm for community extraction in this method models the neural mechanism of short-term memory recall from long-term memory. The benefit of practical use of the proposed method exemplifies that exploring the real brain is helpful for creating new information-processing technologies.

Keywords: Citation network · Community · Brain · Memory recall · Neural mechanism

1 Introduction

A major problem overwhelming modern scientists and engineers is a tremendous number of scientific and technological documents such as academic papers or patents published daily. It is often the case that what one might have to read far exceeds what one can read. It is therefore crucial to prioritize documents in order to find out what one really has to read. Doing this manually, however, requires a huge amount of labor. The purpose of this study is to propose a computational method to remedy this situation; by use of this method, one can readily find a list of documents to read in order of priority.

This method makes use of citation relations between academic papers or patents (the present study focuses on patents). Citation is a good measure to estimate the impact of a technological achievement described in a given patent [1]; citation of a patent in another patent means that the technological achievement described in the

G.A. Di Caro and G. Theraulaz (Eds.): BIONETICS 2012, LNICST 134, pp. 269–282, 2014.
DOI: 10.1007/978-3-319-06944-9_19, © Institute for Computer Sciences, Social Informatics
and Telecommunications Engineering 2014

latter is under the influence or control of that in the former. Some patents have been cited many times; this means that these patents have had broad impact upon subsequent activities of research and development.

In the proposed method, we consider a network of a large number of patents that are connected by citation links (for instance, if patent A is cited by patent B, there is a link between A and B, see Fig. 1). We then extract a 'community' of patents from the entire network (Fig. 2, *right*) [2, 3]. In the literature of network science [4], the term 'community' means a group of nodes that are densely connected within the group but are less connected with nodes outside the group. Community structure is a hallmark of a variety of real world networks such as the World Wide Web/internet, social networks, protein interaction/gene regulatory networks and so forth [4]. Hence we suppose that the citation network of patents is covered by several communities and each community represents a specific topic in technology.

In the proposed method, extraction of a community from a citation network is done by using the computational algorithm previously proposed by the present author [5], which models the mechanism of memory recall in the brain (Fig. 2, *left*). This algorithm calculates the ranking of individual patents in a manner reflecting user's specific interest in some technological topic; the community is hence defined as a set of patents that are highly ranked.

To evaluate the proposed method, we compare its performance of community extraction to that of a benchmark method. Then we examine application of this method to a real citation network of Japan patents and demonstrate that relevant patents are automatically extracted and prioritized.

In the previous study [5] we also demonstrated extraction of relevant documents from a citation network of academic papers. The novelty of the present study lies in the following points: The present study is the first to propose that memory recall in the brain will be compared to community extraction from a network; then we evaluate the efficiency of community extraction by use of the algorithm modeling the mechanism of memory recall in the brain, which itself was proposed in the previous study but was not examined as a process for community extraction there. Furthermore, whereas citation data of only a limited extent of academic papers was available in the previous study, we have been able to exploit citation data of a full extent of Japan patents in the present study, which has enabled us to examine the proposed method in more practical settings.

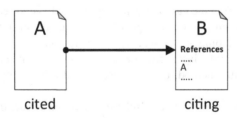

Fig. 1. Patent A is cited by patent B. The arrow line represents the citation relation between them. The arrow head is directed from the cited to citing patents in order to symbolize that the latter is under the influence of the former.

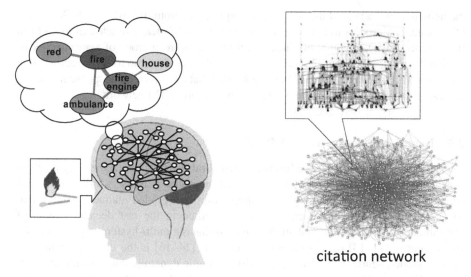

citation network

Fig. 2. *Left*, STM recall from LTM in the brain. *Right*, Extraction of a community from a citation network.

2   Methods

2.1   Citation Network and Spreading Activation

Consider a citation network of patents (Fig. 2, *right*); individual nodes represent individual patents, and individual links represent citation relations between patents. Let $\mathbf{A} = (A_{nm})$ $(n, m = 1, \cdots, N)$ be the adjacency matrix defining the citation network of N patents. If patent m cites patent n, $A_{nm} = 1$; otherwise $A_{nm} = 0$.

We assume that each node has an instantaneous value of 'activity'. Activities spread along links from nodes to nodes in the citation network; at each link, the activity propagates from the citing patent to the cited patent. This process is called 'spreading activation' [6, 7]. We define the 'importance' of a given patent by the value of activity finally acquired by this patent. Spreading activation assigns higher scores of importance to patents that are cited by more numerous and more important patents. The exact dynamics of spreading activation will be given later (see Sect. 2.4).

Spreading activation in the citation network is an analogy to propagation of neuronal activities in the network of neurons in the brain [5, 8]. A node in the citation network can be compared to a single neuron or a population of neurons whose activation encodes a specific 'idea', and a citation link to a synaptic connection between a pair of neurons or a bundle of synaptic connections between a pair of populations of neurons.

2.2   Short-Term Memory Recall from Long-Term Memory in the Brain

In psychology, 'memory' is categorized into 'long-term memory (LTM)' and 'short-term memory (STM)'. The LTM is stored in the brain as an associative network of 'ideas' that the individual has acquired through various experiences from its birth. The STM, on the other hand, is a process to temporally activate a group of ideas extracted from LTM in response to a given situation. For instance, if you see a fire, a group of associated ideas, such as "fire", "red", "house", "fire engine", "ambulance" and so forth, are activated in your brain (Fig. 2, *left*) [6]. Hence we hypothesize that STM recall in response to a given situation is a process of community extraction from the LTM network.

2.3   Neural Mechanism of STM Recall

Recent progress in neuroscience has revealed neural mechanisms of memory recall in considerable detail. In particular, a type of neuronal activity whose magnitude depends on the transient cue signal or stimulation in a graded manner (graded neuronal activity [9, 10]) has been extensively studied in the last decade; results of experimental and computational studies suggest that the multi-hysteretic property of a single neuron [11–13] or a population of neurons [14–16] is the underlying mechanism of graded neuronal activity. The multi-hysteretic dynamics can give the attractor that continuously depends on the initial state; this well accounts for the continuous dependence of the activity on the transient cue signal [11–16]. Extending these findings, we have hypothesized that graded neuronal activity is the neural substrate of cue-dependent recall of STM [8].

2.4   Extracting a Community from a Citation Network
by Multi-hysteretic Dynamics

Our discussion so far has postulated the two hypotheses: STM recall in response to a give situation is a process of community extraction from the LTM network; multi-hysteretic dynamics is the neural mechanism underlying STM recall. Combining these hypotheses, we are led to the idea of extracting a community from a citation network by spreading activation governed by multi-hysteretic dynamics.

Let $x_m(t)$ denote the activity (output) of node (patent) m at time t. The input to node n is given by $I_n(t) \equiv \sum_{m=1}^{N} T_{nm} x_m(t)$, where $T_{nm} = A_{nm} / \sum_{n'=1}^{N} A_{n'm}$. We assume the multi-hysteretic input/output (I/O) relationship for each node, by which the input $I_n(t)$ is converted to the output $x_n(t+1)$ according to the following rule:

$$\text{if}\quad x_n(t) < I_1, \ x_n(t+1) = I_n(t)/\alpha; \tag{1a}$$

$$\text{if}\quad I_1 \leq x_n(t) \leq I_2, \ x_n(t+1) = x_n(t); \tag{1b}$$

$$\text{if}\quad I_2 < x_n(t), \ x_n(t+1) = \alpha I_n(t). \tag{1c}$$

In the above, α is a parameter whose value ranges from 0 to 1 and controls the magnitude of the hysteresis; $I_1 = \alpha I_n(t)$ and $I_2 = I_n(t)/\alpha$. The term 'multi-hysteresis' comes from the fact that the I/O relationship defined above is obtained by multiple stack of bistable (i.e. single-hysteretic) units with each corresponding to a dendritic compartment of a single neuron [11–13] or a single neuron itself [14–16] (see Fig. 2 of [5]).

Because of the multi-hysteretic property, spreading activation done by recursive calculation according to the rule (1a) eventually converges to the steady state, say $\vec{x}(\infty) \equiv \lim_{t\to\infty} \vec{x}(t)$, where the activation pattern of the network continuously depends on the initial activation pattern [8]. Also, nodes that are highly activated in the steady state tend to be mutually connected because activation of one node is supported by activation of others that send links to that node. Thus, a group of nodes highly activated in the steady state tend to form a community. More precisely, a community is represented by vector $\vec{x}(\infty)$ with element $x_n(\infty)$ representing the level of belongingness of node n to that community.

It is worth mentioning the relation between spreading activation governed by the multi-hysteretic dynamics and the PageRank algorithm [17]. The PageRank algorithm is an efficient method to define the importance of individual nodes (typically, web pages) of a network (World Wide Web). This algorithm also uses spreading activation but of the linear dynamics (Markov-chain dynamics)

$$x_n(t+1) = \sum_{m=1}^{N} T_{nm} x_m(t) \ . \tag{2}$$

Note that in the limit $\alpha \to 1$, our algorithm defined by the rule (1) becomes identical to the PageRank algorithm (2). The essential difference between our algorithm and the PageRank algorithm lies in that: The former gives the steady state that depends on the initial state, whereas the latter gives a unique steady state irrespective of the initial state. The PageRank algorithm defines the importance of individual nodes only from the link structure of a network. In contrast, the importance of individual nodes defined by our algorithm can reflect user's specific interest represented by the initial state of a network activation pattern (see Sect. 2.5).

2.5   Seed Patents

In the proposed method, the user is required to express the technological topic that he/ she wants to explore by 'seed patents', which are a set of patents judged by the user to be putatively relevant to the topic [5]. Since the user's in-advance knowledge about the topic might be imperfect, seeds patents might lack some patents that are truly relevant to the topic or include irrelevant ones. Nevertheless the user does not need to be too accurate about the choice of seed patents because, as we shall see, truly relevant patents (namely, highly ranked members of the community) will be retrieved whereas irrelevant ones will be removed in the course of spreading activation.

Let $\vec{\tau}$ be a vector representing a set of seed patents, whose elements are defined as

$$\tau_n = 1 \quad \text{if patent n is a seed patent;} \tag{3a}$$

$$\tau_n = 0 \quad \text{otherwise.} \tag{3b}$$

The initial state of a network activation pattern for spreading activation governed by multi-hysteretic dynamics is defined as

$$\vec{x}(0) = \vec{\tau} \tag{4}$$

2.6   Evaluation of the Performance of Community Extraction

To evaluate the performance of community extraction from a citation network of patents by the proposed method, we carry out a comparison experiment taking the method based on the personalized PageRank (PPR) algorithm as a benchmark [17–19]. The PPR algorithm, unlike the original PageRank algorithm, gives a steady state that depends on $\vec{\tau}$. The PPR algorithm is defined by the formula

$$x_n(t+1) = \rho \sum_{m=1}^{N} T_{nm}x_m(t) + (1 - \rho)\tau_n \quad (0 \leq \rho \leq 1) \tag{5}$$

The second term in the right-hand side functions as a continuous force that biases $\vec{x}(t)$ towards $\vec{\tau}$. Therefore, $\vec{x}(\infty)$ obtained by the PPR algorithm depends on $\vec{\tau}$. Thus, the PPR algorithm can also be considered as a process of extracting a community.

The comparison experiment is conducted using a 'community restoration' problem set as follows. We adopt a mathematical model of a citation network, proposed by Klemn and Eguiluz (KE model network) [20, 21]. The KE model network used in this experiment is defined with the notations in [20] as follows: $N = 100$, $m = 5$ and $\mu = 0.05$. This network might be small compared with real citation networks. However, the community structure of such a relatively small network can be extensively known, and we can thoroughly make use of this knowledge to quantitatively perform comparison experiment.

By the use of the Markov-chain clustering algorithm proposed by the present author [22], the KE network is decomposed into $K = 5$ overlapping communities in such a way that

$$\vec{x}^{(PR)} = \sum_{k=1}^{K} \pi_k \vec{x}^{(k)} \quad \left(\sum_{k=1}^{K} \pi_k = 1\right) \tag{6}$$

Here, $\vec{x}^{(PR)}$ represents the steady state given by the original PageRank algorithm (namely, Markov-chain Eq. (2)), which satisfies $\sum_{n=1}^{N} x_n^{(PR)} = 1$; $\vec{x}^{(k)}$ represents community k, which also satisfies $\sum_{n=1}^{N} x_n^{(k)} = 1$.

Here we assume that a set of seed patents $\vec{\tau}$ provided by the user is a community deteriorated from the 'correct' community $\vec{x}^{(k)}$ corresponding to the topic that the user wants to explore. This deterioration reflects imperfectness of user's in-advance knowledge on the topic. Therefore, extraction of the correct community can be viewed as restoration of a deteriorated community. Hence we can evaluate the efficiency of community extraction by measuring how accurately the correct community is restored from a deteriorated community, namely, a set of seed patents.

A deteriorated community $\vec{\tau}$ is generated from the 'correct' community $\vec{x}^{(k)}$ using a probabilistic model, as follows. Note that components of $\vec{\tau}$ corresponding to seed patents are equal to unity and all the other components are zero. Therefore, probabilistic generation of $\vec{\tau}$, namely, choice of seed patents by the user will be most appropriately modelled by the multinomial distribution

$$p(\vec{\tau}) = \prod_{n=1}^{N} x_n^{(k)\tau_n} \tag{7}$$

In the comparison experiment, we assume that $\vec{\tau}$ has $L = 10$ non-zero components.

How accurately $\vec{x}(\infty)$, obtained either by our algorithm or by the PPR algorithm, restores $\vec{x}^{(k)}$ can be quantified by the correlation coefficient between $\vec{x}(\infty)$ and $\vec{x}^{(k)}$:

$$
r\left(\vec{x}(\infty),\ \vec{x}^{(k)}\right) = \frac{\sum_{n=1}^{N}\left(x_n(\infty) - \overline{x(\infty)}\right)\left(x_n^{(k)} - \overline{x^{(k)}}\right)}{\sqrt{\sum_{n=1}^{N}\left(x_n(\infty) - \overline{x(\infty)}\right)^2}\sqrt{\sum_{n=1}^{N}\left(x_n^{(k)} - \overline{x^{(k)}}\right)^2}} \tag{8}
$$

where $\overline{x(\infty)} \equiv \sum_{n=1}^{N} x_n(\infty)/N$ and $\overline{x^{(k)}} \equiv \sum_{n=1}^{N} x_n^{(k)}/N$. The larger r, the more accurate restoration; especially when $r = 1$, restoration is perfect.

2.7 Evaluation of Robustness

Choice of seed patents by a user must accompany some ambiguities reflecting imperfectness of his/her knowledge. The extracted community should not be significantly altered when seed patents are slightly changed. We therefore examine the robustness of community extraction either by our algorithm or by the PPR algorithm.

To this end we choose a seed pattern, which will be denoted by $\vec{\tau}^{[0]}$. Then we generate serial sets of 'degraded' seed patterns $\vec{\tau}^{[1]}$, $\vec{\tau}^{[2]}$, $\cdots$, as follows: If $\tau_n^{[l]} = 1$, $\tau_n^{[l+1]} = 0$ with probability $p_d = 0.2$ and $\tau_n^{[l+1]} = 1$ with probability $1 - p_d$; if $\tau_n^{[l]} = 1$ and $\tau_n^{[l+1]} = 0$, a zero component of $\vec{\tau}^{[l]}$ (say, $\tau_{n'}^{[l]} = 0$) is randomly selected and $\tau_{n'}^{[l+1]} = 1$. Let $\vec{x}^{[l]}(\infty)$ ($l = 0,\ 1,\ \cdots$) denote the steady state for $\vec{\tau}^{[l]}$ obtained either by our algorithm or by the PPR algorithm. Robustness of the steady state against changes in the seed pattern can be evaluated by measuring the similarity between $\vec{x}^{[0]}(\infty)$ and $\vec{x}^{[l]}(\infty)$ ($l = 1,\ 2,\ \cdots$). For experiment using the model network, we quantify this similarity by the correlation coefficient $\gamma\left(\vec{x}^{[0]}(\infty),\ \vec{x}^{[l]}(\infty)\right)$.

We also compare the robustness of community extraction between the two algorithms using a real citation network of patents. The robustness is evaluated by measuring the similarity between the top $R = 20$ ranking for $\vec{\tau}^{[0]}$ and that for $\vec{\tau}^{[l]}$ ($l = 0,\ 1,\ \cdots$). To quantify this similarity we introduce the following index, which will be referred to as *rank precision*:

$$
\gamma_R = \frac{2}{R(R+1)} \sum_{r=1}^{R} \frac{R - r + 1}{|l_r - r| + 1} \tag{9}
$$

Here, l_r is the rank order for $\vec{\tau}^{[l]}$ of the patent that is ranked the r-th for $\vec{\tau}^{[0]}$. The denominator $|l_R - r| + 1$ represents the difference between the rank orders of the same patent for $\vec{\tau}^{[0]}$ and $\vec{\tau}^{[l]}$; if these rank orders are the same ($l_r = r$), the denominator takes the minimum value. The numerator $R - r + 1$ is the weight assigned to the patent ranked the r-th for $\vec{\tau}^{[0]}$; the higher weight is assigned to the higher rank order. If the patent ranked the r-th for $\vec{\tau}^{[0]}$ is absent in the top R ranking for $\vec{\tau}^{[l]}$, we set $l_r \to \infty$. If the top R rankings for $\vec{\tau}^{[0]}$ and $\vec{\tau}^{[l]}$ are exactly the same, the rank precision takes the maximum value $\gamma_R = 1$; if any of the top R patents for $\vec{\tau}^{[0]}$ is absent in the top R ranking for $\vec{\tau}^{[l]}$, $\gamma_R = 0$. Thus, the rank precision (9) well represents the similarity between the two rankings.

2.8 Bibliographic Data

To demonstrate the practical usefulness of the proposed method, we examined its application to real citation networks. To this end we prepared bibliographic information of Japan patents. This includes for each patent: publication number that we shall use as the identification data (ID) of the patent; inventor(s); title of invention; applicant; year of publication; IDs of cited patents; abstract; claims; and so forth. Among them, only IDs of citing and cited patents are necessary to construct a citation network. It should be noted that citation of patents, unlike citation of academic papers, is given not by authors (inventors) but by examiners who are highly specialized in specific fields of technology.

3 Results

Using the community restoration problem defined in Methods, we compared the efficiency of community extraction between our algorithm and the PPR algorithm taken as a benchmark. How accurately these algorithms restore the correct community, denoted by $\vec{x}^{(k)}$, from a mathematically generated seed pattern (see Methods), denoted by $\vec{\tau}$, was quantified by the correlation coefficient (6). We averaged the correlation coefficient over 100 $\vec{\tau}$'s and the average correlation coefficient $\bar{r}$ was plotted as a function of the unique parameter in each algorithm (α for our algorithm and ρ for the PPR algorithm). The restoration accuracy of our algorithm for $0.7 < \alpha < 1$ is higher than that of the PPR algorithm for the best value ($\rho=0.9$) (Fig. 3A and B). At $\alpha = \rho = 0$ and $\alpha = \rho = 1$, the correlation coefficients for both algorithms are the same because at the former both coincide with the original PageRank algorithm and at the latter both give $\vec{x}(t) = \vec{\tau}$.

For further confirmation, the correlation coefficient was compared for each $\vec{\tau}$ between these algorithms for their best parameter values ($\alpha = 0.9$ for our algorithm and $\rho = 0.9$ for the PPR algorithm). The correlation coefficients themselves largely scatter over $\vec{\tau}$'s, which is consistent with the large standard deviation shown in Fig. 3A and B. Nevertheless, for almost all $\vec{\tau}$'s (97 out of 100) the correlation coefficient given by our algorithm is higher than that given by the PPR algorithm (Fig. 3C). These results demonstrate that community extraction by our algorithm is more efficient than that by the PPR algorithm.

Next we compare the robustness of community extraction between the two algorithms. We calculated the correlation coefficient between $\vec{x}^{[0]}(\infty)$ and $\vec{x}^{[l]}(\infty)$. Here, $\vec{x}^{[0]}(\infty)$ and $\vec{x}^{[l]}(\infty)$ are the steady states obtained by either our algorithm or the PPR algorithm for seed patterns $\vec{\tau}^{[0]}$ and $\vec{\tau}^{[l]}$, respectively. The $\vec{\tau}^{[l]}$ is obtained by l times of degradation from $\vec{\tau}^{[0]}$ (see Methods); that is, the larger l the farther $\vec{\tau}^{[l]}$ is away from

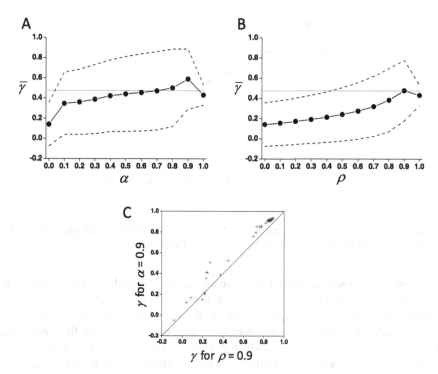

Fig. 3. *A.* The restoration accuracy of our algorithm, which is quantified by the average correlation coefficient $\bar{\gamma}$, is plotted as a function of α, a unique parameter in our algorithm. *B.* The restoration accuracy of the PPR algorithm is plotted as a function of ρ, a unique parameter in the PPR algorithm. The broken lines in *A* and *B* indicate the standard deviation. The horizontal lines in *A* and *B* indicate the value of correlation coefficient obtained by the PPR algorithm for $\rho = 0.9$ that gives the best restoration accuracy. *C.* Each point (+) indicates the correlation coefficient obtained by the PPR algorithm for $\rho = 0.9$ (*x*-coordinate, 'γ for $\rho = 0.9$') and that obtained by our algorithm for $\alpha = 0.9$ (*y*-coordinate, 'γ for $\alpha = 0.9$') for each $\bar{\tau}$.

$\bar{\tau}^{[0]}$. As l increases, the correlation coefficient for the PPR algorithm decreases considerably faster than that for our algorithm (Fig. 4). These results demonstrate that community extraction by our algorithm is more robust than that by the PPR algorithm.

Now we empirically demonstrate the use and the benefit of community extraction by the proposed method applied to a real citation network of Japan patents. We took for example a technological topic expressed by the phrase "two-legged robot". A set of 143 patents with abstracts or claims showing high scores of word matching to this phrase was chosen as seed patents. Table 1 shows the list of the top 20 highest ranking patents in the community extracted by the proposed method. Viewing this table, one can readily know the most important patents in this field of technology. Table 2A shows the list of the top seven applicants defined by the total activities. These applicants are widely acknowledged as the leading companies or organizations in humanoid or two-leg walking robot technology.

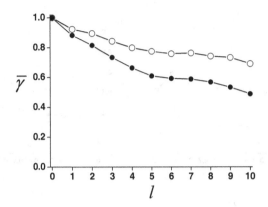

Fig. 4. Robustness of community extraction either by our algorithm (open circle) or by the PPR algorithm (filled circle) against change in the seed pattern.

Figure 5 visualizes citation relations among the top 250 highest ranking patents in the extracted community. Each document icon symbolizes a patent and its size expresses the activity it has acquired, namely, the importance assigned to this patent. Icons are sorted from the top to the bottom in chronological order of the publication dates. Each arrowed line represents the citation relation between two patents. An arrow head is directed from a citied to a citing patent in order to indicate that the latter is under the influence or control of the former. The color of the icons differentiates applicants (red for Honda, blue for Sony, yellow for Toyota and so forth).

Table 1. The top 20 patents in the field of two-leg walking robot technology extracted from a citation network of Japan patents by the proposed method. Bibliographic information is given in Japanese.

rank order	ID (publication number)	Title of invention	applicant	inventor(s)	activity
1	特開平5-337849	脚式移動ロボットの姿勢安定化	本田技研工業株式会社	竹中透	0.1276932
2	特開平11-300661	脚式移動ロボットの制御装置	本田技研工業株式会社	竹中透,長谷川忠明,松本隆志	0.08785171
3	特開平5-305586	脚式移動ロボットの歩行制御装置	本田技研工業株式会社	竹中透	0.08604511
4	特開平11-48170	脚式移動ロボットの制御装置	本田技研工業株式会社	竹中透,河井孝之,長谷川忠明,松	0.08157519
5	特開2003-89083	二足歩行移動体の床反力推定	本田技研工業株式会社	河合雅和,池内廉,加藤久	0.07199144
6	特開平10-217161	脚式移動ロボットの歩容生成装	本田技研工業株式会社	竹中透	0.06745819
7	特開平10-86081	脚式移動ロボットの歩容生成装	本田技研工業株式会社	竹中透	0.06711382
8	特開2002-326173	脚式移動ロボットの目標運動生	本田技研工業株式会社	竹中透,松本隆志,長谷川忠明	0.06695993
9	特開昭62-187671	足底付き多関節二脚歩行ロボッ	株式会社日立製作所,学校法人	佐々木 淑江,大津 誠,吉田 富	0.06408494
10	特開昭62-97006	多関節歩行ロボット制御装置	株式会社日立製作所,学校法人	北川 淑江,吉田 富治,佐々木	0.06174975
11	特開平3-184781	脚式歩行ロボットの足部構造	本田技研工業株式会社	西川正雄,広瀬真人,熊谷智治,阿	0.0593284
12	特開平5-245780	脚式移動ロボットの制御装置	本田技研工業株式会社	吉野龍太郎,高橋英男	0.05889263
13	特開2004-174653	倒立振子制御のゲインを静止時	トヨタ自動車株式会社	山本貴史	0.05406082
14	特開2004-174652	ZMP補償制御のゲインを変化さ	トヨタ自動車株式会社	山本貴史	0.05396421
15	特開2001-150370	脚式移動ロボット及び脚式移動	ソニー株式会社,山口仁一	服部裕一,石田健蔵,山口仁一	0.05226061
16	特開平6-79657	脚式移動ロボットの歩行制御装	本田技研工業株式会社	長谷川忠明,竹中透	0.0475002
17	特開昭61-252081	マスタ・スレーブ・マニピュレータ	工業技術院長	平井 成興,佐藤 知正	0.04725371
18	特開昭55-48596	ロボットの腕機構	株式会社日立製作所	杉本 浩一	0.04722082
19	特開平3-111182	3次元運動機構	新技術事業団	広瀬茂男	0.04636097
20	特開平2-3580	脚移動型歩行機械の制御装置	工業技術院長	曽和俊,服部誠,榊原義宏,細田裕	0.04570828

Table 2. The top seven applicants in the field of two-leg walking robot technology judged by the proposed method (A) and by the PPR algorithm (B).

<table>
<tr><td colspan="2" align="center">A</td><td colspan="2" align="center">B</td></tr>
<tr><td>rank order</td><td>applicant</td><td>rank order</td><td>applicant</td></tr>
<tr><td>1</td><td>Honda</td><td>1</td><td>Honda</td></tr>
<tr><td>2</td><td>Sony</td><td>2</td><td>Sony</td></tr>
<tr><td>3</td><td>AIST</td><td>3</td><td>Toyota</td></tr>
<tr><td>4</td><td>Toyota</td><td>4</td><td>Kansai TLO</td></tr>
<tr><td>5</td><td>Waseda Univ./Hitachi</td><td>5</td><td>Oki</td></tr>
<tr><td>6</td><td>JST</td><td>6</td><td>Fujitsu</td></tr>
<tr><td>7</td><td>Fujitsu</td><td>7</td><td>Bandai</td></tr>
</table>

This visualization documents the well-known trends of two-leg walking robot technology in Japan. The earliest days in this field of technology were led by Waseda University/Hitachi and Agency of Industrial Science and Technology (AIST). Subsequently Honda has obtained a number of important patents and dominated this field. This trend is followed by Sony. Also, it is noteworthy that Toyota has recently entered this field.

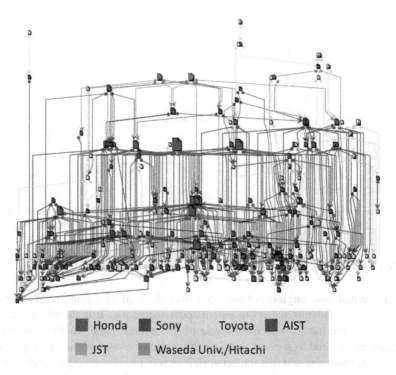

Honda · Sony · Toyota · AIST · JST · Waseda Univ./Hitachi

Fig. 5. Visualization of citation relations between the top 250 patents extracted by the proposed method.

Table 2B shows the top seven applicants extracted by the method based on the PPR algorithm. One finds that AIST and Waseda University/Hitachi, which are widely acknowledged as the leading organizations having marked the beginning of humanoid or walking robot technology in Japan and are extracted in the top seven by our method (Table 2A), are missing in Table 2B. The superiority of the proposed method to the method based on the PPR algorithm demonstrated using the real citation network is consistent with the results of the comparison experiment using a model network (Fig. 3).

For the real citation network we also examined the robustness of community extraction by either method. The similarity between the top $R = 20$ ranking extracted for an original set of seed patents $\vec{\tau}^{[0]}$ (143 patents with abstracts or claims showing high scores of word matching to the phrase "two-legged robot") and that extracted for $\vec{\tau}^{[l]}$ generated from $\vec{\tau}^{[0]}$ with l times of degradation (see Methods) were evaluated by calculating the rank precision γ_R defined by Eq. (9). As l increases, the rank precision given by our method only slightly decreases, whereas that given by the method based on the PPR algorithm drastically decays (Fig. 6). Thus the community extraction from the real citation network by our method is much more robust than that by the method based on the PPR algorithm. These are consistent with the results of evaluation of the robustness using a model network (Fig. 4).

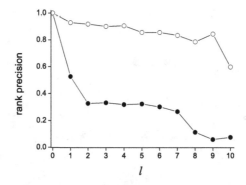

Fig. 6. Robustness of community extraction from the real citation network of patents either by the proposed method (open circle) or by the method based on the PPR algorithm (filled circle).

4   Discussion

Modern scientists and engineers have been overwhelmed by a huge number of documents such as academic papers or patents published daily. It is therefore crucial to find out what one has to read from a pile of documents. Here we have proposed a computational method to support this. For this we have considered a citation network of documents and supposed that the network is covered by communities of documents, with each community corresponding to a specific topic. The proposed method is based on the algorithm that extracts a relevant community from the network. By use of the proposed method the user who wants to explore a specific topic readily obtains a list of documents relevant to this topic as highly ranked members of the extracted community. We have shown the efficiency of the proposed method by comparing it with a benchmark method and demonstrated its benefit and practical usefulness by applying it to the real citation network of Japan patents.

It should be noted that community extraction in the proposed method is 'local' in sense that it extracts a relevant community from the entire network. This is in contrast to standard community extraction in the literature of network science, which mostly aims at 'global' community extraction, namely, finding all the communities covering a network ([4, 23, 24], but see also [2, 3]). Global community extraction can reveal the whole structure of a network but requires a large amount of computation. If it is only necessary to find a relevant community, our local community extraction algorithm can perform this with much less computation.

To be emphasized is that the core algorithm of the proposed method for local community extraction models the mechanism of memory recall in the brain. The STM recall from the associative network of LTM in response to a given situation can be described as a process of local extraction of a community from a network. Thus, this algorithm extracts a relevant community from a network in analogy to STM recall from the LTM network in the brain. The benefit of the proposed method might be so high because it models excellent functions of the real brain. We believe that exploring the real brain will be helpful for creating new information-processing technology.

Acknowledgments. This study was partly supported by KAKENHI (23500379) and KAKENHI (23300061). Bibliographic data of Japan patents used in this study was downloaded from StarPAT, a patent information retrieval system (http://www.scs.co.jp/product/gaiyo/starpat.html).

References

1. Garfield, E.: Citation indexes for science: a new dimension in documentation through association of ideas. Science **122**, 108–111 (1955)
2. Bagrow, J.P., Bollt, E.M.: Local method for detecting communities. Phys. Rev. E **72**, 046108 (2005)
3. Clauset, A.: Finding local community structure in networks. Phys. Rev. E 72, 026132 (2005)
4. Newman, M.E.J.: Communities, modules and large-scale structure in networks. Nat. Phys. **8**, 25–31 (2012)
5. Okamoto, H.: Topic-dependent document ranking: citation network analysis by analogy to memory retrieval in the brain. In: Honkela, T. (ed.) ICANN 2011, Part I. LNCS, vol. 6791, pp. 371–378. Springer, Heidelberg (2011)
6. Collins, A.M., Loftus, E.F.: Spreading-activation theory of semantic processing. Psychol. Rev. **82**, 407–428 (1975)
7. Anderson, J.R., Pirolli, P.L.: Spread of activation. J. Exp. Psychol. Learn. Mem. Cogn. **10**, 791–798 (1984)
8. Tsuboshita, Y., Okamoto, H.: Context-dependent retrieval of information by neural-network dynamics with continuous attractors. Neural Netw. **20**, 705–713 (2007)

9. Romo, R., Brody, C.D., Hernandez, A., Lemus, L..: Neuronal correlates of parametric working memory in the prefrontal cortex. Nature **399**, 470–473 (1999)
10. Aksay, E., Gamkrelidze, G., Seung, H.S., Baker, R., Tank, D.W.: In vivo intracellular recording and perturbation of persistent activity in a neural integrator. Nat. Neurosci. **4**, 184–193 (2001)
11. Egorov, A.V., Hamam, B.N., Fransen, E., Hasselmo, M.E., Alonso, A.A.: Graded persistent activity in entorhinal cortex neurons. Nature **420**, 173–178 (2002)
12. Goldman, M.S., Levine, J.H., Major, G., Tank, D.W., Seung, H.S.: Robust persistent neural activity in a model integrator with multiple hysteretic dendrites per neuron. Cereb. Cortex **13**, 1185–1195 (2003)
13. Loewenstein, Y., Sompolinsky, H.: Temporal integration by calcium dynamics in a model neuron. Nat. Neurosci. **6**, 961–967 (2003)
14. Lisman, J.E., Fellous, J.M., Wang, X.-J.: A role for NMDA-receptor channels in working memory. Nat. Neurosci. **1**, 273–275 (1998)
15. Koulakov, A.A., Raghavachari, S., Kepecs, A., Lisman, J.E.: Model for a robust neural integrator. Nat. Neurosci. **5**, 775–782 (2002)
16. Okamoto, H., Isomura, Y., Takada, M., Fukai, T.: Temporal integration by stochastic recurrent network dynamics with bimodal neurons. J. Neurophysiol. **97**, 3859–3867 (2007)
17. Page, L., Brin, S., Motwani, R., Winograd, T.: The PageRank Citation Ranking: Bringing Order to the Web. Stanford Digital Library Technologies Project (1998). http://ilpubs. stanford.edu:8090/422/
18. Kleinberg, J.: Authoritative sources in a hyperlinked environment. J. ACM **46**, 604–632 (1999)
19. Haveliwala, T.: Topic-sensitive PageRank: a context-sensitive ranking algorithm for web search. IEEE Trans. Knowl. Data Eng. **15**, 784–796 (2003)
20. Klemn, K., Eguiluz, V.M.: Growing scale-free networks with small-world behaviour. Phys. Rev. E **65**, 057102 (2002)
21. Klemn, K., Eguiluz, V.M.: Highly clustered scale-free networks. Phys. Rev. E **65**, 036123 (2002)
22. Okamoto, H.: 9th NETECOSYMP (Okinawa, 2012)
23. Girvan, M., Newman, M.E.J.: Community structure in social and biological networks. Proc. Nat. Acad. Sci. USA **99**, 7821–7826 (2002)
24. Newman, M.E.J., Girvan, M.: Fining and evaluating community structure in networks. Phys. Rev. E **69**, 026113 (2004)

Modeling Epidemic Risk Perception in Networks with Community Structure

Franco Bagnoli[1,2,3]([✉]), Daniel Borkmann[4], Andrea Guazzini[5,6],
Emanuele Massaro[7], and Stefan Rudolph[8]

[1] Department of Energy, University of Florence, Florence, Italy
[2] Center for the Study of Complex Systems, University of Florence, Florence, Italy
[3] National Institute for Nuclear Physics, Florence Section, Florence, Italy
franco.bagnoli@unifi.it
[4] Communication Systems Group, ETH Zurich, Zurich, Switzerland
[5] National Research Council, Institute for Informatics and Telematics, Pisa, Italy
[6] Department of Psychology, University of Florence, Florence, Italy
[7] Department of Computer Science and Systems, University of Florence,
Florence, Italy
[8] Organic Computing Group, University of Augsburg, Augsburg, Germany

Abstract. We study the influence of global, local and community-level risk perception on the extinction probability of a disease in several models of social networks. In particular, we study the infection progression as a susceptible-infected-susceptible (SIS) model on several modular networks, formed by a certain number of random and scale-free communities. We find that in the scale-free networks the progression is faster than in random ones with the same average connectivity degree. For what concerns the role of perception, we find that the knowledge of the infection level in one's own neighborhood is the most effective property in stopping the spreading of a disease, but at the same time the more expensive one in terms of the quantity of required information, thus the cost/effectiveness optimum is a tradeoff between several parameters.

Keywords: Risk perception · SIS model · Complex networks

1 Introduction

Epidemic spreading is one of the most successful and most studied applications in the field of complex networks. The comprehension of the spreading behavior of many diseases, like sexually transmitted diseases (i.e. HIV) or the H1N1 virus, can be studied through computational models in complex networks [4,20]. In addition to "real" viruses, spreading of information or computer malware in technological networks is of interest as well.

The susceptible-infected-susceptible (SIS) model is often used to study the spreading of an infectious agent on a network. In this model an individual is represented as a node, which can be either be "healthy" or "infected". Connections between individuals along which the infection can spread are represented

G.A. Di Caro and G. Theraulaz (Eds.): BIONETICS 2012, LNICST 134, pp. 283–295, 2014.
DOI: 10.1007/978-3-319-06944-9_20, © Institute for Computer Sciences, Social Informatics and Telecommunications Engineering 2014

by links. In each time step a healthy node is infected with a certain probability if it is connected to at least one infected node, otherwise it reverts to a healthy node (parallel evolution).

The study of epidemic spreading is a well-known topic in the field of physics and computer science. The dynamics of infectious diseases has been extensively studied in scale-free networks [2,6,8,24], in small-world networks [19] and in several kind of regular and random graphs.

A general finding is that it is hard to stop an epidemic in scale-free networks with slow tails, at least in the absence of correlations in the network among the infections process and the node characteristics [24]. This effect is essential due to the presence of hubs, which act like strong spreaders. However, by using an appropriate policy for hubs, it is possible to stop epidemics also in scale-free networks [2,9].

This network-aware policy is inspired by the behavior of real human societies, in which selection had lead to the development of strategies used to avoid or reduce infections. However, human societies are not structureless, thus a particular focus must be devoted to the community structures, which are highly important for our social behavior.

Recently, a wave of studies focused the attention on the effect of the community structure in the modelling of epidemic spreading [7,25,26]. However, the focus was only set towards the interaction between the viruses' features and the topology, without considering the important relation between cognitive strategies used by subjects and the structure of their (local) community/neighborhood.

Considering this scenario, an important challenge is the comprehension of the structure of real-world networks [14,15,21]. Given a graph, a community is a group of vertices that is "more linked" within the group than with the rest of the graph. This is clearly a poor definition, and indeed, in a connected graph, there is not a clear distinction between a community and a rest of the graph. In general, there is a continuum of nested communities whose boundaries are somewhat arbitrary: the structure of communities can be seen as a hierarchical dendrogram [22].

It is generally accepted that the presence of a community structure plays a crucial role in the dynamics of complex networks; for this reason, lots of energy has been invested to develop algorithms for the detection of communities in networks [10,12,13]. However, in complex networks, and in particular in social networks, it is very difficult to give a clear definition of a community: nodes often belong to more than just one cluster or module. The problem of overlapping communities was exposed in [23] and recently analyzed in [17]. People usually belong to different communities at the same time, depending on their families, friends, colleagues, etc. For instance, if we want to analyze the spreading of sexual diseases in a social environment, it is important to understand the mechanism that leads people to interact with each other. We can surely detect two distinct groups of people (i.e., communities): heterosexual and homosexual, with bisexual people that act as overlapping vertexes between the two principal communities [1,7,18].

The strategies used to face the infection spreading in a community is itself a complex process (i.e., social problem solving) in which strategies spread (as the epidemics) along the community, and are negotiated and assumed or discarded depending on their social success.

Several factors can affect the social problem solving which is represented by the adoption of a behaviour to reduce the infection risk. Of course, personality factors, previous experiences and the social and economical states of a subject can be considered as influencing variables. Another important variable is represented by the structure of the environment in which the social communities live, because it determines at the same time the speed of the epidemic diffusion and the strategy of the negotiation process; in particular large and more connected communities are often characterized by conservative strategies while small and isolated communities allow more relaxed strategies.

The same strategy can be more or less effective depending on the strategies adopted by the neighbours (community) of the subject. For instance, a subject in a conservative community can adopt a more risky (and presumably profitable) attitude with a certain confidence since he would be protected from the infection because of their neighbours' behaviours. This "parasitic" behavior (like refusing vaccinations) can be tolerated up to a certain level without lowering the community's fitness.

Not only the neighbor's behaviours affect the evolution of the cognitive strategies of a subject, but also the position he has in the network should be a relevant factor. A hub, or a subject with a great social betweenness, is usually more exposed to the infection than a leaf, and as a consequence, the best strategy for him has to be different. In the same way, since the topology of the network (e.g., small world, random) determines variables such as the speed of the spreading, or its pervasiveness, it should also affect the development of the "best strategy".

Moreover, while the negotiation process evolves, the cognitive strategies usually develop within the most intimate community of a subject, thus the behaviour adopted by subjects could be an interesting feature for the community detection problem as well.

The understanding of the effects of the community structure on the epidemic spreading in networks is still an open task. In this paper we investigate the role of risk perception in artificial networks, generated in order to reproduce several types of overlapping community structures.

The rest of this paper is organized as follows: we start by describing a mechanism for generating networks with overlapping community structures in Sect. 2. In Sect. 3, we describe the SIS model adopted to model the risk perception of subjects in those networks. Finally, Sect. 4 contains simulation results from our model with a throughout discussion and future work proposals.

2   The Networks Model

There are n_c different communities with n_v vertices (in this paper we consider only undirected and unweighted graphs); we assume that the probability to

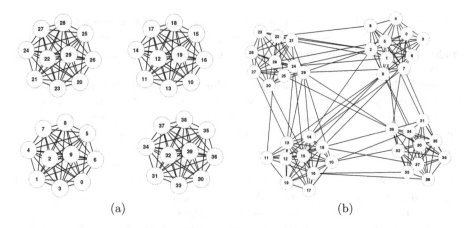

Fig. 1. (a) An example network with 4 different communities composed by 10 vertices: in this case, considering $p_1 = 1$ and $p_2 = 0$, we generate 4 non-interconnected fully connected networks. (b) The same 4 communities with parameters $p_1 = 0.95$ and $p_2 = 0.05$.

have a link between the vertexes in the same community is p_1, while p_2 is the probability to have a link between two nodes belonging to different communities. For instance, with $p_1 = 1$ and $p_2 = 0$, we generate n_c fully connected graphs, with no connections among them as shown in Fig. 1(a). It is possible to use the parameters p_1 and p_2 to control the interaction among different communities, as shown in Fig. 1(b). The algorithm for generating this kind of networks can be summarized as:

1. Define s_1 as number of vertexes in the communities;
2. Define n_c as number of communities;
3. For all the n_c communities create a link between the vertexes on them with probability p_1;
4. For all the vertexes $N = s_1 n_c$ create a link between them and a random vertex of other communities with probability p_2;

Constraining the condition $p_1 = 1 - p_2$, we can reduce the free parameters to just one. The connectivity degree itself depends on the size of the network and on the probabilities p_1 and p_2. In particular, the connectivity function $f(k)$ has a normal distribution from which we could define the mean connectivity $\langle k \rangle$ as

$$\langle k \rangle = (s_1 - 1)p_1 + (n_c - 1)s_1 p_2 \tag{1}$$

with standard deviation $\sigma^2(k) = (s_1 - 1)p_1(1 - p_1) + (n_c - 1)s_1 p_2(1 - p_2)$.

In Fig. 2(a) we show the frequency distribution of the connectivity degree of nodes varying the value of the parameter p_2 for a network composed by $N = 5000$ nodes and $n_c = 5$ communities.

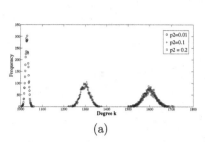
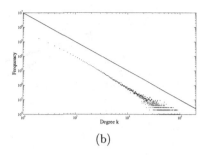

(a)   (b)

Fig. 2. (a) Random networks: in this figure, we show the frequency distribution of the connectivity degree changing the value of the parameter p_2. The circles represent the values for $p_2 = 0.01$, crosses for $p_2 = 0.1$ and eventually squares for $p_2 = 0.2$. Here, $s_1 = 1000$ and $n_c = 5$, thus we have generated networks with 5 communities of 1000 nodes for each. (b) Distribution of connectivity degree for the scale-free network generated with the mechanism described above (dots). The straight line is a power law curve with exponent $\gamma = 2.5$.

It is widely accepted that real-world networks from social networks to computer networks are scale-free networks, whose degree distribution follows a power law, at least asymptotically. In this network, the probability distribution of contacts often exhibits a power-law behavior:

$$P(k) \propto ck^{-\gamma}, \tag{2}$$

with an exponent γ between 2 and 3 [3,11]. For generating networks with this kind of characteristics, we adopt the following mechanism:

1. Start with a fully connected network of m nodes;
2. Add $N - m$ nodes;
3. For each new node add m links;
4. For each of these links choose a node at random from the ones already belonging to the network and attach the link to one of the neighbors of that node, if not already attached.

Through this mechanism we are able to generated scale-free networks with an exponent $\gamma = 2.5$ as shown in Fig. 2(b). There, we show the frequency distribution of the connectivity degree for a network of 10^6 nodes. To generate a community structure with a realistic distribution, we first generate n_c scale-free networks as explained above. Then, for all nodes and all outgoing links, we replace the link pointing inside the community with that connecting a neighbor of a random node in a random community with a probability of $p_2 = 1 - p_1$. Thus, the algorithm can be summarized as:

1. Generate n_c communities as scale-free networks with n_v vertices;
2. For all the vertices, with a probability $p_2 = 1 - p_1$;
   - Delete a random link;

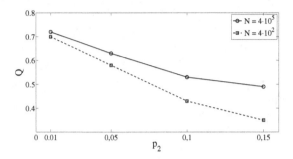

Fig. 3. Different values of modularity (Q) after increasing the mixing parameter p_2 for two networks with $N = 4 \cdot 10^5$ nodes and $N = 4 \cdot 10^2$ nodes.

- Select a random node of another community and create a link with one of its adjacent vertex;

3. End.

In this way, we are able to generate scale-free networks with a well defined community structure. A good measure for the estimation of the strength of the community structure is the so-called modularity [15]. The modularity Q is defined to be:

$$Q = \frac{1}{2} \sum_{vw} \left[A_{vw} - \frac{K_v K_w}{2m} \right] \delta(c_v, c_w), \qquad (3)$$

where A is the adjacency matrix in which $A_{vw} = 1$ if w and v are connected and 0 otherwise. $m = \frac{1}{2} \sum_{vw} A_{vw}$ is the number of edges in the graph, K_i is the connectivity degree of node i and $(K_v K_w)/(2m)$ represents the probability of an edge existing between vertices v and w if connections are made at random but with respecting vertex degrees. $\delta(c_v, c_w)$ is defined as follows:

$$\delta(c_v, c_w) = \sum_{r}^{n_c} \hat{c}_{vr} \hat{c}_{wr} \qquad (4)$$

where $\hat{c}_{ir}$ is 1 if vertex i belongs to group r, and 0 otherwise.

In Fig. 3 we show the values of modularity for two networks that were generated with the same algorithm, but with different sizes. Here, we consider a network with 4 communities: in the first case $s_1 = 10^5$, while in the second case $s_1 = 10^2$. What one can observe in Fig. 3 is that the modularity's behaviour does not change significantly for different network sizes with the same number of communities.

In the case of scale-free networks, the mean connectivity degree $\langle k \rangle$ is fixed a priori when we choose the number of links the new nodes create. In the case of random networks the mean connectivity is given by Eq. 1.

3   The Risk Perception Model

We use the susceptible-infected-susceptible model (SIS) [1,24] for describing an infectious process. In the SIS model, nodes can be in two distinct states: healthy and ill. Let us denote by τ the probability that the infection can spread along a single link. Thus, if node i is susceptible and it has k_i neighbors of which s_n are infected, then, at each time step, node i will become infected with the probability:

$$p(s, k) = [1 - (1 - \tau)^{s_n}]. \tag{5}$$

We model the effect of risk perception considering the global information of the infection level for the whole network, the information about the infected neighbors and the information about the average state of the community. Thus, the risk perception for the individual i is given by:

$$I_i = \exp\left\{-H + J_1\left(\frac{s_{ni}}{k_i}\right) + J_2\left(\frac{s_{ci}}{n_{ci}}\right)\right\}, \tag{6}$$

where $H = J(s/N)$ is the perception about the global network on which s is the total number of infected agents while N is the number of agents in the network. The second term of the Eq. 6 represents the perception about the neighborhood, while the third term represents the perception about the local community of the agent i.

In this model, we assume that people receive information about the network's state through examination of people in the neighborhood. The global information could refer to entities like media while the information about the community could be assumed as *word of mouth*. In this paper, we don't consider the cost that people should pay in order to get these information, but it is clearly an important constraint to consider in future works.

The risk perception I_i, defined in Eq. 6, is assumed to determine the probability that the agents meet someone in its neighbourhood. The algorithm is given by:

1. For all nodes $i = 1, 2, \ldots, N$;
2. For all its neighbors $j = 1, 2, \ldots, k_i$;
3. If $I_i > rand$;
   - i meets j;
   - If j at time $t - 1$ was infected then i becomes ill with probability τ;
4. End.

Then, we propose a gain function defined as the number of meetings in time considering different values of $j = J, J_1, J_2$ and different kind of scale-free and random networks; the gain function $G(j)$ is given by:

$$G(j) = \frac{\sum_{t=1}^{T_e} M_t}{T_e}, \tag{7}$$

in which T_e is the time for the extinction, while M_t is the number of meetings during time. Based on that, we can eventually define a fitness function that considers the probability to extinct the epidemic in the given time. Thus, the fitness function is given by:

$$F_j^T = G(j)P_e(j) \tag{8}$$

It is possible to make a mean-field approximation of this model. Pastor-Satorras and Vespignani defined the mean-field equation for scale-free networks in [24]. In 2010, Kitchovitch and Liò [16] modeled the mean number of infected neighbors $g(k)$ for individuals i with connectivity degree k. In fact, given the probability of receiving an infection by at least one of the infected neighbors (Eq. 5), it is possible to define the rate of change of the fraction of individuals i with degree k at time t by the following:

$$\frac{di_k}{t} = -\gamma + (1 - i_k)g(k), \tag{9}$$

on which γ is the rate of recovery (in our simulations we set $\gamma = 1$).

Then, as shown by Boccaletti et al. [5], for any node, the degree distribution of any of its neighbors is,

$$q_k = \frac{kP(k)}{\langle k \rangle}, \tag{10}$$

hence, it is possible to define the number of infected neighbors as:

$$i_n = \sum_{K_{min}}^{k_{max}} q_k i_k, \tag{11}$$

and it allows to give a definition of $g(k)$ as:

$$g(k) = \sum_{s=0}^{k} \binom{k}{s} p(s, k) i_n^s (1 - i_n)^{k-s}, \tag{12}$$

where $s = s_n$ is the number of infected neighbors.

The temporal behavior of the mean fraction c of infected individuals in the case of a network with fixed connectivity is given by:

$$c' = \sum_{s_n=1}^{k} \binom{k}{s_n} c^{s_n} (1 - c)^{k-s_n} [1 - (1 - \tau)^{s_n}], \tag{13}$$

where $c \equiv c(t)$, $c' \equiv c(t + 1)$ and the sum runs over the number k_{inf} of infected individuals.

4   Results and Discussion

We studied the behavior of our model for different scenarios. In Fig. 5, we show results considering a network of 500 nodes and 5 communities where the initial number of infected agents is $\approx 10\%$ of all agents in the network. We focus on

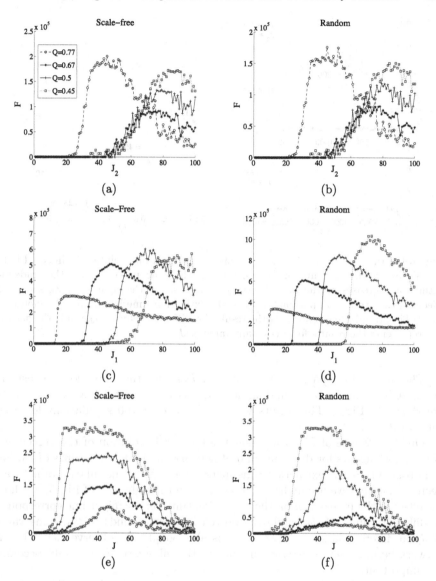

Fig. 4. Effect of the parameters J_2 (a), J_1 (b) and J (c) (*x-axis*) on the fitness function F (*y-axis*) considering different scale-free and random networks with different values of modularity. Results are averaged over 100 simulations for each value of J, J_1 and J_2.

the information about the community (parameter J_2), while we kept $J = J_1 = 1$ fixed. It is very interesting to observe the time necessary for the extinction of the epidemics, with the probability of being infected $\tau = 0.5$ and changing the community structure of the network.

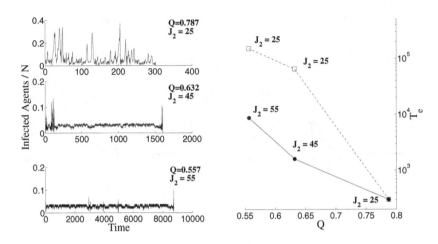

Fig. 5. On the left side of the figure we show the temporal evolution of infected individuals by varying the mixing parameter p_2. The time necessary for the epidemic extinction increases as the modularity Q decreases. On the right side, we show the effects of the precaution parameter J_2 on the extinction time by varying the modularity Q. The straight line represents the results for different value of J_2, while the dashed line represents the results for a constant value of J_2.

The effects of the parameters J_2, J_1 and J on the fitness function F considering different scale-free and random networks with different values of modularity are shown in Fig. 4. The results were averaged over 100 simulations for each value of J, J_1 and J_2.

On the left side of Fig. 5 we show the temporal evolution of the percentage of infectious agents for different kind of networks and different values of J_2. We can observe that the extinction time increases when the modularity of network decreases, even if we use higher values of J_2. On the right side of Fig. 5, we show the effect of the precaution on the extinction time. The straight line corresponds to different values of J_2, while the dotted line corresponds to the same value of J_2 in different kind of networks. It is also possible to observe that when a network becomes *less clustered*, the information about the community becomes less important.

In the case of scale-free networks, the mean connectivity degree $\langle k \rangle$ is related to the number m of links the new nodes create. In the above example, considering $m = 5$, we obtained a mean connectivity degree $\langle k \rangle = 7.8$.

For comparisons, we generated random networks with a mean connectivity degree $\langle k \rangle \in (7,8)$. The first result that we obtained is that the extinction time is larger than in the scale-free case. In Fig. 6, we show the temporal evolution of the infected agents for a random network with modularity $Q = 0.78$ considering $J_2 = 25$ as in the upper plot on the left side of Fig. 5. For the scale-free network the time necessary for the extinction is $T_e \simeq 3 \cdot 10^2$ while for the random one it is $T_e \simeq 3 \cdot 10^3$.

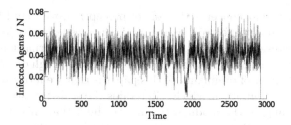

Fig. 6. Percentage of infected agents for a random network of 100 nodes and 5 communities with modularity $Q = 0.78$. Adopting the same parameter used for the simulation reported in Fig. 5, we show how the time for the extinction (approx. 2900 units) of the epidemic is greater than for the scale-free case (i.e., upper plot on the left side of Fig. 5).

Table 1. Critical values for the extinction of the epidemic on case of scale-free networks of 500 nodes and 5 communities considering a maximum threshold time $T_{max} = 1000$, necessary for the extinction of the epidemics.

	Critical Values		
Q (modularity)	J	J_1	J_2
0.78	45	15	25
0.64	40	15	45
0.35	40	20	55

Regarding the effects of the global and local (neighborhood) information, we investigated scale-free networks composed by 500 nodes and 5 communities with a fixed maximum threshold time T_{max}, necessary for the extinction of the epidemics. We assume $T_{max} = 1000$ and separately measure critical values of J, J_1 and J_2. In the Table 1, we show the critical values of the three parameters by changing the modularity Q. As we can observe, the most variable parameter is J_2 while the other two parameters do not appear to change. From this figure, we observe that the information about the fraction of infected neighbors is the most effective for stopping the disease. However, in order to get this piece of information, each node needs to check the status of all its neighbours, a task that can be quite hard and possibly conflicting with privacy. On the other hand, the information on the average infectious level in the community or in the whole population is more easily obtained. Therefore, one needs to add the cost of information into the model in order to decide what the most effective solution for risk perception is.

Summarizing, we have studied the progression and extinction of a disease in a SIS model over modular networks, formed by a certain number of random and scale-free communities. The infection probability is modulated by a risk perception term (modeling the probability of an encounter). This term depends on the global, local and community infection level. We found that in scale-free networks the progression is slower than in random ones with the same average connectivity. For what concerns the role of perception, we found that the local one (information about infected neighbours) is the most effective for stopping the spreading of the disease. However, it is also the piece of information that requires most efforts to be gathered, and therefore it may result a high cost/efficacy ratio.

The main element of originality of this paper is that we introduced a network model based on communities, which still retains the scale-free structure with the possibility of changing the modularity, and we think that this structure (albeit being quite theoretical) is more realistic than standard scale-free networks. The fact the knowledge about own community is more effective than other indicators is surely trivial (and we expected to get this result), but it is also the information that is more expensive to get, at least for the standard data gathering existing today. We would like to quantify the advantage in using this indicator in order to compare its efficiency with respect to its cost (and for doing it we need to include a cost model, that will be done in a future work) and also point to the necessity of gathering this kind of local information, that in a real case may also present problems related to the privacy, but might be of great importance in the case of a pandemic.

In regard to extending the model by inserting a cost model, it should also be taken into account what the best strategies are to avoid the spreading of epidemics in different environments considering agents as intelligent entities capable to change or select the best strategies dynamically in order to minimize the risk and to maximize the economy of the system. In combination to this, we plan to add a more complex model such as the SIR eventually with vaccinations, for which there are important factors like the penetration and the possibility that the modular structure may be exploited to "shield" a community that may remain not exposed – similarly like people that refuse vaccinations, but are "shielded" by a surrounding community that vaccinates.

Acknowledgments. We acknowledge funding from the 7th Framework Programme of the European Union under grant agreement n° 257756 and n° 257906.

References

1. Anderson, R.M., May, R.M.: Infectious Diseases of Humans Dynamics and Control. Oxford University Press, Oxford (1992)
2. Bagnoli, F., Liò, P., Sguanci, L.: Risk perception in epidemic modeling. Phys. Rev. E **76**, 061904 (2007)
3. Barabási, A.L., Albert, R.: Emergence of scaling in random networks. Science **286**, 509–512 (1999)
4. Barrat, A., Barthlemy, M., Vespignani, A.: Dynamical Processes on Complex Networks. Cambridge University Press, New York (2008)
5. Boccaletti, S., Latora, V., Moreno, Y., Chavez, M., Hwang, D.-U.: Complex networks: structure and dynamics. Phys. Rep. Fervier **424**(4–5), 175–308 (2006)
6. Boguñá, M., Pastor-Satorras, R., Vespignani, A.: Absence of epidemic threshold in scale-free networks with degree correlations. Phys. Rev. Lett. **90**, 028701 (2003)

7. Chen, J., Zhang, H., Guan, Z.-H., Li, T.: Epidemic spreading on networks with overlapping community structure. Phys. A: Stat. Mech. Appl. **391**(4), 1848–1854 (2012)
8. Cohen, R., ben Avraham, D., Havlin, S.: Percolation critical exponents in scale-free networks. Phys. Rev. E **66**, 036113 (2002)
9. Dezső, Z., Barabási, A.-L.: Halting viruses in scale-free networks. Phys. Rev. E **65**(5), 055103+ (2002)
10. Van Dongen, S.: Graph clustering via a discrete uncoupling process. SIAM. J. Matrix Anal. Appl. **30**, 121–141 (2009)
11. Dorogovtsev, S.N., Mendes, J.F.F., Samukhin, A.N.: Structure of growing networks with preferential linking. Phys. Rev. Lett. **85**, 4633–4636 (2000)
12. Dorso, C., Medus, A.D.: Community detection in networks. Int. J. Bifurcat. Chaos **20**(2), 361–367 (2010)
13. Fortunato, S.: Community detection in graphs. Phys. Rep. **486**(3–5), 75–174 (2010)
14. Fortunato, S., Castellano, C.: Community Structure in Graphs, December 2007
15. Girvan, M., Newman, M.E.J.: Community structure in social and biological networks. Proc. Natl. Acad. Sci., USA **99**, 7821–7826 (2002)
16. Stephan, K., Pietro, L.: Risk perception and disease spread on social networks. Procedia Comput. Sci. **1**(1), 2339–2348 (2010)
17. Lancichinetti, A., Fortunato, S., Kertész, J.: Detecting the overlapping and hierarchical community structure of complex networks. New J. Phys. **11**, 033015 (2009)
18. Liljeros, F., Edling, C.R., Amaral, L.A., Stanley, E.H., Åberg, Y.: The web of human sexual contacts. Nature **411**(6840), 907–908 (2001)
19. Moore, C., Newman, M.E.J.: Epidemics and percolation in small-world networks. Phys. Rev. E **61**, 5678–5682 (2000)
20. Newman, M.: Networks: An Introduction. Oxford University Press Inc., New York (2010)
21. Newman, M.E.J.: Detecting community structure in networks. Europ. Phys. J. B. **38**, 321–330 (2004)
22. Newman, M.E.J., Girvan, M.: Finding and evaluating community structure in networks. Phys. Rev. E **69**, 026113 (2004)
23. Palla, G., Derény, I., Vickset, T.: Nature **435**, 814 (2005)
24. Pastor-Satorras, R., Vespignani, A.: Epidemic dynamics and endemic states in complex networks. Phys. Rev. E **63**, 066117 (2001)
25. Salathé, M., Jones, J.H.: Dynamics and control of diseases in networks with community structure. PLoS Comput. Biol. **6**(4), e1000736 (2010)
26. Saumell-Mendiola, A., Serrano, M.A., Bogu ná, M.: Epidemic spreading on interconnected networks. arXiv:1202.4087 (2012)

Simulating Language Dynamics
by Means of Concept Reasoning

Gonzalo A. Aranda-Corral[1], Joaquín Borrego-Díaz[2], and Juan Galán-Páez[2(✉)]

[1] Department of Information Technology, Universidad de Huelva,
Crta. Palos de La Frontera S/n., 21819 Palos de La Frontera, Spain
[2] Department of Computer Science and Artificial Intelligence, Universidad de Sevilla,
Avda. Reina Mercedes S/n., 41012 Sevilla, Spain
juangalan@us.es

Abstract. A problem in the phenomenological reconstruction of Complex Systems (CS) is the extraction of the knowledge that elements playing in CS use during its evolution. This problem is important because such a knowledge would allow the researcher to understand the global behavior of the system [1,2]. In this paper an approach to partially solve this problem by means of Formal Concept Analysis (FCA) is described in a particular case, namely Language Dynamics. The main idea lies in the fact that global knowledge in CS is naturally built by local interactions among agents, and FCA could be useful to represent their own knowledge. In this way it is possible to represent the effect of interactions on individual knowledge as well as the dynamics of global knowledge. Experiments in order to show this approach are given using WordNet.

1 Introduction

Complex System (CS) is a broad concept which has specific features but covers very different systems, with an astonishing variety of dynamics. Among them, particularly interesting are those related with human (rational) activities, as for example, organizations, communities and cities. It is usual to study and simulate these kinds of systems by reducing human behavior to simple (but essential) processes. A traditional methodology in CS research is to model these by designing local interactions between nodes (agents) of the CS, then -by means of simulations- global properties are studied (by checking their reliability, accuracy and validity). Several types of agents interactions can be considered: games, communication (messages), competition, etc.

Particularly interesting is the study of Language Dynamics (LD), a rapidly growing field in CS community, that focuses on all processes related with emergence, evolution, change and extinction of languages [11]. For instance, an approximation to the study of self-organization and evolution of the language

Partially supported by TIN2009-09492 project (Spanish Ministry of Science and Innovation) and *Proyecto de excelencia* TIC-6064 of *Junta de Andalucía*, cofinanced with FEDER founds.

G.A. Di Caro and G. Theraulaz (Eds.): BIONETICS 2012, LNICST 134, pp. 296–311, 2014.
DOI: 10.1007/978-3-319-06944-9_21, © Institute for Computer Sciences, Social Informatics and Telecommunications Engineering 2014

and its semantics to consider the community of users as a CS that collectively build the semantics features of their lexicon. Formal Concept Analysis (FCA) [6] aims to collect formal concepts defined by attributes. Thus FCA tools can be applied to enhance models in order to study LD, in which implicit semantic structures in agent's language can be considered.

1.1 Naming Games

A popular approach in LD is to model agents' interaction by means of naming games [19,20]. Naming games were created to explore self-organization in LD (emergence of vocabularies, that is to say, the mapping between words and meanings). Naming games consist on the interaction between two agents, a speaker and a listener. From the basic model, a number of variants for several and specific models can be considered. The aim of the agent community is to achieve a common vocabulary. The minimal naming game is as follows. Each agent has its own context (object/word) and interacts according to the following steps [11]:

1. The speaker selects an object from the current context.
2. The speaker retrieves a word from its inventory associated with the chosen object, or, if its inventory is empty, invents a new word.
3. The speaker transmits the selected word to the listener.
4. If the listener has the word named by the speaker in its inventory and that word is associated with the object chosen by the speaker, the interaction is a success and both players maintain in their inventories only the winning word, deleting all the other words that fitted the same object.
5. If the listener does not have the word named by the speaker in its inventory, or the word is associated to a different object, the interaction is a failure and the listener updates its inventory by adding an association between the new word and the object.

Naming games have been considered both in non-situated and situated models. Situated models place agents in an artificial world, where environmental features as distance between agents or agent's neighborhood can be considered. Situated models are very interesting because the communication among agents does not obey purely random selections: communication takes place between agents which are able to do it (for example, between neighbor ones). Moreover, naming games with spatially distributed agents allow to model the emergence of different language communities by stabilization of the system [18]. This is due to the fact that the "success" of a linguistic innovation is dependent on whether the group, as a whole, has adopted it or not.

Our interest in naming games is based on their *adaptive nature*, that is to say, naming games can produce changes in the lexicon of both, the speaker and listener, as side effect. Thus, agent's lexicon changes during its live within the system. By considering the possible results in each step of the naming game we can list the side effects on agents' lexicon on the game (see [18]).

From FCA point of view, minimal game is a very inspiring interactive method in which it can be applied. For example, by considering *synsets* (sets of cognitive synonyms) from *WordNet*[1] as attributes, which provides an implicit meaning. There exist several variants of the above interaction by considering different levels of reasoning ability of agents.

1.2   Reasons for Selecting FCA (and WordNet)

Classic LD does not consider strong semantic features on agents' interactions. FCA provides a general framework in which semantic features can be added to LD (at object/attribute level). It is interesting to consider a real case study, a language with coherent semantics (also compatible with FCA) as emergence target, to discuss this proposal. In this paper WordNet is selected as the case study but of course it can also be used with any other object-attribute system.

Aim of the Paper. The aim is to describe how the concept lattice associated to the full language emerges (in asymptotic terms) from a community of agents by means of FCA-based semantic interactions. Specifically, the aim is to describe a number of experiments with FCA-based variants of LD approach using *naming games-based models*, by showing models' behaviors. In this way we demonstrate that FCA can enrich LD models, particularly those that focus on vocabulary emergence.

To illustrate the different proposals, WordNet (subsets of) is considered as an universal vocabulary (a global knowledge which the multiagent system aim to achieve). In this way experiments can be confronted with the real structure of a vocabulary (although models presented in this paper -a first approach- are basic and they do not consider every WordNet feature).

The selection of an existent lexical database is not arbitrary. Human languages are surprisingly robust and stable, with persistent categories [14]. In WordNet terms, synsets are categorical classifications that have emerged from human interaction, thus it is a good candidate in order to model its emergence.

Structure of the Paper. In the next section some remarks on FCA and its use in LD modeling are considered. Section 3 presents some variants of naming games by using FCA elements. In order to perform experiments, WordNet lexicon database is selected. Section 4 is devoted to describe a FCA-based version of the naming game. In Sect. 5 we analyse an interesting variant of naming games which uses concept reasoning (actually reasoning with implications of agents' contexts and attribute exploration idea) to perform the interaction. This variant shows a behavior similar than the one in Steels' spatially distributed naming games [18], thus in order to simulate interlingua phenomena, hybrid games are presented in Sect. 6. Section 7 is devoted to discuss experimental results and, in Sect. 8, some conclusions of the work done are given.

---

[1] http://wordnet.princeton.edu

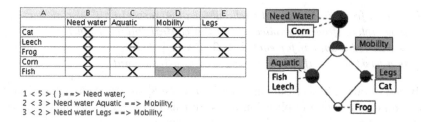

A	B	C	D	E
	Need water	Aquatic	Mobility	Legs
Cat			X	X
Leech	X	X	X	
Frog	X	X	X	X
Corn	X			
Fish	X	X	X	

1 < 5 > { } = = > Need water;
2 < 3 > Need water Aquatic = = > Mobility;
3 < 2 > Need water Legs = = > Mobility;

Fig. 1. Formal context, associated concept lattice and Stem Basis

2 Background: Formal Concept Analysis and Implications

According R. Wille, FCA mathematizes the philosophical understanding of a concept as a unit of thoughts composed of two parts: the extent and the intent [6]. The extent covers all objects belonging to this concept, while the intent comprises all common attributes valid for all the objects under consideration. It also allows the computation of concept hierarchies from data tables. In this section, we succinctly present basic FCA elements, although it is assumed that the reader is familiar with this theory (the fundamental reference is [6]).

A formal context is represented as $M = (O, A, I)$, which consists of two sets, O (objects) and A (attributes) and a relation $I \subseteq O \times A$. Finite contexts can be represented by a 1-0-table (representing I as a Boolean function on $O \times A$). See Fig. 1 for an example of formal context about living beings.

The FCA main goal is the computation of the concept lattice associated with the context. Given $X \subseteq O$ and $Y \subseteq A$ it defines

$$X' := \{a \in A \mid oIa \text{ for all } o \in X\} \text{ and } Y' := \{o \in O \mid oIa \text{ for all } a \in Y\}$$

A (formal) concept is a pair (X, Y) such that $X' = Y$ and $Y' = X$. For example, concepts from formal context about living beings (Fig. 1, left) are depicted in Fig. 1, right. Actually in Fig. 1, each node is a concept, and its intension (or extension) can be formed by the set of attributes (or objects) included along the path to the top (or bottom). E.g. The node tagged with the attribute Legs represents the concept $(\{Legs, Mobility, NeedWater\}, \{Cat, Frog\})$.

2.1 Implications and Basis

Logical expressions in FCA are *implications between attributes*, pair of sets of attributes, written as $Y_1 \rightarrow Y_2$, which is true with respect to $M = (O, A, I)$ according to the following definition. A subset $T \subseteq A$ *respects* $Y_1 \rightarrow Y_2$ if $Y_1 \not\subseteq T$ or $Y_2 \subseteq T$. It says that $Y_1 \rightarrow Y_2$ holds in M ($M \models Y_1 \rightarrow Y_2$) if for all $o \in O$, the set $\{o\}'$ respects $Y_1 \rightarrow Y_2$. See [5,6,17] for more information.

Definition 1. *Let $\mathcal{L}$ be a set of implications and L an implication of M.*

1. *L follows from $\mathcal{L}$ ($\mathcal{L} \models L$) if each subset of A respecting $\mathcal{L}$ also respects L.*
2. *$\mathcal{L}$ is complete if every implication of the context follows from $\mathcal{L}$.*
3. *$\mathcal{L}$ is non-redundant if for each $L \in \mathcal{L}$, $\mathcal{L} \setminus \{L\} \not\models L$.*
4. *$\mathcal{L}$ is a basis for M if it is complete and non-redundant.*

It can obtain a basis from the *pseudo-intents* [8] called *Stem Basis* (SB). SB is only an example of a implication basis. In this paper none specific property of the SB can be used, so it can be replaced by any other basis.

In order to work with formal contexts, stem basis and association rules, the Conexp[2] (cf. [21]) software has been selected. It has been used as a library to build the component which provides implications (and association rules) to the reasoning module of the system we have used in several applications of FCA.

2.2   A Formal Context Associated to WordNet

As it was mentioned before, it is interesting to consider a real and structured language in order to exploit FCA semantic features, thus in this paper experiments are performed on subsets of WordNet system. A Formal context associated to WordNet has been considered by using words as objects of the context and synsets as attributes. Synsets are sets of synonymous words, thus they can be considered as a potential definition (meaning) of each word. The concept lattice associated to this huge lexical database can not be computed, thus small subsets of WordNet have been taken instead, in order to be able to compute FCA elements (concept lattice and SB) in short time. In this way, it will be possible to evaluate the soundness of the proposed models with respect to the FCA elements associated to these subsets.

2.3   Formal Contexts Associated to Agents in Complex Systems

Actually two different scopes of formal contexts should be considered in the experiments. The first one is the formal context associated to the whole language considered, the **global knowledge**, and the second one is the formal context associated to the lexicon that each agent owns, the **individual knowledge**. Starting from an initial lexicon (usually randomly selected) for each agent, they interact using variants of naming games and the result of these interaction transforms theirs contexts (so their formal concepts).

Finally, to compute the collective knowledge, the individual knowledge of each agent is aggregated to obtain a *similarity matrix*. The entries of this matrix is the number of agents owning each pair object-attribute. A pair belongs to the **collective knowledge** only if its value in the similarity matrix is above a certain collective knowledge threshold CK_{th} (which will be detailed later).

Since the aim of the paper is to describe how the concept lattice associated to the full language emerges (in asymptotic terms) from the interaction of a community of agents, it is interesting to study what kind of FCA-based naming

---

[2] http://sourceforge.net/projects/conexp/

games are efficient to achieve this goal, and how these games output stable communities which need other type of interactions (based on attribute exploration idea, for example) to intercommunicate their lexicons.

2.4 Assumptions on the Model

Our approach uses several assumptions -adapted from those enumerated in [9] to a FCA framework- for each model presented in this paper[3]:

- *All agents have same semantic space and pre-existing semantic categories.* This assumption is selected when agents "understand" a pair (*word, synset*) equally to other agents that knew the same pair.
- *Agents are equipped with a symbolic communication ability.* In our case, the ability to reason with Stem basis.
- *Agents can read each other's communicative intentions.* In our case, we simplify this to two performatives (from speech acts): *request* and *answer*.
- *Agents have imitation ability: agents accept information from other agents (credulous agents).*
- *Agents continually detect recurrent patterns.* In our case, agents detect true implications. Our model is not realistic in this assumption, because agents in Sect. 5 selects implications from Stem basis instead of any true implication.
- *Agents have sequencing ability. In our case, sequencing is limited to one-step memory.* We will see that It is sufficient in our model.
- *Agents' behavior is governed by rule competition.* This idea is implemented by selecting, as the knowledge basis of agents, the stem basis (Sect. 5).

The assumptions seem stronger in some cases, compared with other model assumptions as in [10], where minimal abilities for initial agents are supposed.

2.5 Parameters on the Models

Mainly five parameters are considered (see also [7]):

- N is the population size. The values chosen in the experiments are conditioned by the feasibility of the computation of each model.
- δ is the probability for an agent to have within its initial knowledge a pair *lema-synset* (object-attribute) (see section bellow). As general rule, δ is selected in a value range which provides each pair *lema-synset* to appear in at least one agent from the overall population with probability $P = 0.95$.
- Convergence (stabilization) criteria: The convergence test checks whether every existent pair *lema-synset* is present within the *collective knwoledge*.
- The collective knowledge threshold CK_{th} is selected within the range [90 %– 100 %]. This selection has the aim of assuring the convergence to the full language in (almost) every agent.
- The size of the selected subset of WordNet in the experiments is determined by both, the fact that the subsystem must contain complete synsets and by the computational feasibility. The complex concept lattices associated to subsystems of Wordnet are hard to compute [1].

---

[3] Some of them will be weakened in future works.

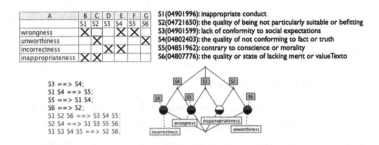

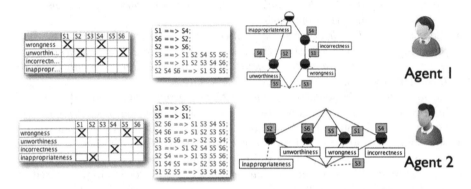

Fig. 2. Example (1st row: context and synsets definitions; 2nd row: stem basis and concept lattice

Fig. 3. Agents for the example

3   Modeling Communication as FCA-based Naming Games

Roughly speaking, the goal of using FCA in naming games is to analyze interactions between agents with a partial knowledge in order to study how a collective knowledge emerges. In these communication games, each interaction is a communicative act where two agents interchange new knowledge. In the experiments performed, the creation steps have been dropped because the aim of these models is to induce the emergence of WordNet.

In order to model communication between agents, the English lexical database *WordNet* has been chosen as the global knowledge. In *WordNet* nouns, verbs, adjectives and adverbs are grouped into sets of cognitive synonyms called *Synsets*, where each of these express a distinct concept. *Synsets* are interlinked by means of conceptual-semantic and lexical relations.

In order to illustrate the process, a tiny WordNet subset has been chosen (see Fig. 2) and the initial knowledge of two agents which will perform the communicative act is shown in Fig. 3.

3.1   Communicative Process Outline

The simulation environment for communication games has been considered as a grid (the world) where agents move freely. In each step, if an agent meets another, a communicative act takes place. Roughly speaking it is as follows.

1. The world is randomly initialized for a given density (*population/gridsize*). Each agent starts with an initial knowledge randomly taken from the global one. To obtain successful communicative games, it is necessary that the union of the initial local knowledge of each agent contains approximately all concepts within the global knowledge.

   As it was already commented, the probability P for each pair within the selected WordNet subset to appear in the initial local knowledge of at least one agent is given by:

   $$P = 1 - (1 - \delta)^N$$

   It is suggested to carry out communication games with at least $P > 0.95$, thus the value δ to be considered depends on the number of agents N in the world (i.e. for $N = 200$ it is suggested that $\delta \geq 0.015$).

2. In each time step, each agent moves randomly to an adjacent cell.
3. Each agent (speaker) chooses randomly a listener agent within the agents in the same cell, in order to start a communicative process (request).
4. After the simulation, the collective knowledge can be measured.

There are different ways of performing the communicative process as well as different ways of measuring the collective knowledge. Those will be depicted in the following sections. Due to the huge size of the *WordNet* database, in order to compute the simulations of communication games, different subsets of *WordNet* have been considered as global knowledge. In order to enrich language dynamics within communication games, only connected subsets of *WordNet* and only formed by full *synsets* have been considered.

3.2   Formal Contexts as Agents' Knowledge in Communication Games

In order to work with FCA-based naming games, the **individual knowledge of an agent** is considered as a local Formal Context, in which *lemas* are *objects* and *synsets* are *attributes*. A relation between an object o_i and an attribute a_i means that that the lema o_i belongs to the synset a_i.

4   Modeling Emergence of WordNet by Intent-Extent Games

The first model is the most direct one due to the relative similarity between a *formal concept* and a *synset*. In this communication game, the communicative act consists on direct interchange of lemmas (objects) and synsets (attributes) between the speaker and the listener.

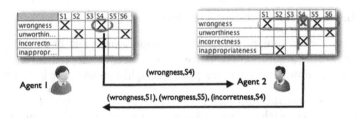

Fig. 4. Intent-extent communicative act

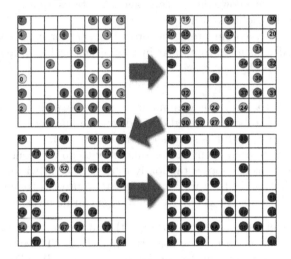

Fig. 5. Evolution of agents knowledge within the world grid

Communicative Act. Each time step, the speaker randomly chooses a pair (o_i, a_j) from its local knowledge (formal context) and sends it (request) to the listener. The answer of the listener will be the two sets $intent(o_i)$ and $extent(a_j)$ (relative to its own formal context). In case the listener does not have any information about o_i or a_i, it will returns an empty set and will add the pair (o_i, a_j) to its local knowledge (see Fig. 4).

Collective Knowledge Emergence. In order to detect and measure the emerging knowledge due to agents interactions in this communication game, the *collective knowledge* is obtained by computing the *similarity matrix*. In each time step the *error rate* between the global and the collective knowledge is measured as the difference between the collective and the global knowledge.

In Fig. 5 four different states of the communicative process for this model are shown. A small example has been chosen in order to show a representation of the world. The circles are agents which randomly move within the grid. The number on the agents shows the number of complete synsets they have within their *individual knowledge*.

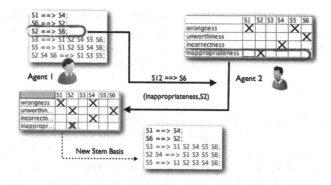

Fig. 6. Communicative act in stem basis game

Convergence Criteria. The communication game ends when the *collective knowledge* emerged from agents interactions is equal to the *global knowledge*. It is worthy to note that the convergence rate of the game highly depends on the collective knowledge threshold CK_{th} considered (see Sect. 7 below).

5 Modeling Emergence of WordNet by Stem Basis Games

This model aims to exploit the power of stem basis (SB) in knowledge detection tasks. In a first approach the communication process goal is the emergence of the collective knowledge by detecting and eliminating inconsistencies within local knowledge of agents. The communicative process in this case concerns to consistency questions. Each agent will contrast its knowledge with others', in order to detect and fix inconsistencies (see Fig. 6).

Communicative Act. The speaker computes the SB of its local formal context, randomly chooses a rule r_i from it and sends it (request) to the listener. If r_i is true within the listeners' local knowledge, it returns a positive answer and finish the communicative act. Otherwise it returns a negative answer and sends to the speaker a counter example (o_i) for r_i, and the speaker adds it to its local knowledge in order to fix the inconsistency. This communicative act is similar to one step of the *attribute exploration* (cf. [6]).

Collective Knowledge Emergence. In this case, as the model works with SB, the collective knowledge has to be considered as the *collective consistent knowledge*, that is to say, the SB corresponding to the collective knowledge have to be consistent with the SB corresponding to the global knowledge. As the rules of the game have changed, in this case, the collective knowledge is measured by its consistency. In order to estimate the soundness of the emerged knowledge, it should be verified whether the true implications within the collective knowledge are entailed by the SB of the original vocabulary. Firstly, the *similarity matrix* is computed as in the former model, and a collective formal context is obtained by filtering de similarity matrix with the aforementioned collective knowledge threshold CK_{th}. Then the collective SB is computed and its consistency is verified against the global SB.

Convergence Criteria. This communication game ends when the game reach the equilibrium, that is to say, when there are no more inconsistencies between agents local knowledge. It is worthy to note that from this model does not emerges the global knowledge (it is not the aim), but a knowledge consistent with the global one. In the following section an hybrid model will be considered in order to get both, consistency and completeness.

6   Modeling Emergence of WordNet by Hybrid Games

In order to obtain a better model for language emergence, an hybrid model is considered. Particularly, to complete the stem basis game, which stabilizes before agents' local knowledge converges to the global knowledge.

The consistency based approach is interesting but does not provide full emergence of collective knowledge. Thus in this hybrid approach the two communicative act types depicted above will be considered, one based on consistency (stem basis interactions) and the second based on direct information exchange (intent-extent interactions). In this model, another question arises, how to merge both types of communicative act?

Communicative Act. In a first approach of merging both types of communicative acts, the simplest solution is to use one communication type or another with a certain probability P (usually $P = 0.5$). This first approach produces a behavior very similar to the intent-extent game, thus it is not very interesting, and another approach should be considered.

Firstly, two new parameters should be considered. One is the time period τ_{eq} necessary to consider that the game as arrived to a equilibrium state, thus, every inconsistency between agents' local knowledge have been suppressed. That is to say, the time period in which agents' local knowledge does not changes. The second parameter is the time period $\tau_{int-ext}$ in which agents interchange information freely (as in the first model) in order to enrich their local knowledge, but leading to new inconsistencies. The process for this second approach is as follows.

Agents normally communicate others as in they do in the stem basis game. If after a time period τ_{eq} the system stays in equilibrium, agents' communication type changes to the one of the intent-extent game during a time period $\tau_{int-ext}$. Then they come back to behave as in the stem basis game in order to solve inconsistencies until the next equilibrium state.

Collective Knowledge Emergence. In this case the both notions of collective knowledge above mentioned should be considered, in order to evaluate both, consistency an knowledge emergence.

Convergence Criteria. The game ends when both objectives are reached, the emergence of consistency and the global knowledge, within the collective agents knowledge.

7   Experiments and Discussion

In order to study the convergence of agents' collective knowledge for each of the aforementioned communication games, many experiments with the different models have been carried out. A connected subset of *WordNet* (as it was mentioned before) has been selected to be used in all experiments.

In Fig. 8 the results of some of those experiments are shown. The figures show the emergence of collective knowledge which tends to the global knowledge (presented as a percentage of the global knowledge). In the cases which correspond, it is also shown the evolution of the implications, associated to the collective knowledge, with respect to those associated to the global one.

The subset of *WordNet* considered for the experiments is of relatively small size (around 400 lema-synset pairs) due to the high computation time of the

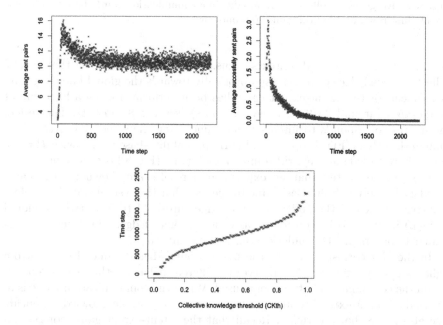

Fig. 7. First row: average number of pairs (lemma-synset) involved in communicative act (left) and the average number of positive (new pair for the listener) ones (right). Second row: convergence rate with respect to the collective knowledge threshold CK_{th}

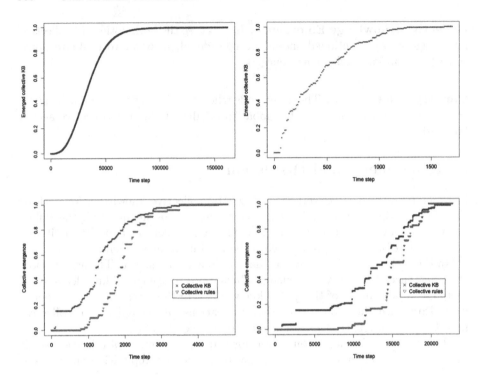

Fig. 8. Convergence of collective Knowledge in communication games based on intent-extent (first row), and hybrid games (second row).

Stem Basis for huge formal contexts. Thus, in order to show the fact that agents' collective knowledge converges asymptotically towards the global knowledge, an experiment for the intent-extent game has been performed using a huge subset of *WordNet* (around 26700 lema-synset pairs) (see Fig. 8 top left). The rest of the figures corresponds to intent-extent game using the same *WordNet* subset than in the others (Fig. 8 top right), hybrid probability-based game with $P = 0.5$ (Fig. 8 bottom left) and hybrid rounds-based game (Fig. 8 bottom right).

It should be noted that no experimental results are shown for the game based exclusively in Stem Basis, due to the fact that the system stabilizes before any concept exceed the collective knowledge threshold (CK_{th}) (which should be high) in order to be considered as collective knowledge. This phenomena is similar to others in LD simulation (language competing).

In the plot corresponding to the hybrid rounds-based game (Fig. 8 bottom right), it is very interesting to observe the different behaviors that arises depending on the communication type being used. When the communication game is in a period of intent-extend communications, intervals of stable knowledge appears (see also Fig. 8, bottom right). Recall that the intent-extent game consists on random knowledge interchanges and introduces *inconsistencies* (counterexamples). In the other hand, when the communication game is in a period of Stem

Basis communications, the knowledge base slightly increases (due to the counterexamples interchange) but the number of collective implications consistent with the global knowledge increases significantly. That is to say, it introduces *consistency* to the knowledge acquired in the previous period.

In order to show the evolution of communicative acts, Fig. 7 depicts the average number of pairs (lemma-synset) that are sent in each time step (1st row, left), and the number of successful ones (1st, right), both for the intent-extent experiment. Finally, Fig. 7 (2nd row) shows the experimental relationship between the collective knowledge threshold (CK_{th}) and average convergence time step. In a nutshell, intend-extend based communication produces collective knowledge quite fast, but it is inconsistent (with respect to the global knowledge) until the whole global knowledge has been learnt. While the communication based on SB produces collective knowledge and don't converges by itself to the global knowledge, but the partial knowledge produced is consistent with the global one.

8   Conclusions and Future Work

In this paper it is shown how to apply FCA in order to enrich LD simulation, in the particular case of naming games in Language Dynamics. FCA provides a solid formal semantic characterization of implicit conceptual structures of Language users. In this way it is possible to analyze the semantic evolution of LD. For instance, a promising research task is to model qualitative *category games* [11].

A consequence of the results of this work is that language convergence seems to be governed by the shared vocabulary (the mapping between words and meaning) instead of the shared language. This conclusion can be justified because the selected δ (which estimates the size of initial vocabulary of agents) is small, and with it the number of pairs $(word, synset)$ initially in each agent. It remains to be investigated vocabulary distributions distinct than the uniform one, for example those that take into account the agents within agent's neighborhood (in order to empower the sharing of vocabulary among close agents).

An interesting research line is to consider weighted distributions on the vocabulary (and concepts [4]), in order to compare local and global knowledge. Also it would be interesting to consider non-uniform distributions of agents within the world (cells). Other distributions (in big sized worlds) would induce the existence of sub-communities with different languages (speaker communities). In the future will also be considered the case where languages get in touch due to communication between speaker communities. This second phenomena is only possible to be simulated if the pair speaker-listener behavior is not driven by means of purely random walks and distributions. In this case, it occurs that this mediates the communication by topological features [18] (see also [13]). The analysis of agent lexicons -in preliminary experiments- reveals that agents will develop stable knowledge within their cluster, but will also develop a second language, an interlingua which is weaker but shared among different clusters [18].

With respect to the use of significative vocabulary subsets (from WordNet or Thesaurus), it is interesting the use of *data weeding* techniques [15] to classify which of them can be explained by CS emergence. The discovering of LD models to explain this would show new ideas on how such vocabularies emerge in the real world. In the case of the concept lattice provided, it is interesting to study the emergence of *concept neighborhoods* [16]. Also, we are investigating other semantic relationships distinct than synsets.

Lastly, it can be sound to use association rules (subsets of Luxemburger basis [12]) instead of Stem Basis. This choice is very related with the idea of lexicon mediated by confidence in the relationship. The reasoning with association rules is more complex, although promising approaches can be exploited [3].

References

1. Aranda-Corral, G.A., Borrego-Díaz, J., Galán-Páez, J.: Scale-free structure in concept lattices associated to complex systems. In: Proceedings of International Conference on Complex Systems, ICCS 2012 (2012) (to appear in IEEE press)
2. Aranda-Corral, G.A., Borrego-Díaz, J., Galán-Páez, J.: Complex concept lattices for simulating human prediction in sport. J. Syst. Sci. Complex. **26**(1), 117–136 (2013). (to appear)
3. Balcázar, J.L.: Redundancy, deduction schemes, and minimum-size bases for association rules. Log. Meth. Comput. Sci. **6**(2), 1–23 (2010)
4. Belohlavek, R., Macko, J.: Selecting important concepts using weights. In: Jäschke, R. (ed.) ICFCA 2011. LNCS (LNAI), vol. 6628, pp. 65–80. Springer, Heidelberg (2011)
5. Bertet, K., Monjardet, B.: The multiple facets of the canonical direct unit implicational basis. Theor. Comput. Sci. **411**(22–24), 2155–2166 (2010)
6. Ganter, B., Wille, R.: Formal Concept Analysis - Mathematical Foundations. Springer, Berlin (1999)
7. Gong, T., Wang, W.S.-Y.: Computational modeling on language emergence: a coevolution model of lexicon, syntax and social structure. Lang. Linguist. **6**, 1–42 (2005)
8. Guigues, J.-L., Duquenne, V.: Familles minimales d'implications informatives resultant d'un tableau de donnees binaires. Math. Sci. Hum. **95**, 5–18 (1986)
9. Ke, J., Holland, J.H.: Language origin from an emergentist perspective. Appl. Linguist. **27**(4), 691–716 (2006)
10. Kirby, S.: Syntax without Natural Selection: How compositionality emerges from vocabulary in a population of learners. In: Knight, C. (ed.) The Evolutionary Emergence of Language: Social Function and the Origins of Linguistic Form, pp. 303–323. Cambridge University Press, Cambridge (2000)
11. Loreto, V., Baronchelli, A., Mukherjee, A., Puglisi, A., Tria, F.: Statistical physics of language dynamics. J. Stat. Mech. Theory Exp. **4**, 1–29 (2011)
12. Luxemburger, M.: Implications partielles dans un contexte. Math. Inf. Sci. Hum. **113**, 35–55 (1991)
13. Patriarca, M., Castelló, X., Uriarte, J.R., Eguiluz, V.M., San, M.: Miguel, modelling two-language competition dynamics. Adv. Complex Syst. **15**(3–4), 1–24 (2012)
14. Mukherjee, A., Tria, F., Baronchelli, A., Puglisi, A., Loreto, V.: Aging in language dynamics. PLoS ONE **6**(2), e16677 (2011). doi:10.1371/journal.pone.0016677

15. Priss, U., Old, L.J.: Data weeding techniques applied to roget's thesaurus. In: Wolff, K.E., Palchunov, D.E., Zagoruiko, N.G., Andelfinger, U. (eds.) KONT 2007 and KPP 2007. LNCS (LNAI), vol. 6581, pp. 150–163. Springer, Heidelberg (2011)
16. Priss, U., Old, L.J.: Concept neighbourhoods in lexical databases. In: Kwuida, L., Sertkaya, B. (eds.) ICFCA 2010. LNCS (LNAI), vol. 5986, pp. 283–295. Springer, Heidelberg (2010)
17. Rudolph, S., Völker, J., Hitzler, P.: Supporting lexical ontology learning by relational exploration. In: Priss, U., Polovina, S., Hill, R. (eds.) ICCS 2007. LNCS (LNAI), vol. 4604, pp. 488–491. Springer, Heidelberg (2007)
18. Steels, L., McIntyre, A.: Spatially distributed naming games. Adv. Complex Syst. 1(4), 301–323 (1999)
19. Steels, L.: A self-organizing spatial vocabulary. Artif. Life 2(3), 319–332 (1995)
20. Steels, L.: Self-organizing vocabularies. In: Artificial Life V: Proceedings of 5th International Workshop on the Synthesis and Simulation of Living Systems, pp. 179–84. MIT Press (1996)
21. Yevtushenko, S.A.: System of data analysis "Concept Explorer". In: Proceedings of 7th National Conference on Artificial Intelligence KII-2000, pp. 127–134 (2000)

Rigorous Punishment Promotes Cooperation in Prisoners' Dilemma Game

Yun Ling[1], Jian Liu[1], Ping Zhu[1(✉)], and Guiyi Wei[1,2]

[1] Networking and Distributed Computing Laboratory,
Zhejiang Gongshang University, 18 Xuezheng St,
Xiasha Higher Education Park, Hangzhou, China
liujiancma@hotmail.com
[2] Wasu Media Network Co. Ltd, Hangzhou, China
{jackyzhu,weigy,yling}@zjgsu.edu.cn

Abstract. In this paper, we introduce a rigorous punishment mechanism into the prisoners' dilemma game. In our model, the punisher punishes the defector with fine β at the cost of γ. Monte-Carlo simulations show the evolution of system is jointly affected by β, γ and system's initial state. We find that when γ is small, the system can evolve into two steady states, i.e., coexisting of cooperators and defectors, and pure punishers. When γ is large, the system can evolve into the only steady state, i.e., coexisting of cooperators and defectors. However, in the middle value of γ, the system can evolve into three steady states, i.e., coexisting of cooperators and defectors, a rock-paper-scissors type of cyclic dominance, and pure cooperators. These results are explained by average total payoff, transition possibility and evolutionary snapshot. We also find the heterogeneity of population distribution can affect cooperation as well.

Keywords: Rigorous punishment · Prisoners' dilemma game · Evolutionary system · Cooperation

1 Introduction

Ranging from biological to social systems, from economic to political activities, cooperative behavior is the heart of many activities and phenomena. Understanding the emergence and persistence of cooperation is a fundamental issue [1]. In particular, there have been a large amount of fruitful interactions between researchers in evolutionary game theory [2,3], especially in the prisoners' dilemma game (PDG) [4]. In PDG, there exists two players, i.e., cooperator (C) and defector (D). The payoffs of players depend on their decisions. For instance, they can get the payoff $R(P)$ while mutual cooperation (defection); if a defector meets a cooperator, the defector can obtain a maximum payoff T and the later can only get a minimal payoff S. The payoffs satisfy $T > R > P > S$ and $2R > T + S$. Because the defector always outperforms the cooperator, the two players will fall into the mutual defection state. However, observations in the real world usually show the opposite. Over the past decades, several mechanisms have been

G.A. Di Caro and G. Theraulaz (Eds.): BIONETICS 2012, LNICST 134, pp. 312–321, 2014.
DOI: 10.1007/978-3-319-06944-9_22, © Institute for Computer Sciences, Social Informatics and Telecommunications Engineering 2014

proposed to explain the altruistic cooperative behavior, such as kin selection [5], direct or indirect reciprocity [6], group selection [7], voluntary participation [8], and so on.

Since Nowak and May [9], spatial games have been given much attention by researchers from many different fields [10–12]. In these works, individuals are located on a spatial network playing with their neighbors. At each round, players interact with their neighbors by choosing cooperation or defection, and get the sum of payoffs. In the next round, each player will update its strategy according to certain rules [9,10,13–17,20]. In such spatial evolutionary game model, cooperation can emerge through the way that cooperators form clusters to resist exploitation by defectors. In this context, the network topology [18,19] plays a key role in the evolution of cooperation, which has been widely studied over the years, e.g., regular networks [20–23], small-world networks [24–27], interdependent networks [28] and scale-free networks [29–31].

In the reality, punishment can maintain cooperation in human societies. For example, polices or other elements of justice system maintain human actions by punishing criminal activities. There are many fruits in this area [32–34]. In these works, fine and punishment cost are considered as two fundamental parameters to determine the stationary distribution of strategies in spatial networks. In particular, Szolnoki [35] found that punishment can promote and stabilize cooperative behavior in the Public Goods Game (PGG). As a natural research extension of N-player interactions, i.e., PGG, in this paper, we are going to investigate the effects of punishment played in PDG.

In present work, we propose a model that incorporates rigorous punishment into PDG and study the model in a lattice network analytically. It is found that the steady-state population could be: (a) a mixture of cooperators and defectors, (b) a mixture of cooperators, defectors and punishers, (c) pure punishers, and (d) pure cooperators, depending on the punishment parameters β and γ. We analyze our findings based on the average total payoff among three of strategies, transition possibility and evolutionary snapshot. We also find that the heterogeneity of population distribution can affect the cooperative behavior as well.

The paper is organized as follows. In Sect. 2, we describe the PDG model with rigorous punishment. The simulation results and discussions are given in Sect. 3. And finally, the paper is concluded in Sect. 4.

2 The Model

We consider an evolutionary PDG with reduced payoff matrix: $T = b$, $R = 1$, $P = S = 0$, where $1 < b < 2$ [36]. To incorporate punishment strategy into PDG, we introduce a third player: the punisher. Thus, in our PDG model with punishment, each player takes one of the three strategies: C (cooperate), D (defect) and P (punish). The extended payoff matrix is given by:

$$
\begin{array}{ccc}
& D & C & P \\
\begin{array}{c} D \\ C \\ P \end{array}
\left(\begin{array}{ccc}
(0,0) & (b,0) & (b-\beta,-\gamma) \\
(0,b) & (1,1) & (1,1-\gamma) \\
(-\gamma,b-\beta) & (1-\gamma,1) & (1-\gamma,1-\gamma)
\end{array} \right)
\end{array}
$$

From the payoff matrix, we can see that defector will get the payoff $b-\beta$ when encountering with the punisher, who gets a payoff of $-\gamma$ correspondingly. The cooperator will get the payoff 1 and the corresponding punisher will receive $1-\gamma$. Here the punisher is considered as a special kind of cooperator who consumes an additional cost of γ while punishing the defectors. Accordingly, the punisher will get $1-\gamma$ when encountering with another punisher.

The PDG is staged in a square lattice with periodic boundary conditions. Initially, a player on site x is designated either as cooperator, defector or punisher at random. To survey the influence of cooperation affected by punishment, we fix the initial proportion of defectors f_D as 0.5. The initial proportion of punisher f_P and cooperator f_C satisfies $f_C + f_P = 0.5$. Then, players play the PDG with their neighbors. At each round, players get the sum payoff P_{S_x}, where S_x is the strategy of player x. Next, player x chooses one of its four nearest neighbors at random, and the chosen player y also acquires its payoff in the same way. Finally, player x imitates the strategy of player y with a probability, controlled by strategy update rule [20]:

$$
\omega(S_x \to S_y) = \frac{1}{1+\exp[\frac{P_{S_x}-P_{S_y}}{\kappa}]} \tag{1}
$$

Where ω is the probability of player x imitating the strategy of player y. Here, $\kappa(0 < \kappa < +\infty)$, characterizes environmental noise, including irrationality and errors. The effect of κ has been well studied in the previous papers [37,38]. According to these works, we set $\kappa = 0.1$.

There follows many interesting questions, including: how the fine β and punishment cost γ jointly affect the cooperative behavior in PDG? how the heterogeneity of population distribution in the initial system impacts the cooperative behavior?

3   Simulation Results and Discussion

To explore the combined influence of punishment and cost on cooperation, we take $N = 100 \times 100$ players and an initial population distribution of $f_D = 0.5$, $f_P = 0.3$ and $f_C = 0.2$. By Monte-Carlo simulations, f_C, f_D, f_P are obtained by averaging over the last 3000 generations of 10000 total generations. Each data is averaged by 10 individual runs.

Figure 1 shows the strategy frequencies of cooperator, defector and punisher when the punishment cost $\gamma = 0.5$. One can clearly find that f_C monotonously increases with β, f_D decreases with β and f_P is emerging when $1.4 < \beta < 1.7$. There are three dynamic evolutions in the range of β. When $\beta < 1.3$, there is only one steady state, i.e., coexisting of cooperators and defectors. When

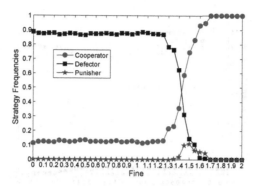

Fig. 1. Strategy frequencies vs fine β for the punishment cost $\gamma = 0.5$, $b = 1.02$, $\kappa = 0.1$, the frequencies of $f_P = 0.3$, $f_D = 0.5$ and $f_C = 0.2$.

$1.3 < \beta < 1.7$, the system is maintained by cyclic dominance among three strategies and the level of cooperation is remarkable increased. When $\beta > 1.7$, pure cooperators dominate the system.

Next, we analyze the underlying mechanism displayed in Fig. 1 by the average payoffs and the transition possibility among three strategies. The average payoffs of cooperators, defectors and punishers are defined as:

$$\overline{P_C} = \frac{\sum_{i=N-M}^{i=N} \sum_{j=1}^{j=N_C^i} P_{i,j}}{\sum_{i=N-M}^{i=N} N_C^i}, j \in C \tag{2}$$

$$\overline{P_D} = \frac{\sum_{i=N-M}^{i=N} \sum_{j=1}^{j=N_D^i} P_{i,j}}{\sum_{i=N-M}^{i=N} N_D^i}, j \in D \tag{3}$$

$$\overline{P_P} = \frac{\sum_{i=N-M}^{i=N} \sum_{j=1}^{j=N_P^i} P_{i,j}}{\sum_{i=N-M}^{i=N} N_P^i}, j \in P \tag{4}$$

Where $P_{i,j}$ denotes for the player $j's$ total payoff in the i generation. $j \in C$ indicates player j is cooperator and similar with $j \in D$ and $j \in P$. N_C^i is the total number of cooperators in generation i and the same as N_D^i, N_P^i. Here, we sum up the last M generations of the total N generations. So the same as $\overline{P_D}$ and $\overline{P_P}$.

As Fig. 2 shows, when $\beta < 1.3$, because punishers always hold a cost of γ, cooperators can easily invade the punishers. Meanwhile, punishers punish the defectors with small value of fine. When the punishers die out in the system, defectors dominate the system. However cooperators can form cooperation clusters, getting higher payoff than defectors. Thus, the cooperators can survive among defectors. Several previous works have shown the emergence of cooperation in lattices network is often induced by formation of cooperator clusters, where cooperators can obtain higher payoff to protect themselves against the

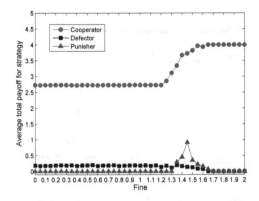

Fig. 2. Average total payoff for $\gamma = 0.5$, $b = 1.02$ and $\kappa = 0.1$, the frequencies of $f_P = 0.3$, $f_D = 0.5$ and $f_C = 0.2$.

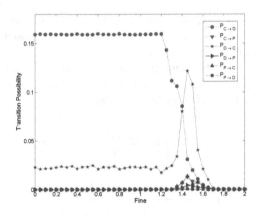

Fig. 3. Transition possibility for $\gamma = 0.5$, $b = 1.02$, $\kappa = 0.1$, the frequencies of $f_P = 0.3$, $f_D = 0.5$ and $f_C = 0.2$, obtained via Monte Carlo simulations of the prisoners' dilemma game on the square lattice network.

invasion of defectors [39]. When $1.3 < \beta < 1.7$, defectors get smaller average total payoff than punishers, they then promote punishers to emerge in the system. As punishers increase, punishers will punish the defector. It will aggravate the decreasing defectors. The cooperators emerge and persist in the system. When β increases, cooperators invade the punishers in the system. Finally, when $\beta > 1.7$, cooperators completely dominate the system.

We can also analysis the strategy transition among cooperators, defectors and punishers. We assume that $P_{C \rightarrow D}$ is the transition possibility where $C \rightarrow D$ denotes cooperator transferring to defector, similarly with $P_{C \rightarrow P}$, $P_{D \rightarrow C}$, $P_{D \rightarrow P}$, $P_{P \rightarrow C}$ and $P_{P \rightarrow D}$. Figure 3, clearly demonstrates the transition possibilities of players for different values of β. As we can see, the system turns into dynamical stable station when $\beta < 1.3$. As β increases, $P_{C \rightarrow D}$ decreases and $P_{D \rightarrow C}$

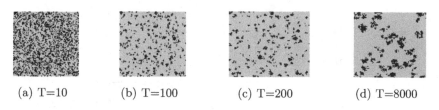

(a) T=10   (b) T=100   (c) T=200   (d) T=8000

Fig. 4. A series of snapshots of distribution of cooperators (blue), defectors (lawn green) and punishers (dark red) on a 100×100 square lattice with $\gamma = 0.5, \beta = 0.5, b = 1.02, \kappa = 0.1$ (Color figure online).

increases. Also, $P_{C \to P}$ and $P_{D \to P}$ are emerging, that is to say, the transition possibility of cooperator transfer to punisher increases. When $\beta > 1.7$, defectors and punishers die out, cooperators dominate the system and punishers give up punishment and turn to cooperation.

To intuitively understand how our mechanism works in the evolution of cooperation, we further investigate the evolution patterns of cooperators, defectors and punishers. Figure 4 shows the evolutionary snapshots when $\gamma = 0.5$, $\beta = 0.5$. Initially, players are randomly distributed in the lattice network from initial population distribution of $f_C = 0.2, f_D = 0.5, f_P = 0.3$. For $T = 100$, we can find that the punishers are easily destroyed and the system soon evolves into the phase where defectors dominate the system. When $T = 8000$, cooperator clusters are shaping and the system evolves into coexisting of defectors and cooperators. However, Fig. 5 shows another different evolutionary snapshot when $\gamma = 0.5$, $\beta = 1.4$. When $T = 100$, the cooperator clusters can resist the invasion of defectors and the frequency of cooperators increases. The evolution system shows a rock-paper-scissors type of cyclic dominance.

To sum up the jointly influence of β and γ on cooperation, we draw the full fine-cost phase diagram presented in Fig. 6. As one can see that there are four system phases in the range of β and γ. When $\gamma < 0.45$ and β is small, the system will enhance the survive of coexisting of defectors and cooperators, then $(C + D)$ phase can occur. As β increases, punishers dominate the system and invade the defectors and cooperators. The system evolves from $(C + D)$ phase to (P) phase. The breaking points of transferring phase depend on the value of γ.

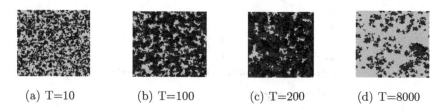

(a) T=10   (b) T=100   (c) T=200   (d) T=8000

Fig. 5. A series of snapshots of distribution of cooperators (blue), defectors (lawn green) and punishers (dark red) on a 100×100 square lattice with $\gamma = 0.5, \beta = 1.4, b = 1.02, \kappa = 0.1$ (Color figure online).

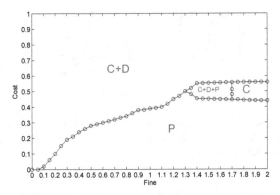

Fig. 6. Full $\beta - \gamma$ phase diagram for $b = 1.02$ and $\kappa = 0.1$, obtained via Monte Carlo simulations of the prisoners' dilemma game on the square lattice.

When $0.45 < \gamma < 0.55$, the system is dominated by defectors and cooperators, i.e., the $(C+D)$ phase in the small value of β. As β increases, the self-organizing patterns with cyclic invasion from $(C+D)$ phase to $(C+D+P)$ phase dominates the system. In this phase, it manifests a new form of cyclic dominance that forms not just between individual strategies but also between strategy alliances in $1.3 < \beta < 1.7$. When β further increases, the system transfers from $(C+D+P)$ phase to the (C) phase, cooperators invade both of defectors and punishers, and completely dominate the system. When $\gamma > 0.55$, the system has the unique $(C+D)$ phase across the full range of β. In other words, at such large values of γ, punishers become inefficient in their contribution of facilitating the evolution of cooperation.

Finally, we visually study the influence of population heterogeneity in cooperation level. Here, f_D is fixed as 0.5. The population heterogeneity is implemented

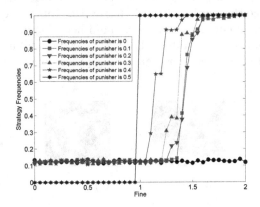

Fig. 7. Sum of strategy frequencies of cooperator and punisher vs fine β for the punishment cost $\gamma = 0.5$, $b = 1.02$ and $\kappa = 0.1$.

by varying f_P as 0, 0.1, 0.2, 0.3, 0.4 and 0.5, under control of $f_P + f_C = 0.5$. Figure 7 shows the relationship of $(f_C + f_P)$ vs β. As we can see that, with increasing of initial proportion of punisher f_P, the sum frequency of $(C + P)$ remarkable increases to guarantee the emergence of cooperation in $1.3 < \beta < 1.7$. As previous work [40] shown, the system's initial states affect the way of evolution station. Our results indicate that a bigger proportion of punisher can effectively increases system's cooperation level. However we have to provide more investment to maintain these punishers.

4 Conclusion

In summary, we have proposed and studied rigorous punishment mechanism in the prisoners' dilemma game. The punishers, who pay an additional cost γ, punish the defectors with fine β. The evolution of system is jointly affected by fine β and cost γ. By simulations, we find that (I) Under small value of cost γ, the system can reach a steady state consisting: (a) a mixture of cooperators and defectors; (b) pure punishers. (II) When cost γ is large, there is only one steady state consisting of pure punishers. (III) However, in the middle value of cost, the systems show the steady states consisting: (a) a mixture of cooperators and defectors; (b) a rock-paper-scissors type of cyclic dominance which is a mixture of cooperators, defectors, and punishers; (c) pure cooperators. We also found that the heterogeneity of population distribution can affect the emergence and persistence of cooperation. Our work indicates rigorous punishment mechanism plays an important role in the evolution of cooperation, and thus it may shed light on understanding the emergence of cooperative behaviors in natural and social systems.

Acknowledgements. This work was supported in part by the National Basic Research Program of China (Grant No. 2012CB315804), the Natural Science Foundation of Zhejiang Province of China under Grants No. Y1110766, the Key Project of Chinese Ministry of Education under Grant No. 210085, the National Development and Reform Commission, China under Special Grants "The Operation System of Multimedia Cloud Based on the Integration of Telecommunications Networks, Cable TV Networks and the Internet", and the Science and Technology Planning Projects of Zhejiang Province, China under Grants No. 2010C13005 and 2011C13006-1. We also thank Runran Liu and Wen-Bo Du for the useful discussions.

References

1. Neumann, V.J., Morgenstern, O.: Theory of Games and Economic Behavior. Princeton University Press, Princeton (1944)
2. Smith, M.J.: Evolution and the Theory of Games. Cambridge University Press, Cambridge (1982)
3. Axelrod, R.: The Evolution of Cooperation. Basic Books, New York (1984)
4. Rapoport, A., Chammah, A.M.: Prisoners Dilemma. University of Michigan Press, Ann Arbor (1965)

5. Hamilton, W.D.: The genetical evolution of social behaviour. II. J. Theor. Biol. **7**(1), 17–52 (1964)

6. Nowak, M.A., Sigmund, K.: Evolution of indirect reciprocity. Nature **437**(7063), 1291–1298 (2005)

7. Traulsen, A., Nowak, M.A.: Evolution of cooperation by multilevel selection. Proc. Natl. Acad. Sci. USA **103**(29), 10952–10955 (2006)

8. Hauert, C., Monte, S.D., Hofbauer, J., Sigmund, K.: Volunteering as red queen mechanism for cooperation in public goods games. Science **296**(5570), 1129–1132 (2002)

9. Nowak, M.A., May, R.M.: Evolutionary games and spatial chaos. Nature **359**, 826–829 (1992)

10. Szabó, G., Fáth, G.: Evolutionary games on graphs. Phys. Rep. **446**, 97–216 (2007)

11. Perc, M., Szolnoki, A.: Coevolutionary games-a mini review. BioSystems **99**, 109–125 (2010)

12. Gross, T., Blasius, B.: Adaptive coevolutionary networks: a review. J. R. Soc. Interface **5**(20), 259–271 (2008)

13. Roca, C.P., Cuesta, J.A., Sánchez, A.: Evolutionary game theory: temporal and spatial effects beyond replicator dynamics. Phys. Life Rev. **6**(4), 208–249 (2009)

14. Wang, W.X., Ren, J., Chen, G.R., Wang, B.H.: Memory-based snowdrift game on networks. Phys. Rev. E **74**(5), 056113 (2006)

15. Wang, Z., Szolnoki, A., Perc, M.: If players are sparse social dilemmas are too: importance of percolation for evolution of cooperation. Sci. Rep. **2**, 369 (2012)

16. Nowak, M.A., Sigmund, K.: A strategy of win-stay, lose-shift that outperforms tit-for-tat in Prisoner's Dilemma. Nature **364**, 56–58 (1993)

17. Wu, Z.X., Xu, Z.J., Huang, Z.G., Wang, S.J., Wang, Y.H.: Evolutionary prisoners dilemma game with dynamic preferential selection. Phys. Rev. E **74**(2), 021107 (2006)

18. Albert, R., Barabási, A.L.: Statistical mechanics of complex networks. Rev. Mod. Phys. **74**(1), 47–97 (2002)

19. Newman, M.E.J.: The structure and function of complex networks. SIAM Rev. **45**(2), 167–256 (2003)

20. Szabó, G., Töke, C.: Evolutionary prisoners dilemma game on a square lattice. Phys. Rev. E **58**(1), 69–73 (1998)

21. Doebeli, M., Knowlton, N.: The evolution of interspecific mutualisms. Proc. Natl. Acad. Sci. USA **95**, 8676–8680 (1998)

22. Hauert, C., Doebeli, M.: Spatial structure often inhibits the evolution of cooperation in the snowdrift game. Nature **428**, 643–646 (2004)

23. Szolnoki, A., Perc, M., Szabó, G.: Phase diagrams for three-strategy evolutionary prisoners dilemma games on regular graphs. Phys. Rev. E **80**(5), 056104 (2009)

24. Abramson, G., Kuperman, M.: Social games in a social network. Phys. Rev. E **63**(3), 030901(R) (2001)

25. Wu, Z.X., Xu, X.J., Chen, Y., Wang, Y.H.: Spatial prisoners dilemma game with volunteering in Newman-Watts small-world networks. Phys. Rev. E **71**(3), 037103 (2005)

26. Ren, J., Wang, W.X., Qi, F.: Randomness enhances cooperation: a resonance-type phenomenon in evolutionary games. Phys. Rev. E **75**(4), 045101(R) (2007)

27. Chen, X.J., Wang, L.: Promotion of cooperation induced by appropriate payoff aspirations in a small-world networked game. Phys. Rev. E **77**(1), 017103 (2008)

28. Wang, Z., Szolnoki, A., Perc, M.: Evolution of public cooperation on interdependent networks: the impact of biased utility functions. EPL **97**, 48001 (2012)

29. Santos, F.C., Pacheco, J.M.: Scale-free networks provide a unifying framework for the emergence of cooperation. Phys. Rev. Lett. **95**, 098104 (2005)

30. Rong, Z.H., Li, X., Wang, X.F.: Roles of mixing patterns in cooperation on a scale-free networked game. Phys. Rev. E **76**(2), 027101 (2007)

31. Du, W.B., Cao, X.B., Zhao, L., Hu, M.B.: Evolutionary games on scale-free networks with a preferential selection mechanism. Physica A **388**, 4509–4514 (2009)

32. Hauert, C., Traulsen, A., Brandt, H., Nowak, M.A., Sigmund, K.: Via freedom to coercion: the emergence of costly punishment. Science **316**(5833), 1905–1907 (2007)

33. Brandt, H., Hauert, C., Sigmund, K.: Punishment and reputation in spatial public goods games. Proc. R. Soc. London B **270**(1519), 1099–1104 (2003)

34. Fehr, E., Gächter, S.: Altruistic punishment in humans. Nature (London) **415**, 137–140 (2002)

35. Szolnoki, A., Szabó, G., Perc, M.: Phase diagrams for the spatial public goods game with pool punishment. Phys. Rev. E **83**(3), 036101 (2011)

36. Nowak, M.A., May, R.M.: The spatial dilemmas of evolution. Int. J. Bifurcat Chaos **3**, 35–78 (1993)

37. Szabó, G., Vukov, J., Szolnoki, A.: Phase diagrams for an evolutionary prisoners dilemma game on two-dimensional lattices. Phys. Rev. E **72**(4), 047107 (2005)

38. Vukov, J., Szabó, G., Szolnoki, A.: Cooperation in the noisy case: prisoners dilemma game on two types of regular random graphs. Phys. Rev. E **73**(6), 067103 (2006)

39. Nowak, M.A., Sigmund, K.: Evolution of indirect reciprocity by image scoring. Nature **393**, 573–577 (1998)

40. Xu, C., Ji, M., Yee, J.Y., Da, F.Z., Hui, P.M.: Costly punishment and cooperation in the evolutionary snowdrift game. Physica A **390**(9), 1607–1614 (2011)

Author Index